GUIDELINES FOR
ASSET INTEGRITY MANAGEMENT

GUIDELINES FOR
ASSET INTEGRITY MANAGEMENT

CENTER FOR CHEMICAL PROCESS SAFETY
of the
AMERICAN INSTITUTE OF CHEMICAL ENGINEERS
New York, NY

WILEY

Published by John Wiley & Sons, Inc., Hoboken, New Jersey.

Published simultaneously in Canada.

For general information on our other products and services or for technical support, please contact our Customer Care Department within the United States at (800) 762-2974, outside the United States at (317) 572-3993 or fax (317) 572-4002.

Wiley also publishes its books in a variety of electronic formats. Some content that appears in print may not be available in electronic formats. For more information about Wiley products, visit our web site at www.wiley.com.

Library of Congress Cataloging-in-Publication Data is available.

ISBN: 978-1-119-01014-2

Printed in the United States of America.

10 9 8 7 6 5 4

CONTENTS

9   Asset Integrity Procedures   211

10   Quality Management   239

11   Equipment Deficiency Management   269

LIST OF FIGURES

LIST OF TABLES

PREFACE

The American Institute of Chemical Engineers (AIChE) has been closely involved with process safety and loss prevention issues in the chemical and allied industries for nearly 50 years. Through its strong ties with process designers, constructors, operators, safety professionals and members of academia, AIChE has enhanced communications and fostered continuous improvement of the industry's high safety standards. AIChE publications and symposia have become information resources for those devoted to process safety and environmental protection.

AIChE created the Center for Chemical Process Safety (CCPS) in 1985 after the chemical disasters in Mexico City, Mexico and Bhopal, India. The CCPS is chartered to develop and disseminate technical information for use in the prevention of major chemical incidents. The Center is supported by more than 150 sponsor companies that provide the necessary funding and professional guidance to its technical committees. The major product of CCPS activities has been a series of guidelines to assist those implementing various elements of a process safety and risk management system. This book is part of that series.

Ensuring the initial and ongoing integrity of process equipment, including instrumentation and safety systems, has become known as *asset integrity management* or *AIM*. Asset Integrity and Reliability is a fundamental component of successful, risk-based process safety programs. However, facilities continue to be challenged to develop and maintain successful AIM programs. CCPS' Technical Steering Committee commissioned these guidelines, as an update and expansion of the previous CCPS document *Guidelines for Mechanical Integrity Systems,* to continue assisting facilities in meeting this challenge. The change in terms from *mechanical integrity* to *asset integrity* reflects international usage, consistent with the elements in CCPS' more recent *Guidelines for Risk Based Process Safety**, and a recognition that a much broader set of assets needs to be properly designed, installed and maintained at process facilities than those requiring "mechanical" integrity.

In addition, the term *equipment* is often associated with an individual piece of equipment such as a pump, compressor or vessel. The term *asset* better reflects how equipment and its associated instrumentation, utilities and connections function together as a system and the integrity of which must be maintained holistically as a system.

---

*American Institute of Chemical Engineers, *Guidelines for Risk Based Process Safety*, Center for Chemical Process Safety, New York, NY, 2007. References in this book are at the end of each chapter.

This document contains approaches and resources for designing, developing, implementing and improving AIM programs. Even the best AIM programs cannot guarantee that incidents will not occur. However, an effective AIM program, integrated with other elements of process safety management, can significantly reduce risks associated with operations involving hazardous materials and energies.

This document only addresses asset integrity for fixed facilities in the process industries. Transportation aspects including pipeline integrity are not covered, although many of the same principles apply.

ACKNOWLEDGMENTS

The Chemical Center for Process Safety (CCPS) thanks all of the members of the Asset Integrity Management (AIM) Subcommittee for providing technical guidance in the preparation of this book. CCPS also expresses its appreciation to the members of the Technical Steering Committee for their advice and support.

The chairman of the AIM Subcommittee was Eric Freiburger of Praxair. Mike Broadribb of Baker Risk was co-chair. The CCPS staff consultant was John F. Murphy. The AIM Subcommittee had the following contributing members:

Andrew Basler, Mallinckrodt	Matt Hedlund, Eastman Chemical
Kevin Blackwell, Honeywell	Tom Sandbrook, Chemours
Russ Davis, Mistras Group	John Traynor, Evonik
Jonas Duarte, Chemtura	Chris Urbanowich, Petrobras

Unwin Company (Columbus, Ohio) prepared this document under contract with CCPS. Robert W. Johnson was the lead contract author, with input from Steven W. Rudy.

Acknowledgement is also given to Robert W. Ormsby, who was the CCPS staff consultant at the beginning of this project; to Tom Folk, who prepared the detailed outline for the AIM Subcommittee; and to ABSG Consulting Inc., which prepared CCPS' precursor to this publication, *Guidelines for Mechanical Integrity Systems*.

CCPS gratefully acknowledges the comments submitted by the following peer reviewers, whose insights and suggestions helped ensure a balanced perspective. It should be noted that the peer review was based on a final draft of this document and, as such, the peer reviewers did not provide comments on the final published manuscript.

Robert A. Bartlett, Pareto Engineering & Management Consulting
James M. Broadribb, Wood Group
Robert A. Bartlett, Pareto Engineering & Management Consulting
Hugh Hemphill, Chevron
Mark Jackson, FM Global
Adrian Sepeda, Process Safety & Risk Management
Razzack Syed, Praxair Inc.
David Thaman, PPG
Terry A. Waldrop, AIG

FILES ON THE WEB

The following files are available to purchasers of *Guidelines for Asset Integrity Management*. They are accessible from the AIChE/CCPS website at http://www.aiche.org/sites/default/files/book-downloads/AIMsupplements.pdf using the password AIM2017. Users of this information are responsible for determining the suitability of these resources to their particular AIM program.

Chapter 8 (AIM Training) Resources

- Sample skills/knowledge list for an electrician
- Sample skills/knowledge list for a mechanic

Chapter 12 (Equipment-Specific) Resources

- Instrumentation and controls:
  - Process control systems
  - Critical alarms and interlocks
  - Chemical monitors and detection systems
  - Conductivity, pH, and other process analyzers
  - Burner management systems

- Rotating equipment:
  - Reciprocating compressors
  - Centrifugal compressors, including specific protection systems (e.g., pressure cutouts)
  - Process fans and blowers
  - Agitators and mixers
  - Electric motors
  - Gas turbines
  - Steam turbines
  - Gearboxes

- Electrical systems:
  - Transformers
  - Motor controls
  - Uninterruptible power supplies (UPSs)
  - Emergency generators
  - Lightning protection
  - Grounding systems

Chapter 15 (AIM Tool) Resources

- Presentation papers related to analysis approaches:
  - *Risk-Based Approach to Mechanical Integrity Success on Implementation*
  - *An Insurer's View of Risk-Based Inspection*
  - *RCM Makes Sense for PSM-Covered Facilities*
  - *Lessons Learned from a Reliability-Centered Maintenance Analysis*

- Resources for performing equipment failure analyses:
  - Additional detailed information on the analysis steps
  - An equipment failure analysis checklist

- Resources for performing root cause analyses:
  - SOURCE™ Investigator's Toolkit
  - Root cause map
  - Causal factor chart and fault tree templates

- AIM program audit resources

- Presentation paper — *Improving Mechanical Integrity in Chemical and Hydrocarbon Processing Facilities - An Insurer's Viewpoint*

1
Introduction

This chapter introduces asset integrity management (AIM), including the scope and objectives of AIM programs throughout a facility life cycle and the relationship of AIM to other process safety elements. The last section in this chapter outlines the structure of this document.

Since a successful AIM program involves leadership, managers, engineers, operating and maintenance personnel, contractors, suppliers and support staff, this document was prepared for a wide range of audiences and potential users. AIM is an integrated product of proper equipment, dependable human performance and effective management systems. Guidelines are given for developing, implementing and continually improving an AIM program that includes these areas of focus. Behind these focus areas needs to be an involved, supportive management. Consequently, this document also includes guidance to those supporting the program.

1.1   BACKGROUND AND SCOPE

For decades, AIM activities, in one form or another, have been a part of industry's efforts to prevent incidents and maintain productivity. Industry initiatives, company initiatives and regulations in various countries have helped both to define AIM program requirements and to accelerate implementation of AIM programs. AIM is already ingrained in the culture of many process plants, as well as facilities in other related industries. AIM activities are essential for process facilities to maintain economic viability.

AIM has been a part of international process safety regulations for many years, including the Seveso Directive and its implementations in Europe (Reference 1-1) as well as Offshore Installation (Safety Case) regulations (Reference 1-2). Since 1992, a major incentive for process industries in the United States to implement AIM programs has been the Occupational Safety and Health Administration (OSHA) process safety management (PSM) standard, 29 CFR 1910.119 (Reference 1-3). This was followed by the Environmental Protection

1

Agency (EPA) risk management program (RMP) rule, 40 CFR 68 (Reference 1-4). These performance-based regulations each contain a mechanical integrity (MI) element that defines the minimum requirements of a program through six sub-elements that address:

- Equipment to be maintained
- Written MI procedures
- MI training
- Inspection and testing
- Equipment deficiencies
- Quality assurance.

Specific requirements are not prescriptively stated in these regulations, but the sub-elements represent time-proven practices for an effective AIM program. The details of each sub-element are left to the discretion of the facility to develop and implement. All PSM- and RMP-covered U.S. facilities in operation since the regulations were issued have been required to audit compliance with these requirements at least every three years. Many of these audits reveal that companies continue to have significant opportunities to improve their AIM programs.

This document was written primarily for process industry facilities. However, most of the content applies to other industries as well. Although this document was written in the United States, a conscious effort has been made to keep the book applicable to facilities worldwide.

1.2   WHAT IS ASSET INTEGRITY MANAGEMENT?

For the purposes of this book, *asset integrity management* (AIM) is a management system for ensuring the integrity of assets throughout the life cycle of the assets. In this context, an *asset* is a process or facility that is involved in the use, storage, manufacturing, handling or transport of chemicals, or the equipment comprising such a process or facility. Examples of assets include off-shore and on-shore extraction and processing equipment; process and auxiliary tanks, vessels and piping systems including their internal components; control systems; safety systems; buildings and other structures; and transport containers. The selection of which assets are "important" is discussed in Section 1.3 and in Chapter 5.

AIM is a product of many activities, usually performed by many people. When these activities are done well, AIM can provide the foundation for a safe, reliable facility that minimizes threats to the workforce, the public and the environment. Effective AIM is also consistent with good business practices.

AIM programs vary according to industry, regulatory requirements, geography and plant culture. However, some characteristics appear to be common to effective AIM programs. For example, they:

- Include activities to ensure that assets are designed, procured, fabricated, installed, operated, inspected, tested and maintained in a manner appropriate for its intended application.

- Clearly designate assets to be included in the program based on defined criteria.

- Prioritize assets to help optimally allocate financial, staffing, storage space and other resources.

- Help plant staff perform planned maintenance and reduce the need for unplanned maintenance.

- Help plant staff recognize when equipment deficiencies occur and include controls to help ensure that equipment deficiencies do not lead to serious incidents.

- Incorporate applicable codes, standards and other recognized and generally accepted good engineering practice (RAGAGEP).

- Help ensure that personnel assigned to perform AIM activities are appropriately trained and have access to appropriate procedures for these activities.

- Develop standard work roles and consistent activities.

- Maintain service documentation and other records to enable consistent performance of AIM activities and to provide accurate asset information to other users, including other process safety and risk management elements.

This document provides guidance for developing an AIM program that includes all of these characteristics.

To present sound guidance for developing and/or improving AIM programs, this document evaluated lessons learned by the process industries. It does not give just one way of managing the integrity of assets, since there are many ways to approach the implementation of an AIM program, and other resources will be needed to develop a full program. Where appropriate, this book gives strengths and weaknesses of different approaches. Company management will need to recognize which approaches best suit their facility and company needs.

Having a successful AIM program is consistent with the business case for process safety. Benefits of AIM programs that can provide greater value for the business include:

- Improved equipment reliability and availability
- Reduced frequency of asset failures that lead to safety and environmental incidents
- Improved product consistency
- Improved maintenance consistency and efficiency
- Reduced unplanned maintenance time and costs
- Reduced operating costs
- Improved spare parts management
- Improved contractor performance
- Compliance with regulatory requirements.

Each of these objectives may have associated costs, such as more detailed procedures, a larger warehouse or improved computer systems, so that companies may need to prioritize their objectives.

One AIM program development approach that is not advocated in this book is to focus only on compliance with regulations. Compliance is one outcome of an AIM program; however, the primary focus of such a program needs to be on the management of risk, in order to deliver the benefits that an AIM program can provide. The requirements for compliance are often vague and subject to misinterpretation. Furthermore, requirements are subject to change via legislated modifications or new interpretations of existing legislation. In addition, a compliance-only program may miss out on many of the benefits of a more holistic approach, such as improved profitability and reduced risks for employees, the facility and the neighboring community. A more holistic approach can help to:

- Present the AIM program as a company priority, rather than just something the company is forced to do; this approach also helps to ensure compliance, since personnel are less likely to take shortcuts.

- Create synergies with equipment and process reliability initiatives that could improve results and/or lower costs.

- Address risks to employees, community and the business.

Therefore, the more holistic approach helps ensure compliance with governing regulations and is often the greater business value than the minimum compliance effort. Although compliance with federal, state and local regulations is often a motivating factor for a facility, following the guidelines in this document can help a facility develop, implement and/or improve an AIM program that:

- is effective in containing and controlling hazardous materials and energies,
- enables the facility to operate reliably and
- is in compliance with regulatory requirements.

1.3   WHAT ASSETS ARE INCLUDED?

One of the key questions when embarking on an AIM program is to identify the facility assets to include in the AIM program. This question is introduced here and further examined in Chapter 5.

As can be seen in the next section on AIM life cycle, an AIM program consists of two major parts:

1.   Properly designing and installing the facility's assets before startup

2.   Maintaining the ongoing integrity of the assets over a lifetime of facility operation.

The question of what assets to include in an AIM program is likely to be different for these two major tasks. For the first part, it would be generally considered as necessary to correctly design and properly install *all* facility assets before startup. However, it is recognized that greater care or rigor is likely to be taken in the design and installation of some assets that are considered as critical to safe or reliable operation of the facility. For example:

- A building that is within an identified potential vapor cloud explosion blast zone (Reference 1-5) and is intended to be blast resistant will require more rigor in its specification, design and construction than a general-purpose maintenance warehouse outside of the blast zone.

- A piping system for transferring a liquefied flammable gas that has the potential for cold embrittlement of common piping materials will require more rigor in its specification, design and installation than standard cooling water piping.

Nevertheless, even the maintenance warehouse and cooling water piping will perform important functions, so their proper design and construction is still warranted.

From a programmatic point of view, the more difficult question pertains to the second major AIM task; namely, maintaining the ongoing integrity and reliability of site processes, including production, utility and support system assets, over a lifetime of facility operation. The question for this task can be rephrased as "What facility assets do we need to maintain, and to what degree of rigor?" A closely related question is "With limited resources, and sometimes needing to deal with unforeseen circumstances, we may not be able to always keep up with every scheduled maintenance task. How do we manage this situation?" These questions have been answered, implicitly or explicitly, in many ways:

Breakdown Maintenance. By this approach, no or minimal inspections, testing and maintenance are performed on facility assets. Repairs are made to equipment only when failure of the equipment is evidenced by a release of material or energy or a failure of the equipment to allow process operations to continue. Although breakdown maintenance requires the least effort in planning, performing, and documenting ITPM activities, it has numerous drawbacks from the perspectives of safety, reliability, maintenance planning and compliance.

Compliance Maintenance. This approach performs only those Inspection, Testing, and Preventive Maintenance (ITPM) tasks required by applicable codes and regulations (and perhaps specific company requirements), at minimal frequencies, with all other maintenance being performed on a breakdown basis.

Risk-Based Maintenance. Analysis tools such as risk-based inspection (RBI) and reliability-centered maintenance (RCM) are used to prioritize maintenance activities and establish ITPM frequencies in such a way as to meet risk-based safety and/or reliability goals. While more planning effort is required than breakdown or compliance maintenance, both on initial and ongoing bases, risk-based approaches can incorporate operating history into the planning process and, in some cases, adjust ITPM plans where data are available to support these adjustments and thereby optimize use of resources. However, risk-based tools are more often used to supplement other approaches rather than to be the primary means of determining what assets are to be maintained.

Prioritized Maintenance. Taking into account the realities of resource limitations and unplanned repairs, this approach identifies some assets as *safety-critical equipment* (SCE) or *critical equipment* and performs no-excuse, no-exception planned maintenance on all such equipment, while conducting necessary ITPMs on all other assets as appropriate but with allowances for task or schedule slippage. The concept of SCE and the selection of assets to receive prioritized maintenance are fully discussed in Chapter 5.

Comprehensive Maintenance of All Assets. This approach seeks to maintain every facility asset according to its pertinent manufacturer's or supplier's recommended tasks and frequencies, as well as by all applicable RAGAGEP and regulatory requirements. While comprehensive maintenance is ideal, it is not often successfully implemented in practice. Some supplier-recommended tasks and frequencies tend to be conservative or even excessive, and company resources rarely are sufficient to plan and perform all specified tasks at the required frequencies, especially when unforeseen circumstances arise.

Some combination of these approaches is often employed. For example, a facility using a risk-based approach is not likely to totally ignore compliance requirements, even when meeting those requirements is not indicated as being necessary to meet risk-based goals. Another facility might perform comprehensive maintenance on some assets and relegate all others to breakdown maintenance.

1.4   AIM LIFE CYCLE

Integrity needs to be built into a facility's assets and planned operation before the facility is started up. This asset integrity is then maintained over time by proper facility operation; by ITPM activities; by managing tasks; and by learning from experience. When failures do occur or deficiencies are detected, integrity is re-established by proper repairs or replacements.

Thus, managing asset integrity begins well before a new process/facility is started up and extends throughout the facility lifetime until final decommissioning. It recognizes that the functional integrity of equipment can degrade over time by corrosion, erosion, fatigue and various other mechanisms (i.e., "aging plants"; see Reference 1-6), so these mechanisms need to be understood, detected, and corrected before containment, control, or the ability to respond to an abnormal or emergency situation is lost. The primary AIM life cycle activities can be summarized as shown in Figure 1-1. These life cycle activities are detailed in Chapter 3.

1.5   RELATIONSHIP TO OTHER PROGRAMS

A practical AIM program will fit within a facility's existing process safety and risk management program, as well as other initiatives such as for reliability and quality improvement. Personnel charged with developing and administering the AIM program can optimize the process by taking advantage of existing programs and by knowing which people and groups of people are responsible for related activities. Table 1-1 illustrates potential interfaces with other facility programs.

AIM starts with a well-designed facility, with clear expectations with respect to expected facility performance. CCPS provides input to design practices and considerations from a process safety perspective in *Guidelines for Engineering Design for Process Safety, Second Edition* (Reference 1-7). Other activities essential to managing asset integrity include the proper procurement, fabrication, construction, and installation of assets, as discussed in Chapter 3.

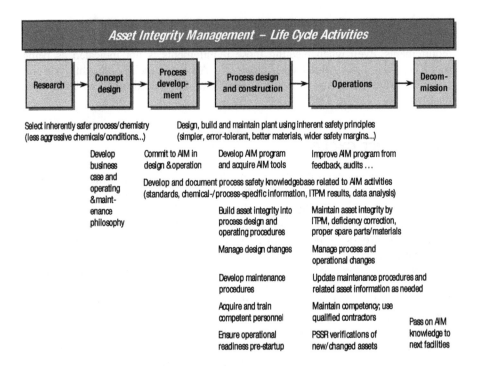

Figure 1-1. Summary of AIM activities throughout a facility life cycle

AIM is also an integral part of an ongoing, risk-based process safety management program. It is part of the "Managing Risks" pillar of CCPS' risk-based process safety model (Reference 1-8). Significant relationships with other process safety elements are shown in Table 1-2. As can be seen from this table, nearly all parts of a risk-based process safety program have some bearing on managing asset integrity.

1.6   RELATIONSHIP TO RAGAGEP

Codes, standards, and practices, which are sometimes termed "recognized and generally accepted good engineering practice" (RAGAGEP), are an important resource for an AIM program. RAGAGEP stems from the selection and application of appropriate engineering, operating, and maintenance knowledge when designing, operating, and maintaining process facilities with the purpose of ensuring safety and preventing process safety incidents.

RAGAGEP involves the application of engineering, operating or maintenance activities derived from engineering knowledge and industry experience based upon the evaluation and analyses of appropriate internal and external standards,

TABLE 1-1. Example AIM Interfaces with Other Facility Programs

Program	Example AIM Interface
Equipment Reliability	• An AIM program is the foundation of a plant's reliability program • Reliability program activities (e.g., vibration monitoring, equipment quality control) contribute to AIM
Occupational Safety; Safe Work Practices (including Hot Work)	• Occupational safety programs help ensure the safe performance of AIM activities • Occupational safety personnel may help maintain the integrity of emergency response equipment
Environmental Control	• Environmental initiatives (e.g., monitoring for fugitive emissions, investigating chemical releases) contribute to AIM
Workforce Involvement	• Employees from various departments have input into the AIM program
Project Management	• Design codes and standards influence AIM activities such as equipment design, inspection and repair • AIM quality management (QM) activities help document that equipment is appropriate for its intended use, including inspection and testing activities during fabrication, construction and commissioning as well as baseline inspections • Process safety information and asset data need to be handed off to facility operator
Purchasing	• Purchasing is involved in the acquisition of new or used equipment and spare parts
Hazard Identification and Risk Analysis	• HIRAs, also known as process hazard analyses (PHAs), can help define the equipment scope for the AIM program and prioritize AIM activities • AIM history can help HIRA teams determine the adequacy of safeguards
Operating Procedures	• Operating procedures may cover AIM-related activities, such as equipment surveillance as part of operator rounds, reporting operating anomalies, recording historical equipment operating data and preparing equipment for maintenance
Training and Performance Assurance	• AIM training in an overview of the process and its hazards can be consistent with the content of the operator training program
Contractor Management	• Inspection and maintenance tasks under the AIM program may dictate skills required of contractors • Because contractors often perform AIM activities, the contractor selection process considers both contractor safety performance and the contractor's work quality
Operational Readiness	• The AIM QA practice to ensure that equipment is fabricated and installed according to design may be fully or partially addressed during a pre-startup safety review
Management of Change (MOC)	• MOC applies to AIM activities and documents (e.g., new or modified equipment, changes to task frequencies and procedures) • The MOC program ensures that AIM issues (e.g., corrosion rates and mechanisms) are considered when evaluating process changes • AIM activities may help establish or dictate a change to safe upper and lower limits • MOC review teams can include process and AIM personnel • The MOC program may be upgraded to help manage equipment deficiencies • Practices for replacing equipment "in kind" are reviewed to ensure that AIM records are not compromised (e.g., inspection records, schedules are updated)
Incident Investigation	• AIM records may be needed by investigation teams • Investigation recommendations may impact AIM activities
Emergency Management	• Emergency response equipment needs to be inspected and maintained
Auditing	• The AIM program will be audited; audit results can help improve the AIM program
Trade Secrets	• Trade secrets needed for AIM activities cannot be withheld

TABLE 1-2. Relationships between AIM Activities and Risk-Based Process Safety Elements

RBPS Pillar	RBPS Element	AIM Activities Related to RBPS Element
Commit to Process Safety	Process Safety Culture	All AIM activities
	Compliance with Standards	Use standards and RAGAGEP
	Process Safety Competency	Involve competent personnel Ensure required inspector and technician certifications
	Workforce Involvement	Develop procedures for critical maintenance activities
	Stakeholder Outreach	
Understand Hazards and Risk	Process Knowledge Management	Identify assets to be included in AIM program Develop ITPM plan Incorporate new knowledge
	Hazard Identification and Risk Analysis	Identify assets to be included in AIM program Develop ITPM plan
Manage Risk	Operating Procedures	Develop procedures for critical maintenance activities
	Safe Work Practices	Develop procedures for critical maintenance activities Plan and perform ITPMs and repairs
	ASSET INTEGRITY AND RELIABILITY	All AIM activities
	Contractor Management	Train employees and contractors Ensure required inspector and technician certifications Audit contractor work on assets
	Training and Performance Assurance	Train employees and contractors
	Management of Change	Update ITPM plan when equipment conditions change Adjust ITPM frequencies, test methods, training and procedures if warranted Keep asset information up to date
	Operational Readiness	Confirm assets as installed meet design specifications Conduct baseline tests and inspections
	Conduct of Operations	Plan and perform ITPMs and repairs Promptly address conditions that can lead to failure Review inspection/test records
	Emergency Management	Plan and perform ITPMs and repairs
Learn from Experience	Incident Investigation	Investigate chronic failures
	Measurement & Metrics	Collect, analyze and archive data
	Auditing	Develop written AIM program
	Management Review and Continuous Improvement	Investigate chronic failures

applicable codes, technical reports, guidance, or recommended practices or documents of a similar nature. RAGAGEP can be derived from singular or multiple sources and will vary based upon individual facility processes, materials, service, and other engineering considerations.

The term *RAGAGEP* comes from U.S. process safety regulations that require its documentation and usage:

- "The employer (owner or operator) shall document that equipment complies with recognized and generally accepted good engineering practices" [OSHA 29 CFR 1910.119(d)(3)(ii) and EPA 40 CFR 68.48(b) and 68.65(d)(2)].

- "Inspection and testing procedures shall follow recognized and generally accepted good engineering practices" [OSHA 29 CFR 1910.119(j)(4)(ii) and EPA 40 CFR 68.56(d) and 68.73(d)(2)].

RAGAGEP gives generally approved ways to perform a specific engineering, inspection or maintenance activity. It may address:

- Equipment design and construction, such as specifying a piping system or fabricating a pressure vessel.

- In-service activities, such as inspecting a storage tank or servicing a relief valve.

- An established work process, such as risk-based inspection (RBI), reliability-centered maintenance (RCM) or specifying and implementing safety instrumented systems (SIS).

RAGAGEP incorporates broad industry experience and technical input and represents the consensus of a relevant organization or technical community. Therefore, it provides a valuable starting point for an AIM program.

In some cases, a country, state, or locality may mandate the use of RAGAGEP. For example, an authority may adopt an NFPA code for its jurisdiction. In addition, many companies internalize standards, often based on RAGAGEP, which are provided by the manufacturer or licensor of a process. Some companies have developed their own internal standards based on company and industry operating experience.

Broad industry experience is not always available for new technologies or for unique or highly specialized processes. In situations where RAGAGEP does not exist, the design of physical facilities or work processes needs to use the best available technology that is relevant to the situation, then take extra care in the design process since the depth of experience may be lacking. This extra care may include additional hazard identification and risk analysis efforts to ensure adequate preventive and mitigative layers of protection are in place to deal with abnormal situations that may develop. Extra inspections, testing and detection methods may also be warranted, during both commissioning and facility operation, until actual operating experience is gained.

To effectively use RAGAGEP, facility management needs to determine which practices are available and then assess the applicability of each practice to its facility. Regardless of the consensus reached to publish RAGAGEP, most standards were not written for a facility's specific equipment, specific chemical application, specific locale or specific operations culture. Some facilities with successful AIM programs are establishing their own data records to help determine (or to validate) the ongoing applicability and use of each standard.

Several chapters of this book address the applicability and use of RAGAGEP in more detail. Descriptions of these practices and approaches are included, such as for determining an inspection interval or technique, but the actual RAGAGEP is not repeated in this book. New and revised codes, standards and recommended practices continue to evolve; therefore, it is advisable for companies to have management systems in place to keep up with the new standards and with changes to existing standards. This is further discussed in Section 2.2.5.

1.7   STRUCTURE OF THIS DOCUMENT

These guidelines begin with four chapters that help set the groundwork for an AIM program. Chapter 2 discusses roles and responsibilities for management and other company personnel and examines the ongoing activities that management undertakes to help ensure AIM program success. Chapter 3 gives an overview of AIM activities from a life cycle perspective. Chapter 4 summarizes asset damage and degradation mechanisms to be understood, evaluated, detected and managed in an AIM program. Chapter 5 reviews considerations a facility may have when defining the equipment to include in its program and prioritizing safety-critical equipment.

Once a basic understanding is gained and the goals, objectives and scope of the AIM program is determined, facility management needs to develop and implement systematic activities related to AIM. These include:

- Inspection, testing, and preventive maintenance
- Training of all affected personnel
- AIM procedures development
- Quality management
- Equipment deficiency resolution
- Auditing of the AIM program.

Each of these activities is addressed in turn in Chapters 6 through 11 and Section 14.2. As illustrated in Table 1-3, Chapters 6 through 11 describe management systems for addressing assets as they are originally designed and installed, as they are maintained over time, and as they are repaired or replaced.

Specific details for these activities depend on facility culture, regulatory obligations and company priorities. Therefore, relatively little prescriptive information is included in this book. Rather, this document presents approaches that have worked in different industries and in facilities of various sizes.

Chapter 6 discusses inspection, testing and preventive maintenance (ITPM). In this document, "preventive maintenance" refers to those activities that are not inspections or tests and that are performed to prevent the failure of equipment within the AIM program. Lubrication of rotating equipment is one example of a PM task meeting this definition. Established approaches for developing test and inspection plans, such as risk-based inspection, are further detailed in Chapter 7.

Chapter 8 covers activities to ensure personnel competency, with the focus on AIM-related training. Chapter 9 addresses the procedures needed for AIM. Quality management activities involving initial design and fabrication as well as ongoing repairs and alterations are discussed in Chapter 10. Chapter 11 covers equipment deficiency recognition and resolution.

TABLE 1-3. Chapters Addressing Management Systems for AIM Activities

Attributes	New Equipment	Inspection and Testing	Preventive Maintenance	Deficiency Correction
Task Definition, Purpose and Documentation Requirements	Chapter 10 (Quality Management)	Chapter 6 (ITPM)	Chapter 6 (ITPM)	Chapter 11 (Deficiency Management)
Acceptance Criteria	Chapter 10 (Quality Management)	Chapter 6 (ITPM) and Chapter 11 (Deficiency Management)	Not applicable	Chapter 10 (Quality Management)
Technical Basis	Chapter 10 (Quality Management)	Chapter 6 (ITPM)	Chapter 6 (ITPM)	Chapter 10 (Quality Management)
Competency	Chapter 8 (Asset Integrity Training)			
Procedures	Chapter 9 (Asset Integrity Procedures)			
Continuous Improvement	Section 14.2 (Program Audits)			

To a large extent, the implementation of AIM activities depends on the specific types of assets involved. Because many codes, standards and other guidelines are written for specific types of equipment, this document contains a section dedicated to the specific approaches applicable to different equipment categories. Chapter 12 is dedicated to the equipment-specific aspects for the management systems covered in Chapters 6 through 11. Activity tables in Chapter 12 and in the electronic resources accompanying this document are presented in a format similar to Table 1-3. Many available codes, standards and practices are listed in this section. Key aspects derived from them, such as time intervals between inspections, are also listed, but the reader is encouraged to consult the referenced documents for more detailed information.

Chapter 13 reviews issues commonly encountered while implementing an AIM program, including budgeting and resources. Frequently, these resources include a computerized maintenance management system (CMMS). Many commercial CMMS packages are available, although some in-house programs are also effective. Chapter 13 includes basic information typically included in any CMMS that is installed as part of an effective AIM program.

The remaining two chapters contain supplemental information related to AIM programs. Chapter 14 discusses performance metrics that apply to AIM program activities, then offers suggestions for continual assessment and improvement of an AIM program. Continuous improvement is needed to ensure that effective AIM programs continue to operate at a high level. Some improvement can be attained simply by asking the right questions (i.e., auditing) and following up to address any identified weaknesses. Performing improvement activities on a regular basis can be expected to result in continuous improvement.

This document closes by providing an overview of other asset management tools that can be used to help make decisions related to AIM activities. Because of the extensive resource requirements for most AIM programs, risk-based decision making can be effectively employed to prioritize resource allocation. Various texts have been written on applicable tools for making these decisions. Chapter 15 includes an overview of many of the tools and references available in these resources.

CHAPTER 1 REFERENCES

1-1   EU Directive 2012/18/EU of 4 July 2012 on the Control of Major-Accident Hazards Involving Dangerous Substances ("Seveso III"), European Parliament and EU Council, 2012.

1-2   HM Government, The Offshore Installations (Safety Representatives and Safety Committees) Regulations 1989, Statutory Instruments, 1989 No. 971, Health and Safety, UK, 1989.

1-3   *Process Safety Management of Highly Hazardous Chemicals*, 29 CFR 1910.119, U.S. Occupational Safety and Health Administration, Washington, DC, 1992.

1-4   *Accidental Release Prevention Requirements: Risk Management Programs*, 49 CFR Part 68 U.S. Environmental Protection Agency, Washington, DC, 1996.

1-5   Center for Chemical Process Safety, *Guidelines for Evaluating Process Plant Buildings for External Explosions, Fires, and Toxic Releases, Second Edition*, American Institute of Chemical Engineers, New York, NY, 2012.

1-6   U.K. Health and Safety Executive, "Managing Ageing Plant: A Summary Guide," Research Report RR823, HSE Books, www.hse.gov.uk, October 2010.

1-7   Center for Chemical Process Safety, *Guidelines for Engineering Design for Process Safety, Second Edition*, American Institute of Chemical Engineers, New York, NY, 2012.

1-8   Center for Chemical Process Safety, *Guidelines for Risk Based Process Safety*, American Institute of Chemical Engineers, New York, NY, 2007.

2
MANAGEMENT RESPONSIBILITY

Many people within a facility's maintenance, operations and engineering organizations will likely be involved with the facility's asset integrity management (AIM) program. An individual's involvement may range from brief encounters to career-length stewardship responsibilities, and the involvement may occur during any or all phases of the assets' life cycle. In successful AIM programs, supervisors and managers emphasize how each person contributes to preventing incidents and improving process reliability. Such an approach is evident when personnel are working within the facility's risk management system and using effective knowledge, skills, resources and procedures associated with the AIM program.

This chapter discusses ways in which supervisors and managers can contribute to the ultimate success of the AIM program through communication and effective application of knowledge, skills and resources. For the purposes of this chapter, supervisors and managers include all personnel with supervisory and/or management responsibilities within engineering, maintenance, operations and related departments, as well as those charged with overall facility leadership.

2.1 LEADERSHIP ROLES AND RESPONSIBILITIES

One of the best ways that the leadership of a facility can help prevent incidents is to provide visible and active involvement in the facility's hazard management system. Key responsibilities of managers and supervisors in the AIM program are:

1. Establishing the direction and scope of the AIM program at the corporate level, beginning with the cultivation of a corporate culture in which managing asset integrity is an essential operating task for the facility.

2. Ensuring that knowledgeable people with the proper training and certifications are performing appropriate activities using effective engineering and decision-making tools and methods.

3. Instilling the expectation that the business plan will be fulfilled only within the safe operating limits of the equipment as dictated by its condition.

4. Ensuring that AIM program activities (e.g., inspections and tests) are being executed and managed on schedule and as planned, inspection/test results are being captured and analyzed, and corrective actions are being appropriately completed.

5. Ensuring that appropriate controls are implemented and maintained within the facility's hazard management system for all related AIM activities.

6. Providing the necessary resources to accomplish all of the above, including the use of third parties where needed.

The primary control mechanisms for these actions are:

- Establishing clear organizational roles, responsibilities and accountabilities for AIM activities, including independence of the testing and inspection organizations, to ensure that site management is receiving current and accurate information on the status of the site AIM program.

- Creating reporting mechanisms for condition of assets, AIM program status, asset failures and integrity-related incidents.

- Ensuring that effective audits of the AIM program and the overall hazard management system are conducted, with the audit results being reviewed by corporate leadership, and with identified gaps being resolved in a deliberate and risk-informed manner while having appropriate interim risk-control measures in place.

Each of these three control mechanisms is described further in the following sections. Similar responsibilities and control mechanisms apply to the leadership of contractor companies, such as those involved in engineering design, construction, inspections, and repairs, with the added necessity of effective interfaces and communications between the operating company and the contractor companies.

2.1.1   Organizational Roles and Responsibilities

AIM is best directed and controlled at the corporate level to ensure consistent implementation and to help establish a positive process safety culture, whereas execution is the operating facility's responsibility. A good practice is to establish an AIM corporate center of excellence, tasked to establish corporate AIM standards and drive efforts to continuously improve the safety and reliability of facility assets.

Process Safety Culture. The first tenet of CCPS' Vision 2020, demonstrating what perfect process safety will look like when it is championed by industry, is a committed culture in which the executives are personally involved, managers drive excellent execution every day, and all employees maintain a sense of vigilance and vulnerability. To create a committed culture, leadership needs to tangibly demonstrate a commitment to process safety from the senior executive team through its line management, so that all employees embrace it and recognize that "it could happen here." Process safety culture is also the first CCPS element of risk-based process safety (Reference 2-1).

Facility Responsibility. The operating facility's responsibility is to ensure that AIM activities are an integral part of the day-to-day operations involving operators, maintenance personnel, inspectors, contractors, engineers, and others involved in designing, specifying, installing, operating and maintaining facility assets. Due to limited resources or other considerations, it may be necessary or even desirable to have multi-skilled maintenance personnel and/or have properly trained operators perform some routine inspections, tests and preventive maintenance tasks such as visual inspections and pump lubrication.

Knowledgeable Personnel. An essential role for a facility's management is to ensure that knowledgeable people are available and assigned to provide the expertise for implementing the AIM program. For example, technical personnel at the facility or available to the facility can direct personnel from various departments to the correct codes, standards and practices for a given application throughout the life cycle of the facility's assets. The life cycle includes process equipment design/engineering, fabrication, procurement, receipt, storage and retrieval, construction and installation, commissioning, operation, inspection, condition monitoring (CM), functional testing, maintenance, repair, modification and alteration, decommissioning, removal and re-use.

Outsourced AIM Activities. Some of the technical aspects of the AIM program can be outsourced if a facility has limited internal expertise. However, it is important that site management maintains ownership of the AIM program. Ownership means that it is the responsibility of facility management to ensure the AIM program is properly designed and maintained. This can be done with proper auditing and management review.

Response to Recommendations. Management personnel have the responsibility to demonstrate further commitment to the AIM program by supporting the recommendations made by technical personnel and by inspectors and auditors. Facility leadership can provide the guidance and direction to ensure that the technical and inspection roles are accepted by everyone within the facility. For example, as custodians of the equipment's condition-based safe operating limits, the technical staff will likely have significant input into decisions regarding whether equipment with known deficiencies will continue to be operated. The technical staff and the inspectors provide the input that enables facility management to manage the facility assets while maintaining safe operations.

Audits and Self-Assessments. Management review of the AIM program is essential to ensure that the program is working properly and producing the desired results. If it is not, management has the responsibility to identify deficiencies, develop corrective actions and ensure that the actions are implemented in a timely manner. Management can combine metrics, incident reports, personal observations, direct questioning, audits results and feedback on various topics to make their assessment.

Independence of the Inspection Function. Persons charged with performing inspections and tests on safety-critical equipment are often part of a more or less formalized inspection authority and/or inspection department. It is good practice to provide some independence to the inspection function such that it is not fully embedded in the maintenance department. For example, the leader of the inspection group may report directly to the facility manager and may be given authority to shut down a unit if an incipient catastrophic failure is detected or a safety-critical system fails inspection.

Importance of the Planning Function. An important but often overlooked role is the maintenance planner. Most large facilities could not function efficiently without this role. Computerized maintenance management systems can facilitate the planning function; however, human judgment is still needed to deal with abnormal situations and resource conflicts.

2.1.2   Roles and Responsibilities Matrix

The roles and responsibilities for program management and implementation can be assigned to personnel in various departments. A comprehensive program document effectively communicates to facility personnel the management systems and roles associated with the AIM program. Many facilities find it a practical necessity to create and maintain a more detailed written AIM procedure in order to sustain all AIM activities. The written program can document roles and responsibilities for various levels of involvement for different aspects of the AIM program. Some people will be directly responsible and accountable for an activity, others may participate in the development or implementation of the activity, and others may need only to be aware of the activity or its results.

A convenient method for displaying these varying roles and responsibilities is in a matrix format. Such a matrix correlates different activities with different job positions by indicating the job position(s) responsible for each activity, as well as the level of participation required of other personnel.

An example matrix showing roles and responsibilities for managing an asset integrity (AI) program is provided in Table 2-1. This matrix lists program activities in the left column and typical job positions by department in the top row. Cells containing letter designators indicate the job position(s) responsible for the activity and the levels of participation for other job positions. This matrix uses the following letter designators:

R   indicates the job position(s) with primary responsibility for the activity

A   indicates the job position(s) to approve the work or decisions involved

S   indicates the job positions that typically support the responsible person(s) and may participate in the performance of the activity

I   indicates the job positions that are likely to be informed of the activity results, may be asked to provide information, or may have minor participation in the activity.

This matrix format will be used to present more specific roles and responsibilities in subsequent chapters of this document. Note that a RASI chart or matrix of this nature can on occasion be expanded to a RASCI chart, with the C standing for consulted.

2.1.3   Reporting and Communication Mechanisms

Reporting mechanisms are necessary to provide appropriate and timely information concerning past performance as well as to alert personnel at various levels within and external to the company that AIM activities are necessary to ensure continued asset integrity.

External Communications.   Communication of AIM-related information to external stakeholders including regulatory authorities can take the form of public progress reports or posted trend information, or be part of a larger document such as a company annual report. Reference 2-2 is an example of an annual public report of safety performance for one industry that includes three AIM-related key performance indicators (KPIs): major releases, findings of noncompliance for verified safety-critical parts, and safety-critical maintenance backlog.

TABLE 2-1. Example AIM Roles and Responsibilities Matrix

R = Primary Responsibility A = Approval S = Support I = Informed — Activity	Facility Leadership / Site Manager	Engineering Manager / Lead (Technical Assurance Role)	Reliability Manager	Maintenance Manager	Maintenance Supervisors	Inspection Manager	Area Superintendent / Unit Manager	Production Supervisors	Process Engineering Manager	EHS Manager	PSM Coordinator
Coordinate AIM program; overall responsibility	A	S	S	R	S	S	S	I	I	I	S
Establish organizational roles	R	S	S	S	I	S	S	I	S	I	I
Develop procedures		R	A	A	S					S	I
Define / maintain RAGAGEP		R	S	A	I	S				I	I
Establish safe design limits	I	A	S				S	S	R	I	I
Train / develop personnel	A	R		R	S					S	I
Establish QA requirements		R	S	A	I		I	I	I		
Provide technical content		R	S	S	I		S	I	S		
Oversee contractors in AIM roles	A			R	S		S				S
Maintain equipment records		R	S	A	S	S					
Manage equipment deficiencies	A	R	S	S	I	S	R	S	S	S	I
Develop metrics	I	R	S	A	S	S	S	S	S	I	I
Report / review metrics	A	R	S	R	I	S	S	I	S	I	I
Establish audit schedule, content	A	S	I	S	I	S	I	I	I	R	R

Note 1: Activities for which responsibility is assigned to more than one job position indicate shared responsibility or responsibility for a specific scope within the activity.

Note 2: At smaller facilities, a single person or position may have more than one "Job Position" function.

Reporting and Communication to Senior Company Management. KPIs and other AIM metrics can also be used effectively to communicate the AIM program performance to company leadership, as a means of determining whether the management system functions are working as intended. Management has the responsibility to both establish and periodically review AIM metrics. Metrics can also be used by the inspection and maintenance organizations to manage the AIM execution tasks. AIM metrics might include (not an exhaustive list):

- A display chart showing which facilities or units have fully implemented specific programs or practices

- The number of asset items included in the AIM program

- The number of asset items with appropriate inspection, testing and preventive maintenance (ITPM) procedures implemented

- The number of inspectors/maintenance employees holding each type of required certification that has expired

- The number (or percent) of overdue ITPM tasks

- The number of deferred repairs, such as known deficiencies that are to be addressed at the next turnaround.

AIM metrics are also discussed in Section 2.2.3 and Chapter 14 of this document. Other ideas for measuring the effectiveness of an AIM program can be found in References 2-3 and 2-4.

Communications to Facility Personnel. Examples of information of interest to personnel working in the facility would include answers to such questions as:

- How well are assets performing their intended functions?

- Will the asset continue to be fit for service tomorrow, next week, next year?

- Are inspections overdue?

- How much scheduled maintenance is on the backlog?

- Are there unresolved safety-critical-equipment deficiencies?

- What information are the measures of asset condition providing?

- Are AIM-related incidents and near misses consistently reported?

- Are qualified people available to perform all required AIM tasks?

- Are engineered safeguards effective?

- Are AIM procedures and standards consistently followed?

- What information are AIM metrics providing?
- With what confidence can AIM data be used?

Management at facilities with effective AIM programs ensures that (1) these and similar questions are being asked periodically and (2) the answers are complete and accurate. Although a facility may have experienced years of incident-free operation, complacency toward procedure adherence and less-than-thorough auditing of the hazard management system, which may have allowed the development of poor maintenance and operating practices, can result in undesirable answers to the above questions.

Management needs to be routinely informed of the answers to questions such as these to help ensure asset integrity. Because every operation and its associated hazards is unique, facility management needs to work with operations personnel to create appropriate questions for each process.

2.1.4   Managing Changes, Abnormal Situations and Incidents

Management can lead an organization in managing change, managing abnormal situations and learning from experience, both on a day-to-day basis and as a result of significant AIM-related incidents. Keeping informed of and ensuring a proper, technically informed response to changes, as well as to abnormal situations such as going outside an established operating envelope due to a process-related upset, can prevent changes and deviations from affecting asset integrity by mechanisms such as thermal fatigue or accelerated corrosion, or from compromising the integrity of the system as originally designed. Identifying and managing both temporary and permanent changes is critical, due to the large number of changes that are made at a typical facility and the many ways in which changes can affect asset integrity. Also important is a process of sharing lessons learned from managing abnormal situations and near misses, both for site personnel and also across the organization.

Management needs to be involved with and informed of the results of AIM-related incident investigations associated with both near misses and actual loss events (References 2-5, 2-6, 2-7). Root cause analysis generally points to management system failures that require management attention, such as refocusing a training program or providing more resources to complete required AIM tasks at their established frequencies.

2.1.5   Auditing

Auditing is another important aspect of any management system. Formal and informal auditing provides a means for facility management to understand the effectiveness of the AIM activities and to uncover weaknesses, gaps and noncompliance in the management system and its implementation.

Some level of AIM auditing activities normally occurs on a continual basis within a process facility. These audit activities may include routine walk-around discussions, periodic stewardship/status reports, regulatory-required audits, periodic technical audits and departmental effectiveness audits. Ongoing and continual auditing practices can help uncover equipment integrity problems or AIM program deficiencies before an incident occurs. Specific details on AIM program auditing are provided in Chapter 14.

2.1.6   Management Review and Continuous Improvement

The final element of the CCPS risk-based process safety model (Reference 2-1) is Management Review and Continuous Improvement. This applies to AIM programs as well as to process safety management. Management review involves the continuous, active oversight of the AIM program activities and their effectiveness in (1) building asset integrity into new and modified facilities, then (2) maintaining asset integrity during the facility's operating lifetime. Having the goal and mindset of AIM program continuous improvement requires leadership in seeking out and implementing better technologies and practices to perform AIM functions, as well as finding and fixing weaknesses in the existing program.

2.2   TECHNICAL ASSURANCE RESPONSIBILITIES

Technical management and supervisory personnel ensure that the AIM program activities are completed in accordance with the requirements of the hazard and risk management system and in a manner that meets the requirements of the specific tasks. Therefore, it is a management responsibility to ensure that qualified personnel are available to perform all tasks. In some cases, qualified personnel can be drawn from company central organizations. However, having staff members on site who are qualified to perform the necessary AIM tasks is often preferable. (Chapter 8 discusses training and qualifying personnel to perform AIM tasks.) The following sections discuss some specific technical assurance responsibilities that are provided in support of an AIM program.

2.2.1   Defining Acceptance Criteria

Management personnel are responsible for defining appropriate acceptance criteria for assets. These criteria include the acceptable integrity operating window (IOW) over the life of the asset in practical terms of operating limits (process variable or measured material limits) and run time before the next condition inspection, functional test, repair or replacement. The upper and lower limits of the IOW are based on both the asset's current condition or functionality assessment and its projected condition or functionality over time (Reference 2-8).   Figure 2-1 illustrates this operating window.

Each AIM activity needs to satisfy acceptance criteria and be consistent with company standards and good engineering practice.   Supervisory and technical management personnel facilitate these activities by employing technical assurance personnel to

1. Develop procedures and other AIM documents with technical content requirements

2. Determine metrics for equipment condition monitoring and AIM program performance and

3. Technically review the results of AIM activities.

In smaller facilities, technical personnel often have multiple responsibilities. Nevertheless, someone will need to be assigned the technical assurance role, even if that is not his or her sole responsibility.

2.2.2   Providing Technical Content

Technical content requirements to be provided include (1) details of how work is to be performed or checked, (2) requirements for materials of construction, (3) applicable codes and standards to be followed, (4) acceptance criteria for

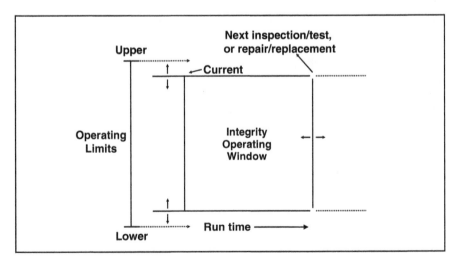

Figure 2-1. Definition of the integrity operating window (IOW)

inspections and tests, and (5) asset-specific inspection techniques. Many companies have a Chief Engineer or Engineering Authority that oversees these issues and approves any departures from codes and standards.

In addition, decisions to operate an asset with a deficiency can be based on data from technical assurance personnel. Such information may include the current and predicted asset condition, repair requirements, on-stream monitoring requirements and/or revised safe operating limits. See Chapter 11 for information on managing equipment deficiencies.

2.2.3   Establishing Metrics

Technical assurance personnel can identify relevant standards of measurement, or metrics, as indicators of asset integrity. A facility's metrics can include measures related to (1) AIM program implementation and schedule adherence, (2) asset condition trends, (3) adherence to procedures and (4) training status. Discipline-specific measures can provide the information needed to communicate the status of asset integrity activities and trends that affect asset integrity (e.g., increase in past-due inspections and tests). Typically, metrics are used to report the status of such AIM concerns as

- AIM program budget

- AIM activities backlog management

- Preventive maintenance (PM) schedule adherence

- Asset deficiency resolution

- Each asset type's inspection and testing schedule adherence

- Recommendation resolution and implementation, such as from inspection and testing activities

- Equipment incident investigation recommendation resolution and implementation

- Equipment operating near design limits or nearing the end of the remaining life

- AIM program audit findings and recommendation resolution and remediation

- AIM procedure use (e.g., Are the procedures current?)

- Craft and technician training and qualifications.

Developing and maintaining AIM metrics requires input from many stake-holders, as can be seen from the "Develop metrics" and "Report/review metrics" line items in Table 2.1 above. To have value, the metrics need to be meaningful for management review. Using metrics is further discussed in Chapter 14.

2.2.4   Ensuring Technical Review

Technical review of AIM activity results is a critical technical assurance role. The volume of data gathered and the potential for rapidly changing asset conditions necessitate timely review of the inspection, testing and preventive maintenance (ITPM) activities and asset repair results. Prompt technical review of such results

1.   Identifies equipment with deficient or near-deficient conditions

2.   Allows rapid communication to affected personnel regarding the unavailability of safeguards

3.   Improves timing and effectiveness of corrective actions and

4.   Helps prevent incorporating incorrect information into equipment history records.

Technical review is primarily the responsibility of facility or corporate technical personnel having background and experience to provide the needed technical assurance. Authorities having jurisdiction over the facility may also be required to perform technical reviews, such as for boiler or pressure vessel inspections.

2.2.5   Incorporating New Knowledge

Technical personnel are generally tasked with keeping up with new technical developments that are relevant to different parts of the AIM program, particularly in the area of technical content. This new knowledge may be obtained from many possible sources, such as new technical reports, new or updated information provided by an equipment supplier or technology licensor, updated codes and standards, new best practices, or industry benchmarking. The obtaining of new knowledge is often gained through technical and professional networks. However, a more systematic approach may be needed to ensure new safety-critical information is found in a timely manner.

Ensuring relevant new knowledge is incorporated into the AIM program is as important as identifying the new information in the first place. This might be successfully achieved on an informal basis; however, a more formal, documented approach may be warranted such as the updating of company best practices as controlled documents, keeping track of which versions of codes and standards are used, and disseminating new knowledge throughout the organization via periodic technical communications or regular meetings of internal technical groups.

CHAPTER 2 REFERENCES

2-1 Center for Chemical Process Safety, *Guidelines for Risk Based Process Safety*, American Institute of Chemical Engineers, New York, NY, 2007, Chapter 12.

2-2 Oil & Gas UK, health & safety reports, The UK Oil and Gas Industry Association Ltd., London, UK, e.g. Publication HS077, Oil & Gas UK Health & Safety Report 2013, http://www.oilandgasuk.co.uk/cmsfiles/modules/publications/pdfs/HS077.pdf.

2-3 Center for Chemical Process Safety, *Guidelines for Process Safety Metrics*, American Institute of Chemical Engineers, New York, NY, 2009, pages 144-145.

2-4 Center for Chemical Process Safety, *Guidelines for Risk Based Process Safety*, American Institute of Chemical Engineers, New York, NY, 2007, pages 359-360.

2-5 Center for Chemical Process Safety, *Guidelines for Investigating Chemical Process Incidents, Second Edition*, American Institute of Chemical Engineers, New York, NY, 2003.

2-6 NFPA 921, *Guide for Fire and Explosion Investigations*, National Fire Protection Association, Quincy, Massachusetts.

2-7 API RP 585, *Pressure Equipment Integrity Incident Investigation,* American Petroleum Institute, Washington, DC.

2-8 API RP 584, *Integrity Operating Windows,* American Petroleum Institute, Washington, DC.

3
AIM LIFE CYCLE

This chapter examines the various activities associated with managing asset integrity from the perspective of the life cycle of a facility (or process). The remaining chapters of this document give detailed guidance for implementing the asset integrity management (AIM) activities throughout the various life cycle stages.

3.1 OVERVIEW

Although the primary activities associated with managing asset integrity are during a facility's operating phase, decisions affecting AIM start at the earliest design stages and AIM does not end until the final decommissioning of facility assets. The activities related to AIM change as a facility's life cycle progresses. These activities can be seen as fulfilling four major objectives:

1. Define the requirements to be achieved by the assets.
2. Design and build integrity into new and modified assets.
3. Maintain the integrity of the assets throughout the facility lifetime.
4. Detect and correct deficiencies and failures that occur during operation.

The sections in this chapter that step through a facility life cycle from an AIM perspective are put on the Chapter 1 life cycle timeline in Figure 3-1 (see Figure 1-1). Key AIM activities at the first life cycle stages, as discussed in Section 3.2, are to consider inherent safety principles and to make strategic AIM program decisions such as expected reliability/availability goals and technical specifications. The primary focus when progressing from design through construction (Sections 3.3 and 3.4) is to design and build integrity into new facility assets. Operational readiness is established during the commissioning phase (Section 3.5).

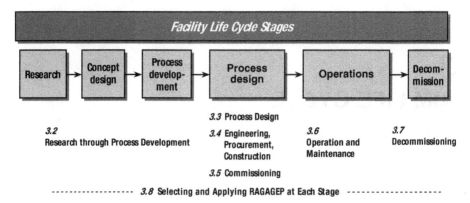

Figure 3-1. Life cycle stages cross-referenced to Chapter 3 sections

After initial startup, the AIM focus (Section 3.6) shifts to maintaining the integrity of the assets throughout the facility lifetime and to detecting and correcting deficiencies and failures. AIM activities associated with the end of facility life are discussed in Section 3.7. Since recognized and generally accepted good engineering practice (RAGAGEP) cuts across all life cycle stages, their selection and application are addressed separately in Section 3.8.

Life Cycle Perspective on Facility Costs and Risks. Much has been learned about evaluating projects and new facilities from a life cycle perspective. A good time to look at life cycle issues is when a new plant or process is first being considered. For example, opting for better controls, higher-grade materials of construction, more reliable rotating equipment, etc. may have a higher initial cost but result in fewer problems when the plant is in operation.

A broad example of this perspective has been experienced in the United States with respect to ethanol plants, the capacity of which has increased from 1.63 to 13.9 billion gallons per year over the years 2000 to 2013 (References 3-1, 3-2). The industry's main focus was getting plants built as fast as possible to meet an increasing demand for ethanol. This led to issues such as most plants not installing cleaning filtration systems (Reference 3-3) and using less-expensive materials of construction. As a result, the industry is now investing in retrofits and upgrades (Reference 3-3), which are generally more expensive than if the same improvements were made in the initial design and construction.

Building improvements into the initial design of a facility instead of waiting until problems show up can also reduce the likelihood of process safety incidents by decreasing the frequency of abnormal situations such as piping leaks and flow deviations, as well as by increasing the reliability of protection layers, and thus decreasing the likelihood of hazardous material releases, fires and/or explosions. Using inherently safer technologies and decreasing the frequency of initiating causes such as pumps stopping and utilities being lost can also decrease or eliminate the need for added-on layers of protection such as safety instrumented systems, which can be expensive to both install and maintain.

Process safety incident risks can be understood to be balance-sheet liabilities, similar to how the cost of raw materials and labor are balance-sheet liabilities, with the difference being in the likelihood of whether the liability will actually be realized within a given period of time (Reference 3-4). For example, if the likelihood of a major fire in a processing unit is a 10% chance per year of operation, with the average total cost of a fire being $1,000,000, then the liability associated with this fire risk is $100,000 per year on an annualized basis. Of course, risk estimates generally have a greater degree of uncertainty than cost estimates.

Other life-cycle issues associated with managing asset integrity include trade-offs between equipment such as certain types of pump seals that may be inexpensive to purchase and install but expensive and time-consuming to maintain, versus equipment that may cost more initially but have minimal maintenance requirements and a lower failure rate. Part of the evaluation of a new facility is making assumptions as to the percentage of time the plant will be in operation, versus being down for maintenance or repairs. A true life cycle analysis of facility costs will look not only at initial costs but also at process safety incident risks, loss-of-production risks and ongoing AIM costs to get a true comparison of alternatives.

3.2   RESEARCH THROUGH PROCESS DEVELOPMENT

Although managing asset integrity is centered on the operating phase of a facility's life cycle, decisions are made at the earliest life cycle stages that will have a profound effect on an AIM program. Opportunities are best taken at the research, development and design stages to choose options that will make the need less demanding for ongoing containment and control of hazardous materials and energies. As the developmental phases progress, AIM philosophies and then technical specifications that address integrity (such as materials of construction, code selection, etc.) need to be established before detailed engineering can progress.

Inherently Safer Design. Inherent safety reviews, as discussed in References 3-5 and 3-6, can be performed early in a facility's life cycle as well as at later stages. Their value at the early research and process development stages cannot be overemphasized. The primary decisions related to inherent safety at these early stages are associated with selection of process materials and chemistry.

The following examples illustrate the AIM impacts of choosing inherently safer options when fundamental decisions are being made regarding the future nature of a process facility:

- Selecting a process chemistry route that has little or no potential for a runaway chemical reaction will reduce the need for multiple layers of protection to keep the reaction under control, to respond when control is lost and to safely relieve a runaway situation.

- Selecting a catalytic reaction path may allow a process to operate closer to ambient conditions, putting less demand on maintaining an elevated pressure/temperature process facility.

- Choosing less corrosive process materials will reduce corrosion potential and the need for more corrosion-resistant materials of construction and/or more rigorous corrosion monitoring.

- Selecting non-chloride-containing process materials may reduce or eliminate certain damage mechanisms such as chloride stress corrosion cracking.

- Opting for materials with a lower toxicity hazard may reduce requirements for lethal service high-integrity containment systems.

- Using noncombustible process materials or liquids with a higher flash point may reduce the need for certain fire protection systems and features such as deluge systems that would need to be inspected and maintained on an ongoing basis.

Of course, other benefits such as lower safety management costs and reduced potential for community impacts can also be realized by selecting inherently safer options.

Establishment of AIM Program Requirements. The best time to define requirements to be achieved by the new or updated facility's assets is before the start of the process design. In the international standard ISO 55000 (Reference 3-7) on asset management, this is termed the *strategic asset management plan.* It is the documented information that specifies how organizational objectives are to be converted into asset management objectives, the approach for developing asset management plans, and the role of the asset management system in supporting achievement of the asset management objectives. For example, one organizational objective for a new facility may be to achieve a given production rate of final product. An AIM objective necessary to achieve this organizational objective might be an overall availability of assets as a system for the given production rate to be achieved. This and other objectives would express the up-front expectations that will need to be fulfilled.

Activities at this point include developing the organization's AIM philosophies and top-level program documentation, followed by technical specifications such as selection of applicable codes and standards and materials of construction. Once these AIM program requirements are established, the detailed engineering can proceed.

3.3   PROCESS DESIGN

Inherent safety reviews and consideration of inherent safety principles can continue during process design. The primary decisions related to inherent safety during process design are associated with reducing hazardous material inventories, simplifying the process equipment and designing to operate closer to ambient conditions. The following examples illustrate the impact of choosing inherently safer options at the process design stage:

- Reducing the number of required storage vessels of hazardous raw materials, intermediates and/or finished products reduces the number of vessels and associated piping and instrumentation that needs to be inspected and maintained on an ongoing basis.

- Selecting a plug flow reactor rather than a continuous stirred tank reactor eliminates the need to maintain an agitator system and its associated seals and bearings.

- Choosing a pump design or a raw material supply pressure that is not capable of overpressurizing downstream process equipment reduces emergency relief inspection and maintenance requirements.

- Designing a piping system with all-welded construction eliminates the potential for leaking flange connections.

- Shortening flow paths reduces the length of piping systems needing to be inspected and maintained.

- Specifying a higher-grade material of construction (e.g., stainless steel instead of carbon steel) can significantly decrease corrosion potential.

- Using gravity flow or pressure transfer of liquids eliminates transfer pump maintenance requirements.

Note that many of these options relate to "second-order" inherent safety (Reference 3-5), in that options such as all-welded construction do not eliminate or reduce the underlying hazards (e.g., flammability and/or toxicity of the process material that is inside the primary containment system) but they do permanently and inseparably reduce the likelihood of process safety incidents. Other decisions related to second-order inherent safety are likely to be made during the process design stage, such as those involving the location and layout of the facility.

Reliability in Design. The process design can have a profound effect on the reliability of facility assets. The design stage is the primary opportunity a facility has to "build in" reliability to assets. After startup, during the operating phase, AIM is focused on preserving this fundamental designed-in reliability.

Building reliability into facility assets at the design stage can have additional benefits besides improved reliability once the facility is in operation. Along with following inherent safety principles, they both can reduce the need for added-on safety devices and features such as safety instrumented systems (Reference 3-8). From a risk-based perspective, reducing hazards can reduce potential loss event impacts (i.e., the severity side of the risk equation) and having more reliable assets can reduce the frequency of initiating causes that could lead to loss events (i.e., the likelihood side of the risk equation). If the combined likelihood and severity is sufficiently reduced by these means in the facility design, then fewer layers of protection will be required to meet risk goals.

Reliable designs start with competent and creative engineering. Where possible, designs employ features that have been tested and proven. As a result, designs benefit from lessons learned so that mistakes are not repeated. Many of these proven designs then become the foundation for codes and standards. Table 3-1 contains a few widely used design codes, but is by no means an exhaustive list. This topic is further discussed in Section 3.8 with respect to applying RAGAGEP throughout a facility's life cycle.

TABLE 3-1. Typical Design Code Applications

Application	Design Code or Standard
Boilers (power)	ASME Boiler and Pressure Vessel Code (BPVC), Section I, Power Boilers
Electrical Systems	National Fire Protection Association (NFPA) 70
Instrumentation	International Electrotechnical Commission (IEC) 61511 / ANSI/ISA-84.00.01 (IEC 61511 Mod) and others
Piping (process)	ASME B31.3
Piping (power)	ASME B31.1
Pressure Vessels	ASME BPVC, Section VIII, Pressure Vessels
Pumps	API 610, ASME B73.1, ASME B73.2, others
Storage Tanks	API 620, API 650, and Underwriters Laboratories Inc. (UL) 142

Building reliability into the design of a facility's assets is not likely to stop at meeting applicable codes and standards. Special equipment and new technologies will require more care in their design and engineering, since there may be insufficient industry experience for applicable standards to have been developed. Risk-based studies during the design stages may identify assets that could warrant overdesign (such as specifying a higher pressure rating than minimally required) or selection of superior equipment to meet projected reliability goals. Company experience may dictate the use of specific configurations to address prior equipment failures. An example of the latter might be a company choosing to use a special design or material of construction for certain compressor parts as a result of a previous root cause failure analysis. These same types of considerations can also apply to building process safety into the facility design, meeting risk goals and choosing to use more reliable equipment to avoid the need for extra layers of protection such as by safety instrumented systems.

Process Safety in Design. CCPS' *Guidelines for Engineering Design for Process Safety* (Reference 3-9) gives guidance on the many engineering design aspects directly related to process safety. These include, but are not limited to:

- Plant siting and layout
- Civil / structural / support design
- Process utilities and support systems
- Materials of construction and corrosion control
- Insulation
- Equipment-specific design issues (vessels, dryers, fired equipment, etc.)
- Piping systems and components
- Material handling and warehousing
- Design of the basic process control system
- Instrumented systems to respond to process deviations
- Emergency relief and depressuring systems design
- Effluent treatment and flare systems
- Deflagration / detonation arresters
- Ignition source control
- Fire protection systems
- Special mitigation systems (explosion suppression, water curtains, etc.)
- Emergency response equipment and alarm systems.

Of course, much overlap occurs when building both reliability and process safety into the design of a new facility.

Different facilities may have other process safety features, besides those listed above. For example, offshore oil facilities may have safety-related assets such as mooring systems, buoyancy and ballast control, dynamic positioning and blowout prevention.

Many of these design elements related to process safety will be identified and managed within the AIM program as safety-critical equipment. They may need to not only be relied upon to prevent or mitigate a major incident, but also to survive an initial fire or explosion and still perform its critical function. The selection of safety-critical equipment is detailed in Chapter 5.

Design Documentation. At the process design stage, AIM documentation builds on the strategic program objectives and AIM philosophy developed before beginning the process design (Section 3.2 above).

All facilities need accurate and complete design documentation for many purposes. It forms part of the knowledge base to be used for performing process hazard analyses, for developing standard operating and maintenance procedures, for upgrading facilities and managing change, and for supporting ongoing AIM activities such as generating baseline test data, performing preventive maintenance, and correcting deficiencies and failures. When the process design and engineering is being performed by an organization other than the operating company, the potential exists for some of this critical information not to be turned over at the end of the project. Detailed contractual requirements for design documentation as part of the project scope and the use of design turnover checklists are two ways of increasing the likelihood of all design documentation being turned over to the operating company.

Design documentation can begin with a reference to specific codes and standards applicable to various equipment types. Another source of specifications for many processes is the design manuals or other information in the original engineering and purchasing records. In creating its own specifications, a facility using the original manuals and/or codes and standards can consider supplementing that information with (1) lessons learned that are specific to that facility (e.g., those documented during failure analysis investigations) and (2) updated or new information that may have become available after the original construction.

Maintaining proper design specifications is important. Documenting the design specifications and developing the supporting drawings and data sheets (i.e., process safety information) are often required by regulations. Applying codes, standards and specifications correctly and having appropriate checks in place help produce a robust design. Companies use a variety of methods, including safety and design reviews (e.g., piping and instrumentation diagram [P&ID] reviews, relief system reviews and various hazard reviews) to help ensure the quality of the designs. CCPS' *Guidelines for Hazard Evaluation Procedures* (Reference 3-6) includes descriptions of many hazard review techniques. This reference also includes an extensive hazard evaluation checklist that companies can use to aid their design efforts. During the design stage, facility personnel can establish a quality assurance (QA) plan for evaluating the design at the different design phases, as well as a QA plan to be used during fabrication and construction (as described in Section 3.4). Appendix 3A provides suggestions that can be incorporated into a plan for performing reviews during different phases of design (Reference 3-10).

Not only does design documentation need to be developed, it also needs to be kept up to date and readily accessible. Updating design information may be necessitated by facility changes, by lessons learned from experience and failure analyses and by changes in the underlying codes and standards. These and other documentation issues are discussed in CCPS' *Guidelines for Process Safety Documentation* (Reference 3-11).

The challenge of keeping specifications current is compounded following company mergers and acquisitions. Merged companies can be hampered by the use of "legacy specifications," leaving engineers with multiple versions of similar specifications. A comparable problem occurs for companies that rely on vendor specifications when different vendors were used to build different process units. Companies can assign a person or team to address these issues and to be responsible for producing and maintaining a definitive set of specifications.

3.4   ENGINEERING, PROCUREMENT AND CONSTRUCTION

The primary AIM focus during the engineering, procurement and construction life cycle stages is to ensure the process and its associated assets are properly designed to ensure a safe and reliable operating facility, then constructed in a manner that is consistent with the design specifications and readies the facility for managing asset integrity when started up and operating. The intended result is that the final as-built plant, including instrumentation, controls and supporting facilities, fully meets appropriate design specifications and has the asset information documented in such a way as it can be effectively used for ongoing AIM.

The engineering and design of a new or modified facility has a significant impact on the integrity of the operating facility. Reference 3-9, as described in the preceding section, addresses process safety in the engineering phase as well as the initial process design.

AIM activities during the construction phase are primarily quality assurance responsibilities involving vendor selection, procurement, fabrication, construction, and installation, with inspections along the way that include factory acceptance tests, welding tests and positive material identification. Formal reviews might include subject-matter-expert verification of proper materials of construction and corrosion studies. Various issues related to quality management during engineering, procurement and construction are addressed in Chapter 10.

Life-cycle thinking can prove important in engineering, procurement and construction. Selecting a material of construction that is less susceptible to corrosion by the process materials and anticipated impurities may be more expensive when purchased but, in the long run, be much more economical by significantly reducing containment integrity problems, replacement costs and loss-of-containment liabilities. Likewise, selection of equipment suppliers or fabrication contractors on a lowest-cost basis alone might reduce initial costs but increase long-term costs and liabilities by having less reliable equipment subject to earlier failures. Startup of the facility might also be delayed if positive material identification (PMI) tests find wrong materials of construction or factory acceptance tests or QA/QC checks indicate substandard construction.

Used Equipment. Unforeseen problems can arise when purchasing used equipment. The purchasing facility may be provided design documentation and some data about previous equipment service and/or repairs. Often, however, little or none of this information is available, and some of the information may be incorrect or incomplete. Facilities that purchase used equipment may consider developing specific procedures for doing so. These procedures may include many of the same considerations listed in Section 3.7 for asset re-use, such as recommissioning procedures. In addition, procedures for obtaining used equipment need to identify methods for securing and maintaining equipment file information.

The purchasing and re-use of used equipment may need to consider the following:

- A potential damage/corrosion assessment for vessels in hazardous or severe service vessels (this type of assessment may not be required for vessel installed in benign or non-critical service)

- A gap or fitness-for-service assessment to determine if the used equipment meets current applicable standards and local jurisdictional requirements

- A full PMI study on the equipment, documenting the materials of construction to ensure they are suitable for the intended process

- A thorough review via an authorized inspection agency using suitable NDE methods, to detect areas with damage or corrosion

- A hydrotest per applicable standards, such as ASME, for re-commissioned vessels

- Removal of any insulation to check for CUI damage.

3.5   COMMISSIONING

The commissioning stage of a facility's life cycle involves the final pre-paration activities involving newly constructed or modified assets in making the transition to an operating facility. Commissioning involves not only the physical assets (including auxiliary equipment and functions) but also the operating and maintenance personnel and the facility documentation and written procedures.

Commissioning is a planned, deliberate sequence of steps that may have certain "hold points" to ensure everything is prepared, documented, consistent with the intended design, and working properly (such as the functionality of instrumented protective systems) before proceeding. Commissioning checklists are often employed to ensure all planned commissioning steps are completed.

Commissioning is not necessarily all performed at one time. It may be necessary to perform line flushing, loop checks, baseline inspections, etc. that are sometimes considered "precommissioning" activities. It may also be necessary to commission sections of a facility in a staged manner by completing operational readiness activities and starting up one part of the process at a time, such as commissioning utility systems while construction is being completed on the main process. If commissioning is performed in a staged manner, attention needs to be paid to the installed assets as the commissioning proceeds, since the stages may not proceed as quickly as planned. This might include such necessities as the oiling and greasing of rotating equipment, hand-rotating standby equipment, and maintaining a clean and dry nitrogen purge until introduction of actual process materials.

When the design and/or construction of a new or modified facility involves one or more third parties such as an engineering and construction firm, part of commissioning is ensuring complete design, fabrication and operating documentation is turned over to the facility in a way that is usable. This is essential to documented process knowledge management as well as to the management of asset integrity. Key documentation includes, but is not limited to:

- Layout, isometric and construction drawings

- Building design information including ventilation system design

- Piping and instrumentation diagrams (P&IDs)

- Control and electrical systems documentation

- Equipment files, including dimensions and materials of construction

- Equipment manufacturers' documentation, including installation and operating instructions, recommended preventive maintenance activities and activity intervals, and recommended spare parts

- Emergency relief system design and design basis documentation

- Hazard identification and evaluation reports, if performed by the supplier.

Another important aspect of the commissioning stage is the obtaining of baseline data that can then be used in the ongoing management of asset integrity after startup.   If baseline data, such as actual piping and vessel thickness measurements, are not obtained at this time, then the availability of predicted values such as remaining life calculations will be delayed by an entire test interval, which may be several years for some equipment.   Baseline data measurements need not be made right at startup; actual data can be obtained during equipment fabrication instead for some types of equipment.   Taking and documenting baseline readings may be required or recommended by code, and particularly for re-used vessels when changing ownership or service.

During commissioning, AIM-related issues are likely to be encountered or uncovered and will require careful treatment in the same way as if the facility was in operation.   For example, some assets may not perform as intended or may fail during an equipment check-out sequence.   Such deficiencies will need to be successfully managed and may require proper replacement of equipment or parts. Some assets may need calibration or adjustment, while others such as a furnace lining may require conditioning, passivation, etc.

Operational Readiness Review.   One of the final opportunities to identify integrity issues before introducing hazardous materials into a process is during an operational readiness review, also known as a pre-startup safety review (PSSR). Facilities may consider requiring a QA review as part of operational readiness activities.   During the QA review, the installed equipment can be compared to the design documentation, and any project-specified installation requirements can be verified.   Upon completion of the QA review, further steps are to

1.   Document any discrepancies between design and installation

2.   Evaluate whether each discrepancy is tolerable (this evaluation can be similar to a change review process)

3.   Make necessary corrections prior to equipment startup

4.   Document the as-built condition and closure of any identified items.

3.6   OPERATION AND MAINTENANCE

Managing asset integrity during the operational phase of a facility's life cycle consists of several activities:

- Initially starting up the facility in a manner such that the integrity of the facility assets is not compromised or weakened

- Operating the facility within its intended design parameters

- Recognizing deviations from the intended integrity operating window and properly responding to these deviations

- Performing regular preventive/predictive maintenance tasks on schedule and tracking/trending results

- Monitoring the condition of assets by inspections, tests, condition monitoring and performance monitoring as required

- Executing transient operations such as shutdowns and restarts in a manner such that asset integrity is not compromised or weakened

- Successfully managing changes to the facility, including not only changes to equipment but also to utilities and support facilities, personnel, technology, procedures, chemicals and feedstocks

- Detecting latent safeguard deficiencies and failures by inspections and functional tests

- Detecting primary containment system deterioration by inspections and tests

- Correcting deficiencies and failures as they arise using proper spare and replacement parts and restoring the system to its fully functional state.

These activities form the heart of asset integrity management and are the subject of Chapters 5 through 13 of this document. Many AIM responsibilities are performed in support of the above-listed activities, such as the need to develop AIM procedures, train and equip personnel to properly perform AIM activities, analyze inspection and test data, determine failure root causes, manage the maintenance workload, maintain documentation and perform various oversight, auditing and quality assurance functions.

Managing Change. A significant number of plants may be required to continue to operate beyond their intended life, or may be required to change service such as handling a different raw material. Subtle changes and personnel changes can also affect AIM.

Aging Plants. Many of the AIM activities during the operating and maintenance life cycle stage are needed because of the potential for degradation of assets over time due to age-related mechanisms such as corrosion, erosion, fatigue and embrittlement. This degradation potential has been termed "aging plants" (or "ageing plants") and has received considerable attention in some industries such as nuclear power and offshore oil and gas. Publications are available from the U.K. Health and Safety Executive that are intended to bring a focus to aging plant considerations for onshore process facilities as well (see References 3-12 and 3-13). A discussion of asset failure modes and damage mechanisms is given in Chapter 4 of this document.

Nearing End of Operational Phase. During late-life operation and maintenance, preparation may need to be started for the decommissioning of the facility. This may include a review of all critical assets, along with their reliability and associated maintenance requirements.

Often, a risk-based approach includes a review of both preventive and corrective maintenance associated with each asset and system that has strategic input to decommissioning planning. This may involve an assessment of the reliability of critical assets that are required to be functional during decommissioning.

3.7   DECOMMISSIONING

Decommissioning may be an AIM concern if re-use of assets is intended or possible. It may also be an AIM concern if decommissioning a large facility or one that would otherwise take an extended amount of time.

Potential for Re-use of Assets. Any equipment that is not removed and disposed of upon decommissioning might potentially be re-used. "Mothballed" units and "boneyards" present opportunities for saving money. However, they also present significant AIM challenges.

Facilities that recognize these challenges may consider establishing decommissioning and recommissioning procedures. A decommissioning procedure can consider depressurization and cleaning of equipment, additional measures for equipment preservation, and any ongoing inspections and/or preventive maintenance (PM) activities that need to be performed to maintain assets in a state of readiness or near-readiness. In addition, design and inspection documentation needs to be retained. Equipment on its way to the boneyard or warehouse needs to be labeled or tagged. Similarly, units that are mothballed for re-use at a later date (e.g., seasonally operated equipment) need to have procedures to ensure that liquids are drained, systems are purged and other measures are taken to help preserve equipment life (e.g., maintaining a proper atmosphere to prevent corrosion) and protect the safety of personnel and contractors from hazardous materials and energies and unsafe atmospheres inside equipment.

Some facilities provide quality assurance (QA) for reusing equipment through recommissioning procedures. Recommissioning procedures may include a change-of-service approval process. Variables to consider in such an approval process include (1) the length of time the equipment was out of service and (2) the extent to which ongoing inspections and/or PMs were performed. Generally, recommissioning involves inspections and other equipment checks to verify that the used equipment is suitable for the new service. If appropriate, pressure equipment may be rerated according to the practices listed in Section 10.7.

Decommissioning of Larger Facilities. Some facilities may take many years to decommission. For example, a large offshore oil platform typically takes up to five years from cessation of production to the time the platform is physically removed, and even longer for subsea assets to be removed. During this time, different decommissioning phases can have a direct impact on asset integrity management. Each facility is different; however, a decommissioning program will typically need to consider at least four phases of varying durations:

1. *Late-life operation and maintenance.* Normal operation, while deinventorying raw materials, intermediates and products as much as possible.

2. *Cessation of production.* Facility shut down; remaining process materials contained.

3. *Cleanout.* Positive isolation of inputs and outputs (such as pipelines), remaining process materials deinventoried and purged out; process equipment engineered down and cleaned.

4. *Removal.* Reverse of installation process, requiring global isolation of systems to allow efficient separation of complex systems.

The above decommissioning phases account for the gradual engineering down of systems and isolation of various material and energy sources.

3.8   RAGAGEP SELECTION AND APPLICATION AT EACH STAGE

RAGAGEP is an acronym for "recognized and generally accepted good engineering practice," a term that is found in two places in the U.S. Occupational Safety and Health Administration's Process Safety Management (PSM) Standard, 29 CFR 1910.119. As was detailed in Section 1.6 above, RAGAGEP stems from the selection and application of appropriate engineering, operating, and maintenance knowledge when designing, operating and maintaining chemical facilities with the purpose of ensuring safety and preventing process safety incidents. It involves the application of engineering, operating or maintenance activities derived from engineering knowledge and industry experience based upon the evaluation and analyses of appropriate internal and external standards, applicable codes, technical reports, guidance, or recommended practices or documents of a similar nature. RAGAGEP can be derived from singular or multiple sources and will vary based upon individual facility processes, materials, service, and other engineering considerations.

Facilities have different issues that are relevant to the use of codes, standards and practices at the various life cycle stages. Table 3-2 summarizes some of the relevant issues.

The selection of RAGAGEP at each stage starts with a knowledge and awareness of codes, standards and practices available from various relevant organizations. This is an essential part of process safety knowledge that needs to be maintained over time. Industry organizations may be helpful in this regard. See Reference 3-14 for an example of a public document listing possible RAGAGEP for natural gas processing plants.

With a knowledge and awareness of RAGAGEP, engineers involved in the design of a new facility or the operation of an existing one can then determine which of those codes, standards and practices apply to the pertinent facility and life cycle stage. This also involves being familiar with the various codes, standards and practices, as well as the purpose and relevance of each one, to the facility in question. Documentation of what RAGAGEP has been employed in the design of a facility is an important aspect of process safety knowledge.

TABLE 3-2. Issues Relevant to Use of RAGAGEP at Different Life Cycle Stages

Life Cycle Stage	Issues Relevant to Use of RAGAGEP
Research and Process Development	Identification of hazards, use of inherently safer technologies
Process Design and Engineering	Good engineering practices with respect to selecting materials of construction and designing assets to meet service requirements, evaluation of the adequacy of designed safeguards, design of protective systems to meet asset integrity level requirements
Procurement and Construction	Inspections, quality assurance of new and re-used assets, fitness for service
Commissioning	Quality assurance, operational readiness
Operation and Maintenance	Aging plants, detection of deterioration and failures, inspection and testing protocols, inspector qualifications, inspection and testing intervals, preventive maintenance activities, quality assurance of spare parts and replacements in kind, management of change
Decommissioning	Aging plants, fitness for service and operational readiness (in case of possible re-use of mothballed or decommissioned equipment)

CHAPTER 3 REFERENCES

3-1   Lichts, F.O., "Industry Statistics: 2010 World Fuel Ethanol Production". Renewable Fuels Association, retrieved 2011-04-30, in en.wikipedia.org/wiki/Ethanol_fuel_in_the_United_States#U.S._government.

3-2   "U.S. Fuel Ethanol Plant Production Capacity," U.S. Energy Information Administration, www.eia.gov/petroleum/ethanolcapacity, May 20, 2013.

3-3   Jessen, H., "CIP Matters," *Ethanol Producer Magazine*, www.ethanolproducer.com/articles/9570/cip-matters.

3-4   Johnson, R.W., "Risk Management by Risk Magnitudes," *Chemical Health and Safety 5*(5), September/October 1998.

3-5   American Institute of Chemical Engineers, *Inherently Safer Chemical Processes: A Life Cycle Approach, Second Edition*, Center for Chemical Process Safety, New York, NY, 2008.

3-6   American Institute of Chemical Engineers, *Guidelines for Hazard Evaluation Procedures, Third Edition*, Center for Chemical Process Safety, New York, NY, 2008.

3-7   ISO 55000, Asset Management – Overview, Principles and Terminology, International Organization for Standardization, Geneva, Switzerland.

3-8   Broadribb, M.P. and M.R. Currie, "HAZOP/LOPA/SIL: Be Careful What You Ask For!" *6th Global Congress on Process Safety,* San Antonio, Texas, March 2010.

3-9   American Institute of Chemical Engineers, *Guidelines for Engineering Design for Process Safety, Second Edition*, Center for Chemical Process Safety, New York, NY, 2012.

3-10   Casada, M., R. Montgomery, and D. Walker, *Reliability-focused Design: Inherently More Reliable Processes Through Superior Engineering Design*, presented at the International Conference and Workshop on Reliability and Risk Management, San Antonio, TX, 1998.

3-11   American Institute of Chemical Engineers, *Guidelines for Process Safety Documentation*, Center for Chemical Process Safety, New York, NY, 1995.

3-12   Wintle, J. et al., "Plant Ageing: Management of Equipment Containing Hazardous Fluids or Pressure," U.K. Health and Safety Executive Research Report RR509, HSE Books, 2006, www.hse.gov.uk.

3-13   Horrocks, P. et al., "Managing Ageing Plant: A Summary Guide," U.K. Health and Safety Executive Research Report RR823, HSE Books, August 2010, www.hse.gov.uk.

3-14   U.S. Environmental Protection Agency, Interim Chemical Accident Prevention Advisory, Design of LPG Installations at Natural Gas Processing Plants, EPA 540-F-14-001, January 2014, http://www2.epa.gov/rmp/interim-chemical-accident-prevention-advisory-design-lpg-installations-natural-gas-processing.

Additional Chapter 3 Resources *(consult the latest version of each document)*

API 510, *Pressure Vessel Inspection Code: Maintenance Inspection, Rating, Repair and Alteration*, American Petroleum Institute, Washington, DC.

API 570, *Piping Inspection Code: Inspection, Repair, Alteration, and Rerating of In-service Piping*, American Petroleum Institute, Washington, DC.

API 610/ISO 13709, *Centrifugal Pumps for Petroleum, Petrochemical and Natural Gas Industries*, American Petroleum Institute, Washington, DC.

API 620, *Design and Construction of Large, Welded, Low-pressure Storage Tanks*, American Petroleum Institute, Washington, DC.

API 650, *Welded Steel Tanks for Oil Storage*, American Petroleum Institute, Washington, DC.

API 653, *Tank Inspection, Repair, Alteration, and Reconstruction*, American Petroleum Institute, Washington, DC.

American Society of Mechanical Engineers, *International Boiler and Pressure Vessel Code*, New York, NY.

ASME B31.3, *Process Piping*, American Society of Mechanical Engineers, New York, NY.

ASME B73.1, *Specification for Horizontal End Suction Centrifugal Pumps for Chemical Process*, American Society of Mechanical Engineers, New York, NY.

ASME B73.2, *Specifications for Vertical In-line Centrifugal Pumps for Chemical Process*, American Society of Mechanical Engineers, New York, NY.

ASTM E 1476, *Standard Guide for Metals Identification, Grade Verification, and Sorting*, ASTM International, West Conshohocken, PA.

IEC 61511, *Functional Safety: Safety Instrumented Systems for the Process Industry Sector - Part 1: Framework, Definitions, System, Hardware and Software Requirements*, International Electrotechnical Commission, Geneva, Switzerland. Also by the same title ANSI/ISA-84.00.01 Part 1 (IEC 61511-1 Mod), International Society for Automation, Research Triangle Park, NC.

National Board Inspection Code, National Board of Boiler and Pressure Vessel Inspectors, Columbus, OH.

NFPA 70, *National Electrical Code*, National Fire Protection Association, Quincy, MA.

Pipe Fabrication Institute, *Standard for Positive Material Identification of Piping Components Using Portable X-Ray Emission Type Equipment*, New York, NY.

UL 142, *Steel Aboveground Tanks for Flammable and Combustible Liquids*, Underwriters Laboratories Inc., Northbrook, IL.

APPENDIX 3A. DESIGN REVIEW SUGGESTIONS

The design process for a significant project has several steps (phases), so that AIM life cycle activities may include different design reviews at each phase.

Evaluation Phase. Evaluation is the beginning phase of the design process. At this stage, some general information is available on the needs/requirements for a new process and the range of technologies that may be employed to satisfy those needs/requirements. The evaluation phase determines (1) whether this project should be pursued (relative to other projects competing for the same resources), (2) the specific objectives for the project, and (3) what fundamental direction (such as primary technological choices) the project should take. Analysis during this phase focuses on determining the feasibility of, and the risks associated with, the project. Much of the effort at the evaluation phase is in selecting which projects to pursue from among many competing alternatives — a decision in which the total cost of ownership (and hence, integrity-related characteristics) ought to play a key role.

Conceptual Design Phase. Conceptual design is an intermediate phase of the design process. During this phase, a development team investigates and proposes the "best" type of process configuration for the project. Choices about general types of processing equipment, interconnections between equipment, ranges of expected operating conditions, and operating environments are made during this phase; however, specific selection of components and process parameters will not be determined until later stages of the design process. Analysis during this phase focuses on determining how overall performance goals/objectives translate into goals/objectives for individual systems. This analysis provides an assessment (at a high level) of whether required performance is realistically achievable and what modifications/improvements would need to be made to attain overall performance goals/objectives. In addition, various design concepts are compared to determine which option(s) warrants further development based on a variety of factors, including project risk and expected life-cycle costs. Several analysis tools (e.g., What-If Analysis, Relative Ranking, Pareto analysis) can be applied at this phase. Analyses at this stage strongly affect (support) equipment selection and configuration decisions. Inherent safety reviews are also important at the early design stages, to evaluate possible process configuration alternatives that will allow the process to operate with reduced hazards such as using less hazardous process materials, reducing inventories and operating closer to ambient conditions.

Preliminary Design Phase. Preliminary design is another intermediate phase of the design process. During this phase, a development team expands on the work of the conceptual design phase by (1) finalizing choices about types and numbers of various equipment items, (2) producing the drawings that define the configuration of the equipment, (3) optimizing process parameter choices, and (4) evaluating impacts of tie-ins to plant services (e.g., air systems, electrical systems). Specific preplanning for how the process will be operated (such as operating procedure development) and how the process will be maintained (such as planned maintenance task/frequency definition) occurs in this phase. Analysis during this phase focuses on determining how individual system goals/objectives translate into component goals/objectives. Analysis also provides an assessment (at more detailed levels) of whether required performance of the individual systems is realistically achievable and what modifications/improvements to the individual systems or components are necessary. Optimization of system performance characteristics is completed during this phase and needs to consider a number of factors, such as costs, quality of products, and reliability-related characteristics (e.g., average system availability, system capability). Some analysis tools commonly used during this phase include:

- What-If Analysis focusing on major items

- Relative Ranking to evaluate key component reliabilities for competing designs

- Block diagram analysis to assess overall system reliability characteristics (e.g., overall equipment effectiveness) based on estimated or allocated individual system reliability characteristics.

Analyses during this phase also support decisions regarding equipment selection and configuration, but generally go further to begin characterizing overall system performance. Issues important for reliable maintenance and operation, as well as manufacturing/assembling the process, begin to surface in these analyses.

Detailed Design Phase. Detailed design is the final formal phase of the design process. During this phase, final choices are made regarding the specific equipment in the process (and associated vendors) and the layout of equipment. This phase could also include

1. Completing the final design
2. Preparing equipment files

3.   Developing and commissioning operating procedures/instructions

4.   Developing and scheduling inspection, testing and preventive maintenance (ITPM) tasks for the process

5.   Selecting a spare parts stocking strategy for the process.

The aim of the detailed design phase is having a complete specification of equipment that can be successfully fabricated/installed, operated and maintained. Analysis during this phase focuses on ensuring that equipment selection and configuration allow systems to meet design goals/objectives. An assessment is performed at the equipment level to determine whether required performance of the equipment (and ultimately the individual systems and overall system) is realistically achievable and, if not, what modifications and/or improvements are necessary. Optimization of equipment selection (based on factors such as equipment cost, quality of products, equipment reliability-related characteristics, etc.) is completed during this phase. In addition to equipment selection and configuration, analyses in this phase serve to identify:

• Major loss contributors for key systems (or components)

• Critical parameters for reliable fabrication/construction/manufacturing

• Important operating limits and startup criteria

• Appropriate planned maintenance tasks

• Necessary spare parts/materials stores.

The detailed design phase typically requires more analyses and tends to use a larger variety of analysis techniques than previous phases. Simpler techniques (e.g., checklists, What-If Analysis) provide appropriate amounts of resolution to assess certain systems and particular equipment reliability. A Hazard and Operability (HAZOP) Study is often performed at the final design stage for many processing operations involving significant hazards and more complex controls. Other systems, due to various factors such as redundancy or increased hazard, may require a more in-depth, quantitative analysis technique such as Fault Tree Analysis (FTA) or Common-Cause Failure Analysis. More detailed and more complex analyses can be applied to areas of significant risk or areas in which significant uncertainties exist regarding the level of risk. A guideline for analysis selection is to not perform more analyses than are necessary for decision making. Chapter 15 discusses tools for risk-based decision making.

4

FAILURE MODES AND MECHANISMS

Before taking steps to develop and implement an AIM program, it is important to understand some key concepts and terminology related to the kinds of asset failures that an AIM program seeks to prevent, as well as what tools are available to identify and analyze potential failures. The objective of this chapter is to introduce key concepts related to asset failures, failure modes, damage mechanisms and hazard analysis tools as they relate to AIM programs.

4.1   INTRODUCTION

As soon as a new facility is started up (and even before initial startup), the new and refurbished assets that comprise the facility are subject to many different damage and degradation mechanisms such as impact, wear, stress, fatigue and corrosion. These damage mechanisms, some of which may be a result of improperly designing equipment, specifying the wrong material of construction for a given service, or operating a process outside its intended limits, can lead to asset failures and their resulting effects on the system and its surroundings.

   Most damage mechanisms, failures, and failure effects can be anticipated on the basis of prior experience with the same or similar assets. They can be systematically analyzed for their likelihood of occurrence as well as the severity of their effects. An analysis of this type is useful in making design improvements and providing a foundation for operating and maintaining the facility in such a way as to reduce risks and direct limited resources where they will do the most good, including knowing what inspection, testing and preventive maintenance tasks to perform and how often. The concepts and tools for identifying and analyzing potential asset failures and applying the results to managing asset integrity are described in the following sections, with references given to additional available resources.

4.2   EQUIPMENT FUNCTIONS AND FUNCTIONAL FAILURE

An understanding of failure modes and mechanisms starts with an understanding of *failure* and its opposite, which is the successful functioning of an asset. The definition of *failure* that is used in this document is:

Failure
Loss of ability to perform as required

So, to recognize failure of a given asset, it is essential to know two things: first, what **required performance** is for the asset; second, what constitutes **loss of ability** to perform as required.

4.2.1   Required Performance

The required performance of an asset is its successful functioning while in service, achieving its operational intent as part of a larger system or process. For example, a gas detector might have required performance that would include the detection of a particular gas or range of gases at a given location or covering a given area, over a specified concentration range, within a defined response time. All of these parameters can be definitively or quantitatively specified as to what is considered required performance. This is known as a *functional specification*.

Assets often have more than one function. For example, a hydrocarbon inlet separator may have a primary function of separating the gas phase from the liquid phase of an incoming two-phase stream. However, it may also have a secondary function of removing entrained water from the liquid hydrocarbon phase. The required performance of an asset is defined in terms of both its primary and secondary functions. This is true for instrumentation and controls as well as for other types of assets. A distributed control system may have as its primary function to provide basic process control; it may also provide safety functions such as depressuring or alerting operations to an abnormal situation.

4.2.2   Loss of Ability

If required (successful) performance of an asset is properly defined, then *failure* is readily understood as not achieving that required performance while the asset is in service. For example, if part of the functional specification of a gas detector is that its response time will be no more than 10 seconds, then if the response time is found to exceed 10 seconds, the gas detector is failing to meet its functional specification and can be said to be in a *failed* state.

4.2.3   Observed Performance: Revealed vs Unrevealed Failure

Many assets are readily or rapidly observed as exhibiting a loss of ability to perform as required. For example, if a feed pump fails such that it is no longer pumping process material to an operating unit, the failure will be readily observed by its effect on multiple operating parameters such as flows, pressures, levels and temperatures. The failure might also be observed more directly by observing a difference in or lack of pump sounds, lack of shaft rotation, etc. A failure that is readily apparent by its effect on the system or surroundings, and characterized by having a relatively short duration before either being detected or causing a loss event (i.e., the failure condition is manifested within seconds to hours), is known as a *revealed failure* (Reference 4-1).

On the other hand, other assets, particularly those that work on an intermittent or on-demand basis such as a standby pump, a safety shutdown or a relief device, may be in a nonfunctional state but the loss of ability to perform as required would not be observed until either (a) it was called upon to work and failed to do so, or (b) it was tested or inspected and found to not meet its functional specification. Such failures, where a component could be in a nonfunctional state for an extended time (often months or years) before being detected, are known as *latent* or *unrevealed failures*.

4.2.4   Failures vs Incipient Failures

Part of an AIM program is recognizing, responding to and correcting failures as they occur in an operating facility. However, the larger part of an AIM program is taking actions to avoid failures. The likelihood of asset failures can be reduced by various quality assurance and preventive maintenance activities. It can also be reduced by conducting tests and inspections that may identify potential failures by various mechanisms before they result in actual failures.

Some asset failures occur suddenly and without warning. Most failures develop more gradually over time. If *incipient failures*, as defined below, can be detected in the early stages and corrected in time, then actual failures may be avoided.

Incipient failure

**An imperfection in the state or condition of hardware
such that a degraded or catastrophic failure can be expected to result
if corrective action is not taken**

These concepts can be illustrated by considering a pressure vessel. Its required performance might be to provide leak-tight containment of certain process materials within given pressure and temperature limits. Other parameters might also be included in the functional specification of the vessel, such as minimum throughput or residence time. Vessel failure, by definition, would be if there was loss of ability to contain the process material. If observed performance was degraded but did not yet reach the functional specification limit, such as due to a gradually increasing flow restriction that reduced throughput but still met the minimum-throughput required performance, this might be considered an *incipient failure*. If the flow restriction can be detected and corrected before a loss of ability to perform as required, as defined by the functional specification, then a failure can be avoided.

In the example above, a restriction that reduces vessel throughput can be directly observed and identified by flow conditions. Detection of some incipient failures can only be <u>indirectly</u> observed and inferred. For the vessel to provide leak-tight containment, observation of a leak or rupture in the pressure boundary of the vessel would constitute vessel failure. If one way to get to the point of leak or rupture is by gradual thinning of the vessel wall by erosion, then an incipient failure can be detected by taking periodic measurements of the vessel wall thickness where erosion is occurring. Note that "vessel wall thickness" is not part of the "required performance" of the vessel. However, a reduction in wall thickness is a means of detecting an imperfection in the state or condition of the vessel.

4.2.5   Failures vs Deviations

It can be seen that "required performance" as described above is very similar to the concept of a Hazard and Operability (HAZOP) Study *design intent*, which describes the function and parameter limits that define the boundaries of normal operation for part of a process. The HAZOP Study systematically looks for *deviations* from the design intent, which is essentially the same as looking for potential ways to have unacceptable differences between required and observed performance. The main divergence between these concepts is that design intent may be associated with more than one equipment piece, such as a piping run that includes a pump, valves, filter and heat exchanger, whereas failures are generally observed on a single-component basis.

4.2.6   Documentation of Functional Specifications

Defining "required performance" for an asset starts with the process design, which also involves knowing an operating context for how the asset will be employed and what its intended function will be. For vessels, this often means determining the equipment limits such as the maximum allowable working pressure (MAWP) at a given maximum temperature, as well as how much vacuum the vessel is designed to withstand. The required performance of a centrifugal pump may be defined by its pump curve as well as pressure and temperature limits. This information is typically kept in equipment files, which are generated as the life cycle proceeds from engineering design to equipment selection, vendor selection and procurement. The intended functions and required performance of safety systems can be both documented in equipment files and summarized using lists and/or matrices. In the context of risk-based process safety management, this design and equipment information is part of process knowledge management.

Having documented functional specifications is essential to an AIM program as part of knowing the required performance of assets and identifying failures and incipient failures. A facility can also benefit from the common use of this information in AIM programs, including reliability analyses, process hazard analyses and risk studies. This is further discussed later in this chapter in the context of Failure Modes and Effects Analyses (FMEA) and risk concepts.

4.3   FAILURE MODES

A proper understanding of asset failures includes not only an ability to recognize failures and incipient failures but also to distinguish *how* assets can fail, since most assets can fail in more than one way. The ways in which an asset can fail are called its *failure modes*. The following definition will be used in this document:

Failure mode

A symptom or condition by which a failure is observed

A failure mode might be identified as loss of function, spurious operation (function without demand), an out-of-tolerance condition or a simple physical characteristic such as a leak observed during inspection

---

Some of the most common failure modes for different types of process equipment are listed in Table 4-1. Note that the "primary containment" failure modes apply to any component that is part of the primary containment system, so are not listed again under equipment types such as pumps and heat exchangers. More detailed failure mode taxonomies are available as part of CCPS' Process Equipment Reliability Database (PERD) project (Reference 4-2) and from ISO 14224 (Reference 4-3).

Other types of assets will have different failure modes than those listed in Table 4-1. Examples of these failure modes are given in Table 4-2. Reference 4-4 is an example of a specialized compilation of failure modes for offshore floating storage unit turret and swivel systems.

TABLE 4-1. Example Process Equipment Failure Modes

Failure Mode	Description
Primary containment (vessel, tank, piping, pump casing, heat exchanger shell, etc.)	
Rupture or split	Catastrophic failure of the pressure/containment boundary
Leak to surroundings	Release of hazardous material from the primary containment system (less than rupture or split)
Piping system and fluid transfer (pipe, filter, heat exchanger tubes, vessel internals, etc.)	
Block or plug	Impede fluid flow into, through or from component, reducing flow below the minimum specified as required performance
Rotating and conveying equipment (pumps, compressors, conveyors, agitators, etc.)	
Fail off	Stop performing intended function (e.g., moving liquid or solids; compressing gas; mixing vessel contents)
Overspeed	Exceed upper limit of specified rotational speed
Heat exchangers	
Tube rupture	Catastrophic failure of internal fluid/fluid boundary
Tube leak	Leak across internal fluid/fluid boundary
Fouling	Reduction in heat transfer capability to below limit specified as minimum required performance
Check valves; excess flow valves; block valves	
Fail to respond	Fail stuck; fail to transfer to new position; fail to block flow when flow reverses, or fail to provide a tight seal upon closure
Instrumentation, controls, detectors (all three modes do not apply to every type)	
Fail low/no output or closed	Fail to the closed position or fail giving an output signal below the lower limit for the current process state or input signal
Fail high output or open	Fail to the open position or fail giving an output signal above the upper limit for the current process state or input signal
Fail to respond	Fail stuck; fail to transfer to new position or output based on changing process state or signal input
Relief devices (safety relief valve, rupture disk, etc.)	
Open prematurely	Create open path from the process at a pressure/temperature condition below minimum specified opening conditions
Fail to open on demand	Fail to create an open path from the process at a pressure/temperature condition exceeding the upper limit of the set pressure tolerance range
Flame arresters; detonation arresters	
Fail to arrest	Fail to stop propagating flame front or detonation
Flares; thermal oxidizers	
Fail to perform	Fail to convert incoming stream that falls within design limits to products meeting required conversion/composition parameters

TABLE 4-2. Example Failure Modes for Some Other Asset Types

Failure Mode	Description
Buildings / structures	
Structural failure	Collapse due to design error; maloperation (e.g., overloading); loss of structural integrity due to an external force such as windstorm, snow loading or earthquake
Degradation	Deterioration of building structural or utility components to the point of loss of structural integrity, loss of function or loss of personnel protection (including failure to stop ingress of smoke or toxic vapors)
Secondary containment	
Failure to contain	No secondary containment upon loss of primary containment due to degraded structural integrity, improper design (e.g., inadequate strength), or maloperation (e.g., drain left open)
Partial failure	Only part of secondary containment function achieved due to inadequate sizing or design (e.g., sloshing over sides of dike)
Electrical switchgear / substations	
Power loss	Mechanical failure of switchgear, wiring, etc.; overload; shorting
Degraded function	Transformer degradation of oil or internal components, oxidation/ corrosion of contacts, insulation breakdown, etc.
Fire safety equipment (fixed, portable)	
Fail to function on demand	Failure modes will depend on specific type of equipment; e.g., portable fire extinguishers not charged, discharged, not pressurized, plugged, valve fails to operate; detector fails to respond (including not being powered, miscalibration, sensor failure); deluge fails to function due to inadequate water supply, valve failure, piping failure, plugging or improper design; fire pump fails to operate
Fireproofing / insulation	
Loss of function	Deterioration over time to the point of not being present or not having adequate structural integrity when needed; loss of insulation capability e.g. due to being liquid-soaked

4.3.1   Mechanical Failures vs Operational or Maintenance Errors

It is tempting to call some things asset failures that are not really failures. For example, a manual block valve might be inadvertently closed, resulting in a no-flow condition when flow through the valve is the required condition. Although the flow rate deviates from the design intent, the manual block valve has worked according to its design specification as it closed when its valve handle was turned to the closed position. Such a condition is an operational error and not an asset failure. The same is true for controller set point and rate/quantity input errors.

A failure mode sometimes listed for rotating equipment is "motor wired backwards" or similar, resulting in the rotating equipment operating in the reverse direction from the required condition. However, this is not an asset failure but rather an installation or maintenance error. Likewise, a relief device opening at its set pressure as a result of increasing process pressure is not a failure, since the relief device is functioning as designed.

These types of operational or maintenance errors are generally detectable by different means than by inspection and testing AIM activities that are looking for asset failure and degradation mechanisms. Typical means of identifying such errors are by installation and maintenance checklists, run tests and operational readiness reviews.

4.3.2   Common Cause Failure Identification

When identifying failure modes of system components, one consideration is that failure modes of multiple components can often have a common cause. This can come from many sources, such as a set of valves all coming from the same supplier, piping insulation all installed by the same contractor, a series of pumps all being maintained by the same mechanic, or a high temperature deviation when operating the process causing thermal stress on many system components. These causes can also come from process materials having the wrong composition or impurities that result, for example, in internal corrosion of process equipment or chloride stress corrosion cracking.

4.3.3   Fail-Safe Direction

A distinction needs to be made between *failure mode*, defined above as "a symptom or condition by which a failure is observed," and *fail-safe direction*. Uses of the term "fail-safe" can be potentially ambiguous or misleading. Three examples of how this term may be used are:

- Some electrical components in instrumentation such as transmitters may be designed to burn out or fail preferentially by design in a certain direction. This would be a proper use of the term "fail-safe" only if the preferential failure direction will bring the system to a safe state.

- It may be intended to mean the state to which a component is designed to move to if its input signal is lost. In this case, the component is not failing at all, but rather changing state upon loss of signal. It may be more proper to designate such a component as, e.g., signal-to-open or signal-to-close than fail-safe, fail-closed or fail-open. Whether a given failure mode is the fail-safe direction or the fail-dangerous direction will depend on the process of which the component is a part.

- It may also mean the state to which the process is intended to go if the output signal from the component is lost. For example, if a combustion system loses signal from its combustible gas analyzer or flame sensor, then its logic solver would be programmed to automatically shut down the process or shut off fuel to the combustion system.

4.3.4 Less-Common Failure Modes

Table 4-1 includes some of the most common failure modes for typical process system components. Less-common failure modes are also possible. For example, a less-common failure mode for rotating equipment might be for it to fail on; i.e., fail to stop when turned off. Managing asset integrity generally involves taking actions to prevent failure modes that are reasonably likely to occur during the lifetime of a facility. Less-common failure modes may also need to be addressed if consequences of failure are severe. For example, rupture of a pressure vessel might be considered very unlikely to occur during the lifetime of a facility, but if failure consequences are severe, taking steps to detect degradation and incipient failure conditions and maintain equipment so as to avoid catastrophic vessel failures is certainly an appropriate part of the facility's AIM program.

4.3.5 Level of Detail for Failure Modes

For most assets, failure modes are identified and evaluated at the component level. (By contrast, damage mechanisms, as described in the next section, might be found at the system or circuit level as well as at the component level.) Typical components for process operations were listed in Table 4-1. They can also be identified as the individual equipment items identified on P&IDs.

However, more-critical equipment and complex equipment items, such as a recycle gas compressor in a catalytic cracking process, may warrant a greater level of detail to match a suitable failure management approach. The objective in selecting an appropriate level of detail is to understand failure modes in sufficient detail that the information can be used when determining AIM activities for inspecting, testing and maintaining the asset.

4.4 DAMAGE MECHANISMS

The next step in understanding asset failures is to know the various means by which assets are damaged or degraded to the point of failure. These underlying asset failure causes are termed *damage mechanisms*. Identifying and inspecting for indications of damage mechanisms can be used to predict future failures.

Damage mechanism

A mechanical, chemical, physical or other process that results in asset degradation

---

Some damage mechanisms can occur before an asset is ever put into service. Some damage mechanisms occur more or less gradually while in service; other damage mechanisms are sudden and episodic. Common types of damage mechanisms are given in Table 4-3. Specific damage mechanisms are defined in Table A-1 of ASME PCC-3 (Reference 4-5) and detailed for refining industry fixed assets in API RP 571 (Reference 4-6). Additional damage mechanisms are possible for assets that are not processing equipment, such as electrical power systems, protective structures, and secondary containment systems. Electronic assets can be damaged by heat, corrosion, chemicals, mechanical damage, magnetic fields or static electricity.

Some of the damage mechanisms in Table 4-3 are well understood, whereas others are less commonly encountered or more difficult to conceptualize. For example, five different damage mechanisms all involve hydrogen attack (References 4-5 and 4-6):

- *High-temperature hydrogen attack (HTHA)* is decarburization and cracking caused by hydrogen at elevated temperatures and pressures diffusing into steel and reacting with carbides in the steel to form methane (CH_4), which, since it cannot diffuse through the steel, accumulates and eventually causes intergranular fissuring.

- *Hydrogen embrittlement* is a loss in ductility of high-strength steels due to the penetration of atomic hydrogen, which can lead to brittle cracking. Hydrogen embrittlement can occur during manufacturing, welding or from services that can charge hydrogen into the steel in an aqueous, corrosive or a gaseous environment.

- *Hydrogen blistering* may form on the surface or within the wall thickness of a pipe or pressure vessel by hydrogen atoms generated by corrosion diffusing into the steel and collecting at a discontinuity in the steel such as an inclusion or lamination. The hydrogen atoms combine to form hydrogen (H_2) molecules that are too large to diffuse out, so the pressure builds to the point where local deformation occurs, forming a blister.

TABLE 4-3. Example Damage Mechanisms

Mechanism Type	Description / Examples
Prior to service, or introduced during repair or modification	
Incorrect material(s) of construction for service	Wrong selection of materials for intended service or materials used for fabrication not the intended materials (generally manifested during commissioning or in service by another damage mechanism such as accelerated corrosion)
Fabrication or installation defects	Weld or casting porosity/voids; various types of weld defects; inadequate tempering; misalignment; not wired/installed correctly
Mechanical damage	Asset physically damaged during fabrication, transit or installation
During commissioning	
Improper asset commissioning	Inadequate passivation, rapid heat-up, etc. (generally manifested in service by another damage mechanism such as accelerated corrosion or cracking)
In-service deterioration or aging mechanisms	
Corrosion, chemical attack	Chemical conversion of material of construction resulting in loss of mechanical strength
Erosion, abrasion, wear, fretting, galling	Physical wearing away of material of construction resulting in loss of mechanical strength
Fatigue, softening, weathering	Vibration or thermal cycling or aging of material of construction or fireproofing resulting in loss of mechanical strength
Embrittlement	Damage to metal or other material of construction resulting in loss of ductility, cracking
Stress corrosion cracking	Cracking at grain boundaries as a result of exposure to chlorides, ammonia, carbonates, sulfides or other specific chemical types
De/carburization, graphitization	Metallurgical damage by carburization, decarburization or graphitization
Creep	Plastic deformation over time under stress and temperature
Hydrogen attack	High-temperature hydrogen attack; hydrogen embrittlement; hydrogen blistering; hydrogen-induced cracking *(see text)*
Mechanical damage	Asset physically damaged by external impact or vibration
As a result of operating or environmental condition(s) beyond intended limits	
Pressure excursion leading to yield, rupture or buckling/collapse	Exceed or drop below design pressure limits to a sufficient degree as to result in asset damage (single event or repeatedly)
Temperature excursion leading to e.g. melting, creep or embrittlement	Exceed or drop below design temperature limits or exposure to hot or cold material (e.g., molten slag) to a sufficient degree as to result in asset damage (single event or repeatedly)
Thermal shock	Differential thermal strains exceed strength of material
Chemical attack	Exposure to unintended and aggressive process material, impurity or environmental contaminant
Loss of foundational support	Loss of support for buildings or equipment due to subsidence or flooding; subsidence affecting cross-country pipelines

- *Hydrogen-induced cracking (HIC)* is the formation of hydrogen blisters at different depths from the surface of the steel, in the middle of the plate or near a weld, followed by the development of cracks that link together neighboring or adjacent blisters that are at slightly different depths (planes). Interconnecting cracks between the blisters often have a stair-step appearance; hence, HIC is sometimes referred to as "stepwise cracking."

- *Stress-oriented hydrogen-induced cracking (SOHIC)* is similar to HIC but potentially more damaging. High levels of residual or applied stress drive the formation of arrays of cracks stacked on top of each other, resulting in a through-thickness crack that is perpendicular to the surface.

Techniques for identifying and addressing different damage mechanisms are the subject of later chapters in this document.

4.5   FAILURE EFFECTS

The effects of an asset failure are the failure consequences. The consequences will differ significantly depending on whether the failure is *revealed* or *unrevealed*, as described in Section 4.2.

4.5.1   Effects of Unrevealed Failures

Unrevealed (latent) failures generally have no immediate effect on a process or system. An unrevealed failure will not be manifested until it is either discovered directly by inspection or functional testing, indirectly in some manner, or it is discovered when the failed component is called upon to function and fails to perform its intended function. Two examples will be used to illustrate this point.

If a high-high level switch is installed on a storage tank to provide an output signal (e.g., used to activate an alarm and/or shut down incoming feeds) when the liquid level reaches the high-high level point, and the high-high level switch fails stuck one day during routine operation of the storage system, there is no immediate effect on the system. The switch failure will only become known when either the switch is effectively tested or inspected (the nature of the test/inspection depending on the switch technology employed, which may include on-line diagnostics) or the liquid level in the tank exceeds the high-high level point and continues to rise with no warning or shutoff of feeds.

Another example is a passive safeguard such as a secondary containment system. Secondary containment can deteriorate over time to the point where it no longer can provide its required performance upon loss of primary containment. The deterioration may not be obvious or may be noticed with no response. However, the primary system can continue to operate indefinitely and the failure of the secondary containment system may not become known until after loss of primary containment occurs and the secondary containment is needed.

4.5.2   Effects of Revealed Failures

Revealed failures have an immediate or near-term effect on a process or system. In general, the process or system cannot continue to operate or function normally with a revealed-failure component in the failed state. Revealed failures will often be manifested by an effect on process variables such as flow, pressure, temperature or level.

Revealed failures are manifested as *initiating events* for event sequences that can lead to loss events with associated loss and harm impacts. Enabling conditions and/or safeguard failures may also need to be realized in order for the loss event and its impacts to result. Two examples follow.

If the metal wall of a pipe carrying a hazardous slurry under pressure is thinned over time by erosion (the *damage mechanism*) to the point of rupture (the *failure mode*), then the effects of this revealed failure will be an immediate release of the hazardous slurry to its surroundings (the *loss event*). The *impacts* of the loss event may include injury to nearby personnel, loss of process material, interruption in production and/or possible environmental damage. In this scenario, the pipe rupture is an initiating event and there are no preventive safeguards intervening between the pipe rupture and the hazardous slurry release, although there may be some mitigative safeguards such as secondary containment or PPE pertinent to this scenario that may reduce the release impacts.

Not all revealed failures lead immediately to loss events. As a second example, consider a cooling control failure on an exothermic (heat-releasing) chemical reaction system. The specific failure mode may be a cooling water control valve failing closed. This will have an immediate effect on the reaction process variables such as temperature, pressure and reaction rate. However, other layers of protection such as operator response, independent safety shutdown and adequately sized emergency relief protection may need to all fail for the cooling water control valve failure to result in a thermal runaway reaction and vessel rupture explosion.

4.6   RISK

Risk is a key concept in managing the integrity of assets. Risk estimates are often used as inputs to the management decision process when allocating resources and selecting between alternatives. The fundamental approach is to allocate limited resources where they will do the most good in reducing risks. Directing resources toward reducing risks that are already very low or trivial is not as beneficial as directing them toward reducing higher risks. A secondary approach is to help decide how far to reduce known risks; i.e., until the estimated risks are low enough to be considered "tolerable" or "as low as reasonably practicable" (ALARP). Some fundamental risk concepts, as they relate to managing asset integrity, are presented in Appendix 4A.

Non-Risk-Based Decisions. Not all AIM decisions need to be made on the basis of managing estimated risks. Two alternatives for making AIM decisions are (1) on the basis of good practice or regulatory requirement and (2) on the basis of maximum possible impact.

Good-practice and regulatory-requirement decisions are made by comparing some aspect of managing asset integrity with industry norms and/or regulatory requirements that have developed over many years, often as a result of learnings from previous incidents. For example, fire protection equipment is fairly standardized as a result of many years of industry experience with many thousands of fixed and portable systems worldwide, so that a risk estimate is not needed to decide how often to inspect fire extinguishers, check hydrants, or run-test diesel firewater pumps.

Decisions made on the basis of maximum possible impact foresee the possibility of a loss event having unacceptably large impacts, so that regardless of how low the estimated likelihood might be, something is judged to be necessary to reduce the possible impacts, modify the process to be inherently safer, or abandon the risk. For example, if a consequence analysis indicates a given heavier-than-air flammable release scenario could drift into a nearby community and result in a fire/explosion event with devastating impacts, it may be decided to change the process to reduce the size of a potential release, install a barrier between the process and the community, change to a non-flammable process material or discontinue the operation. This decision would be solely on the basis of intolerable impacts and not attempt to take the likelihood of occurrence into account.

4.7   ANALYSIS

Systematic approaches are commonly used that aid the identifying and addressing of failure modes and mechanisms. Reference 4-7 summarizes many of these approaches with worked examples, including Failure Modes and Effects Analysis (FMEA), Fault Tree Analysis (FTA) and Event Tree Analysis (ETA). Reference 4-7 also contains a chapter on deciding on the most appropriate method for a given situation. Guidelines for using fully quantitative risk analysis techniques are given in Reference 4-8. Expert decision making may also be employed, but would not be as systematic as the above-mentioned approaches.

Failure Modes and Effects Analysis. One approach that can be particularly useful for AIM purposes is Failure Modes and Effects Analysis (FMEA). The basic FMEA approach involves determining the physical boundaries and operating context of the analysis, selecting a level of resolution (e.g., to study failures at the subsystem, product, component or parts level), systematically identifying and examining all failure modes of one item at a time within the study boundaries, then evaluating the immediate and ultimate effects of each failure mode on the system as a whole. Functional/reliability block diagrams may need to be created as the analysis proceeds to evaluate failure effects. The analysis is documented in a consistent, tabular format.

FMEAs can be used for several AIM-related purposes, such as:

- Evaluating the adequacy of existing safeguards, in the context of a hazard evaluation

- Identifying safety-critical equipment

- Providing input to the determination of ITPM tasks and their respective frequencies

- Designing out failures or redesigning the process, particularly where single-point failures are found that could lead to mission-critical effects

- Evaluating the maintainability of system components

- Discussing other issues such as where redundancy may be warranted or what spare parts may need to be on hand.

An example FMEA is shown in Table 4-4. Many FMEAs include additional documentation in the FMEA worksheet, such as how each failure mode would be detected. A few of the more common variations on the basic FMEA methodology include the following:

- *FMECA* (Failure Modes, Effects and Criticality Analysis) categorizes the severity (criticality) of the failure effects.

- *FMEDA* (Failure Modes, Effects and Diagnostic Analysis) extends FMEA to examine frequency and detectability of failure modes, and is primarily used for analyzing and certifying instruments such as for safety instrumented systems.

- *DFMEA* (Potential Failure Mode and Effects Analysis in Design).

- *PFMEA* (Potential Failure Mode and Effects Analysis in Manufacturing and Assembly Processes).

The use of FMEA in the context of developing ITPM test and inspection plans is discussed in Section 7.5 and Section 15.3 (in conjunction with RCM). Procedures and requirements for conducting FMEAs for various purposes can be found in References 4-7, 4-9, 4-10 and 4-11.

TABLE 4-4. Example FMEA *(from Reference 4-7)*

AREA: HCl Distillation Column TEAM: ...					DRAWING: E-708, Rev. F
#	Component	Failure Mode	Effects	Safeguards	Findings, Actions
1	Temperature control valve (TCV-201)	Fails open	Increased heating of the HCl column; potential column overpressure and release of HCl	Multiple TIs on column High temperature alarm and shutdown on column Excess overhead condensing capacity (spare condenser)	Add a high pressure alarm Develop an emergency checklist for operators to follow on high temperatures
		Fails closed	HCl column cools down; no effect of concern	---	
		Leaks externally	Loss of steam; no effect of concern	---	
		Leaks internally	Increased heating of the HCl column; however, at a slow rate Potential column overpressure and release of HCl	Multiple TIs on column High temperature alarm and shutdown on column Excess overhead condensing capacity (spare condenser) Operators have ample time to diagnose and manually isolate TCV	
2	Temperature transmitter (TT-201)/ temperature controller (TIC-201)	False high output	HCl column cools down; no effects of concern	---	
		False low output	Increased heating of the HCl column; potential column overpressure and release of HCl	Multiple TIs on column Excess overhead condensing capacity (spare condenser)	Same as above Move the high and low temperature alarms to another column transmitter
		No signal change	Possible increase in column temperature due to minor process upsets; slight chance of an overpressurization and release of HCl	Multiple TIs on column Excess overhead condensing capacity (spare condenser) Ample time to diagnose problem and manually control column temperature	Same as above Move the high and low temperature alarms to another column transmitter

4.8   ITPM TASK ASSIGNMENTS

The primary maintenance issues related to addressing failure modes and damage mechanisms are properly performing established inspection, testing and preventive maintenance (ITPM) activities on schedule, as well as successfully dealing with asset deficiencies as they arise. Correcting deficiencies may involve asset repair or replacement along with the associated operational readiness steps, and may also involve managing change when situations arise such as the unavailability of replacements in kind. General considerations related to these ITPM tasks are addressed in Chapter 6 of this document. Considerations related to specific types of equipment are addressed in Chapter 12.

In addition to these maintenance tasks, technical input is generally required to properly address failure modes and mechanisms. When an inspection finds an unusual result such as a higher-than-expected corrosion rate, some research or troubleshooting may be needed to understand and address such issues.

4.9   OPERATIONAL ISSUES

The main operational issues related to addressing failure modes and mechanisms are (1) operating a facility consistently according to standard operating procedures within established limits, and (2) managing changes to all aspects of the facility and operation that could affect asset integrity. If these two key elements of process safety—conduct of operations and change management—are not properly established and implemented, then the best inspection, testing and preventive maintenance program may not be effective in maintaining asset integrity.

Reference 4-12 discusses both conduct of operations and management of change as key elements of risk-based process safety management. The recommended practice API RP 584 (Reference 4-13) also defines and discusses the AIM importance of operating a process within an established integrity operating window (IOW) as illustrated in Section 2.2.1 of this document.

Not only is it important to avoid operational deviations outside established limits, it is also essential to identify and evaluate the potential consequences when an operational deviation does occur that may result in either near-term or long-term damage to facility assets. This may result in the need to change how often ITPMs are to be performed and/or to require additional tests or inspections. For example, the one-time or repetitive over-firing of furnace tubes may require earlier tube inspection or replacement, or the inadvertent introduction of a corrosive material into a process may require additional purging/cleaning or more frequent corrosion testing to be performed. For this reason, operating personnel need to be encouraged to report and document process deviations when they occur—something that may require a culture change within an organization.

4.10 OTHER RELATED ACTIVITIES

Other activities related to addressing failure modes and damage mechanisms include the managing of items that may affect or be affected by damage mechanisms. For example, providing and maintaining cathodic protection, sacrificial anodes, coatings and/or chemical treatment may be essential to avoiding specific corrosion mechanisms for various assets. Such items need to be included in regularly scheduled inspection and replacement activities in order to be maintained over time.

The maintaining of adequate spares on-hand may be in part determined by possible damage mechanisms. An analysis such as by FMEA as described above may point to the need for having spares on hand or even installed (such as a hot standby pump) for critical components that may be difficult to obtain from a supplier and/or to replace without a major process interruption. A successful preventive maintenance program may reduce the need for spare parts by reducing the frequency of in-service failures or by extending service life. These and related quality management issues are discussed in Chapter 10.

CHAPTER 4 REFERENCES

4-1   Center for Chemical Process Safety, *Guidelines for Enabling Conditions and Conditional Modifiers in Layer of Protection Analysis*, Appendix A, American Institute of Chemical Engineers, New York, NY, 2013.

4-2   Center for Chemical Process Safety, Process Equipment Reliability Database (PERD) taxonomies, American Institute of Chemical Engineers, New York, NY, www.aiche.org/ccps/resources/perd/taxonomies.

4-3   ISO 14224, *Petroleum, Petrochemical and Natural Gas Industries - Collection and Exchange of Reliability and Maintenance Data for Equipment*, International Organization for Standardization, Geneva, Switzerland.

4-4   OTO 2001/073, *Failure Modes, Reliability and Integrity of Floating Storage Unit (FPSO, FSU) Turret and Swivel Systems*, UK Health & Safety Executive, www.hse.gov.uk/research/otohtm/2001/oto01073.htm.

4-5   ASME PCC-3, *Inspection Planning Using Risk-Based Methods,* American Society of Mechanical Engineers, New York, NY.

4-6   API RP 571, *Damage Mechanisms Affecting Fixed Equipment in the Refining Industry*, American Petroleum Institute, Washington, DC.

4-7   Center for Chemical Process Safety, *Guidelines for Hazard Evaluation Procedures, Third Edition*, American Institute of Chemical Engineers, New York, NY, 2008.

4-8   Center for Chemical Process Safety, *Guidelines for Chemical Process Quantitative Risk Analysis, Second Edition*, American Institute of Chemical Engineers, New York, NY, 1999.

4-9   MIL-STD-1629A, *Procedures for Performing a Failure Mode, Effects and Criticality Analysis*, U.S. Department of Defense, Washington, DC.

4-10 IEC 60812, *Analysis Techniques for System Reliability – Procedure for Failure Mode and Effects Analysis (FMEA)*, International Electrotechnical Commission, Geneva, Switzerland.

4-11 SAE Standard J1739, *Potential Failure Mode and Effects Analysis in Design (Design FMEA), Potential Failure Mode and Effects Analysis in Manufacturing and Assembly Processes (Process FMEA)*, SAE International, Warrendale, Pennsylvania

4-12 Center for Chemical Process Safety, *Guidelines for Risk Based Process Safety*, American Institute of Chemical Engineers, New York, NY, 2007.

4-13 API RP 584, *Integrity Operating Windows,* American Petroleum Institute, Washington, DC.

4-14 Center for Chemical Process Safety, *Guidelines for Developing Quantitative Safety Risk Criteria*, American Institute of Chemical Engineers, New York, NY, 2009.

4-15 Center for Chemical Process Safety, "Process Safety Leading and Lagging Metrics …You Don't Improve What You Don't Measure," American Institute of Chemical Engineers, New York, NY, 2011, www.aiche.org/ccps/resources/overview/process-safety-metrics/recommended-process-safety-metrics.

4-16 Moosemiller, M., "Development of Algorithms for Predicting Ignition Probabilities and Explosion Frequencies," *5th Global Congress on Process Safety,* Tampa, Florida, 2009.

4-17 Buncefield Major Incident Investigation Board, *The Buncefield Incident 11 December 2005: The Final Report of the Major Incident Investigation Board,* 2008.

4-18 Johnson, D.M., "The Potential for Vapour Cloud Explosions – Lessons from the Buncefield Accident," *J. Loss Prevention in the Process Industries 23,* 2009, pp. 921-927.

4-19 Center for Chemical Process Safety, *Guidelines for Process Equipment Reliability Data, with Data Tables*, American Institute of Chemical Engineers, New York, NY, 1989.

4-20 ANSI/API RP 754, *Process Safety Performance Indicators for the Refining and Petrochemical Industries*, American Petroleum Institute, Washington, DC.

Additional Chapter 4 Resources *(consult the latest version of each document)*

API 510, *Pressure Vessel Inspection Code: Maintenance Inspection, Rating, Repair and Alteration*, American Petroleum Institute, Washington, DC.

API 570, *Piping Inspection Code: Inspection, Repair, Alteration, and Rerating of In-service Piping*, American Petroleum Institute, Washington, DC.

API RP 579-1, *Fitness-for-Service, 2nd Edition*, American Petroleum Institute, Washington, DC.

API RP 941, *Steels for Hydrogen Service at Elevated Temperatures and Pressures in Petroleum Refineries and Petrochemical Plants*, American Petroleum Institute, Washington, DC.

API Std 653, *Tank Inspection, Repair, Alteration, and Reconstruction,* American Petroleum Institute, Washington, DC.

ASM International, *J. Failure Analysis and Prevention* (bimonthly publication), Springer.

Becker, W.T. and Shipley, R.J., *ASM Handbook,* Volume 11 - Failure Analysis and Prevention. ASM International, Materials Park, Ohio, 2002.

Bloch, H.P. and F.K. Geitner, *Machinery Failure Analysis and Troubleshooting, Fourth Edition: Practical Machinery Management for Process Plants*, Butterworth-Heinemann, Oxford, UK, 2012.

BS 7910, *Guide to Methods for Assessing the Acceptability of Flaws in Metallic Structures*, British Standards Institute, London, UK, 2013.

COMAH Guidance on the technical measures for the prevention, control and mitigation of major accidents for various systems/unit operations, UK Health & Safety Executive, www.hse.gov.uk/comah/sragtech/systemsindex.htm.

McEvily, A.J. and J. Kasivitamnuay, *Metal Failures: Mechanisms, Analysis, Prevention*, Wiley-Interscience, Hoboken, NJ, 2013.

Moubray, J., *Reliability-Centered Maintenance, Second Edition*, Industrial Press Inc., New York, NY, 1997.

NACE Corrosion Engineer's Reference Book, Third Edition, NACE International, Houston, Texas, 2002.

WRC 488, *Damage Mechanisms Affecting Fixed Equipment in the Pulp and Paper Industry*, Welding Research Council, Shaker Heights, Ohio, 2004.

WRC 489, *Damage Mechanisms Affecting Fixed Equipment in the Refining Industry*, Welding Research Council, Shaker Heights, Ohio, 2004.

WRC 490, *Damage Mechanisms Affecting Fixed Equipment in the Fossil Electric Industry*, Welding Research Council, Shaker Heights, Ohio, 2004.

Wulpi, D.J., *Understanding How Components Fail, 3rd Edition*, ASM International, Materials Park, Ohio, 2013.

APPENDIX 4A. RISK CONCEPTS RELATED TO AIM

Definitions. In the realm of process safety, *risk* is defined as a measure of human injury, environmental damage and/or economic loss in terms of incident likelihood and the magnitude of losses or injuries. In equation form, this translates to:

Risk = Loss Event Frequency • Loss Event Impact

A *loss event* is an irreversible physical event that has the potential for loss and harm impacts (Reference 4-7). Examples include a release of a hazardous material, ignition of flammable vapors or ignitable dust cloud, or rupture of a tank or vessel. An incident might involve more than one loss event, such as a flammable liquid release (first loss event) followed by ignition of a flash fire and pool fire (second loss event) that heats up an adjacent vessel and its contents to the point of rupture (third loss event). Each of these three loss events would have an associated frequency and impact.

The *frequency* (likelihood) is an estimated probability of the loss event occurring per unit of time, with the unit of time generally taken as a year. For example, a "hundred-year flood" for a given location has a frequency of 0.01 per year, which means there is estimated to be a 1% probability of a flooding event occurring at that location per calendar year.

The *impact* (severity) is a measure of the ultimate loss and harm of a loss event. Impact may be expressed in terms of numbers of injuries and/or fatalities, extent of environmental damage, and/or magnitude of losses such as property damage, material loss, lost production, market share loss and recovery costs.

Risks are additive at all levels. Hence, the risk of a fire in a given process unit is the sum of the risks from all fire scenarios associated with the process unit. The total business interruption risk for a facility is the sum of the business interruption risks in each of the process units. For this reason, risks need to be managed at all levels in an organization (Reference 4-14).

AIM-Related Impacts. Since hazardous materials and energies are handled at process facilities on an industrial scale, the impacts of AIM-related loss events such as fires, explosions and hazardous releases can be quite significant. Many different categorizations of impacts and severity levels are in common use. The CCPS document "Process Safety Leading and Lagging Metrics" (Reference 4-15) uses the following impact categories to indicate the severity of a process safety incident:

- Safety/Human Health Impacts
- Fire or Explosion (including overpressure) direct costs
- Potential Chemical Impact
- Community/Environmental Impact.

ANSI/API RP 754 (Reference 4-20) has similar impact categories.

Safety/human health impacts from AIM-related loss events are generally acute injuries that can range from a single minor injury to multiple on-site and/or off-site fatalities. The human health impacts can be realized from many different mechanisms, including direct contact with a released process material; exposure to fire thermal radiation, hot combustion gases and/or direct flame; or explosion blast effects including blast overpressure, shrapnel and flying debris, building collapse, or being thrown against a hard/sharp surface or knocked off an elevated location.

Fire or explosion *direct costs* are costs to the company of repairs or replacement, cleanup, material disposal, environmental remediation and emergency response. Direct costs do not include indirect costs such as business opportunity, business interruption and feedstock/product losses, loss of profits due to equipment outages, costs of obtaining or operating temporary facilities, or costs of obtaining replacement products to meet customer demand. Direct costs do not include the cost of the failed component leading to loss of primary containment, if the component is not further damaged by the fire or explosion (Reference 4-15). Note that losses can be secondary damage from follow-on effects as well as damage from the primary loss event. Note also that even though the CCPS metric for fire/explosion impact does not include the indirect costs such as those listed above, these indirect costs are nevertheless genuine impacts and may need to be included in risk estimates and decisions.

Potential chemical impacts can range from small, contained, localized releases to massive releases with the potential for a vapor cloud explosion and/or off-site health effects. *Community/environmental impacts* can range from environmental effects requiring only short-term remediation to those requiring long-term, very costly remediation; these impacts can also include community shelter-in-place or evacuation and/or extensive media coverage.

Loss Event Frequencies. The frequency (probability per unit time) of a specific loss event occurring is estimated based on the scenario events necessary to get to the specific loss event. This can range from a simple loss-of-primary-containment (LOPC) event to a complex sequence of events with multiple safeguards failing and/or multiple conditions needing to be present. The overall loss event frequency is determined by combining the initiating cause frequency with the requisite failure-on-demand and conditional probabilities. The following examples will illustrate this point.

For a simple LOPC event, the loss of primary containment *is* the loss event (irreversible physical event with potential for loss and harm), so the loss event frequency is equal to the estimated likelihood of the LOPC event. For example, if a pressure vessel operating within its integrity operating window (IOW) suddenly and catastrophically fails, the release of pressure and vessel contents as well as any flying debris and follow-on effects is the irreversible physical event, and no other events, failures or conditions are necessary for negative impacts to be realized. If the estimated likelihood of such a LOPC event is taken, for example, as 10^{-5}/year (Reference 4-16), then that estimate is also the loss event frequency used to calculate scenario risk.

More commonly, the occurrence of a loss event requires the combination of an initiating event along with the failure of one or more safeguards and/or the presence of one or more conditions such as the presence of an ignition source. For example, the first loss event of a gasoline (petrol) storage tank overfilling and spillage in the Buncefield, UK incident of 11 December 2005 (Reference 4-17) required the tank level gauge failing to indicate the proper level while the tank was being filled, plus failure of the tank overfill protection system. The second loss event of a damaging vapor cloud explosion apparently required additional conditions to exist at the time of the tank overfilling; namely, the composition of the gasoline being a 'winter mix' incorporating about 10% butane, the overfill occurring at nighttime under very low wind speed conditions thus minimizing dispersion, the presence of an ignition source, the presence of undergrowth and trees providing sufficient congestion to accelerate the flame velocity, and a transition to detonation in part of the remaining vapor cloud (Reference 4-18). Additional loss events occurred as domino effects damaged nearby tanks and the ensuing fire spread.

For these more complex scenarios, the loss event frequency is calculated by combining the frequency of one event (the *initiating cause* or *initiating event*) with dimensionless probabilities for all of the other failures and conditions. The non-initiating failure probabilities are either simple probabilities of failing to respond and so keeping the scenario from continuing (such as an operator failing to respond in time to an alarm condition) or are probabilities of failure on demand (PFDs) of standby systems. The latter, which are generally associated with unrevealed (latent) asset failures, can be calculated for individual repairable components having constant failure rates as:

$$PFD = 1 - \exp(-\lambda\tau)$$

where λ is the fail-dangerous or fail-to-respond failure frequency and τ is the failure duration (i.e., the average time to detect and correct the failure). Assuming random failures, on the average, the failure duration is equal to one half of the functional testing/inspection interval. If a safeguarding layer of protection involves multiple components, the PFD of the safeguard is basically the sum of PFDs of the individual components making up the layer of protection (such as the sensor, logic solver and final control element of a simple safety instrumented system).

Thus, it can be seen from the above that asset failure frequencies, along with test/inspection intervals for repairable components, are the key determinants in assessing the loss event frequency for a scenario. The asset failure frequencies can be LOPC events, initiating events and/or safeguard component failures. These failure frequencies are best determined by actual in-service operating experience when sufficient relevant experience is available to estimate the failure frequencies. Development of failure frequencies from experiential data, as well as discussion of failure taxonomies and presentation of some generic failure rate data for when sufficient experience is not available, are all given in Reference 4-19. ISO 14224 (Reference 4-3), which is also issued with identical content as API Standard 689, defines a taxonomy of failure modes and gives a standardized approach to the collection of reliability and maintenance data, with a focus on reliability performance. Reference 4-16 gives a basic compilation of LOPC frequencies typical of oil and chemical process operations. Other failure data sources, as well as a full discussion of frequency determination and risk calculations, are given in Reference 4-8.

5
ASSET SELECTION AND CRITICALITY DETERMINATION

Early in the development of an asset integrity management (AIM) program, facility personnel need to establish boundaries and develop a list of the assets to include in the program. This chapter discusses some criteria to consider when identifying the assets to include in an AIM program, as well as to possibly determine a criticality level for each asset. Addressing these items will help to ensure that (1) the program includes all desired assets, (2) the assets are prioritized for AIM activities consistent with the program's objectives and philosophy, and (3) the basis for the program is consistently understood. Some of the steps that facilities take to accomplish this include:

- Reviewing the program objectives and philosophy

- Establishing the asset selection criteria, including defining which types of assets to include or exclude from the AIM program as well as possibly determining a criticality for each asset or asset type

- Defining the level of detail, such as whether assets are to be included individually or only as part of a system

- Documenting the asset selections and criticality determinations.

This effort produces an asset list that can serve as the basis for establishing inspection and testing plans, record-filing systems, spare parts and materials lists and other asset-specific elements of the AIM program.

5.1 PROGRAM OBJECTIVES AND PHILOSOPHY

Reviewing the AIM program objectives before establishing the asset selection scope will ensure the asset selection is consistent with the philosophy and intent of the program. If the facility must comply with regulatory requirements, then the jurisdiction may dictate program boundary requirements including specific assets that must be included in the AIM program. However, be aware that regulations may be performance-based, and interpretations of the requirements may change.

As discussed earlier, company management is generally motivated to develop AIM programs for reasons extending beyond regulatory compliance. Reducing the likelihood of process safety, occupational safety and/or environmental incidents is a common incentive for including additional assets in the program. Other motivations for including additional assets include reliability improvements, such as downtime reduction and/or extended asset life, and product quality improvements. Note that these motivations may also suggest the need for additional inspection, testing and quality assurance tasks.

Various alternatives or supplements to compliance-based asset selection are employed by industry, with selection criteria varying from company to company depending on program objectives and philosophy. Alternatives to a compliance-based approach include the following or any combination thereof:

- *Consequence-based* asset selection, which considers the potential severity of consequences (impacts) if an asset fails when considering whether to include it in the AIM program or when assigning a criticality to individual assets, to types or categories of assets, or to systems that include multiple assets. For example, if failure of a level control system has the potential to result in a tank being overfilled and resulting in a toxic release or major fire, then the level control system would be included in the AIM program and perhaps assigned a criticality level (as discussed in Section 5.4) based on the potential toxic release or major fire impacts. A consequence-based approach may also include preventive safeguards such as overflow prevention systems and/or mitigative safeguards such as secondary containment, water curtains or fire protection systems if they provide a layer of protection against incidents with impacts above a threshold severity level.

- *Risk-based* asset selection, which is similar to a consequence-based approach but considers both the potential impacts as well as likelihood of occurrence when selecting assets for the AIM program and/or assigning criticalities.

- *Prescriptive* asset selection, which is generally based on corporate-level directives indicating mandatory inclusion of certain asset types regardless of potential consequences or level of risk. For example, a company may decide that assets such as particular classes of pressure vessels, all safety instrumented systems and specific types of relief systems are always to be included in the AIM programs at all of the company's facilities.

It should be noted that selection of assets for an AIM program is not merely a one-time effort. The asset list, as well as any criticality determinations (as discussed in Section 5.4), will change as new and modified assets are installed or systems are decommissioned. (See Reference 5-1 for a full discussion of change management systems.) In addition, depending on the selection criteria, assets might be added or removed from the AIM program or their criticality level changed as their condition or service changes over time. For example, if significant corrosion is detected in a vessel or piping system to the extent that is starting to affect its corrosion allowance, the criticality level of the vessel or piping system might be elevated. Other possibilities may include changes based on operating and inspection history, hazard identification and risk analysis updates, near misses, incident investigations, new standards and new technical knowledge.

Some facilities have found that they can improve program credibility by not trying to be overly ambitious in asset selection. This credibility is important for gaining cooperation in achieving compliance and progressing toward further objectives. Successful AIM programs are often implemented with a relatively small initial scope (such as a pilot test) that can be expanded or intensified after AIM becomes part of the facility's culture. However, following through with program expansion when practical is obviously important.

5.2   ASSET SELECTION CRITERIA AND PRINCIPLES

This section identifies rationale for establishing criteria for selecting what assets to include in an AIM program. Defining the types of assets to include may involve interpretation of regulations, clarification of other program objectives, statement of performance objectives, and often some case-by-case decision making.

While portions of this exercise may involve uncertainties, establishing and documenting clear criteria that explain the rationale for including and excluding assets in the program is important to the program's success. (Note: Many times the rationale for excluding assets is at least as useful as the rationale for including them.) Clearly defining criteria may not prevent asset selection decisions from being challenged (such as by internal or third-party auditors); however, consistent application of clear asset selection principles is more readily defensible than inconsistent application of vague guidelines, even when the clear principles are less ambitious. Documentation of the selection criteria, the process used to define an asset list using the criteria, and the appropriate roles and responsibilities are discussed in Sections 5.5 and 5.6. An example of asset criteria documentation is provided in Appendix 5A.

Fixed Equipment, including Containment and Relief Systems. Because a common objective in the process industries is to contain hazardous materials and energies, AIM programs almost always include (1) pressure vessels; (2) atmospheric and low-pressure storage tanks; (3) piping and piping components including hangers/supports, bolting, valves, heat exchangers, in-line filters, eductors and venturis; and (4) the vent systems and pressure relief devices (such as pressure safety valves, rupture disks, pressure-vacuum vent valves and weighted hatches) that protect these vessels, tanks and piping systems. Listing pressure vessels, tanks, and relief devices is usually straightforward. Adding piping to the asset list can be more difficult; this is discussed in Section 5.4 on determining the right level of detail.

In addition to pressure vessels, tanks and piping systems, secondary containment components such as dikes, curbs, sumps and other waste collection systems are often considered for inclusion. Frequently, environmental, safety and regulatory considerations drive companies to include secondary containment. Fireproofing on structural components and insulation on tanks, particularly when the relief design takes credit for insulation, are additional passive protection systems that may be considered for inclusion in the AIM program. Functional piping components such as filters, eductors, and venturis may have similar performance objectives.

Rotating Equipment. The following are considerations regarding rotating and reciprocating assets:

- If they contain hazardous materials, maintaining containment is a primary function.

- If ensuring process flow is an objective, then the drivers (e.g., turbines, motors) may need to be included.

- If loss of function of non-containment items such as agitators, conveyors, blowers and fans has process and/or personnel safety implications, these items may also need to be included.

- Containment and/or functionality of non-process rotating and reciprocating items, such as in cooling water systems, steam systems, refrigeration systems and power distribution systems may also be important, resulting in the inclusion of some non-process assets as well.

- Lube oil and seal fluid systems will likely have the same requirements as the rotating or reciprocating assets with which they are associated.

Instrumentation. Instrumentation is an important consideration in most AIM programs. Identifying which instruments to include and determining the particular activities desired (e.g., functional tests, QA verifications) can be problematic. In addition, an instrument may be included solely due to its inherent need to contain the process rather than the ultimate use of the process information transmitted.

Some instrumentation provides a process control function, maintaining the process within its integrity operating window. If a particular control loop is important from a process safety perspective, then ITPM activities need to include all the components in that control loop.

Other instrumentation is intended to warn of or respond to abnormal situations in the process. If this instrumentation is considered an essential layer of protection against a process safety incident (or part of an essential layer of protection, such as a critical alarm intended to alert an operator to take corrective action, or an operator-actuated emergency shutdown system), it will also need to be fully maintained. Such instrumentation may also be important for business purposes, such as for monitoring fugitive emissions or providing an essential backup function.

Instrumentation may also be used for testing and/or calibration purposes. These instruments will have their own requirements for inspection, testing and maintenance.

Still other instrumentation may provide a last line of defense against a significant loss event. Examples include emergency blowdown systems and safety instrumented systems. These last-resort instrumented systems are usually a primary focus of AIM activities involving instrumentation.

A variety of methods exist for developing instrumentation lists for AIM programs (sometimes called "critical instrument" lists). Some facilities include all instrumentation associated with any other piece of covered equipment, some have asked the facility operations department(s) for instrument lists, while others have extracted instrument lists from the safeguards documented in hazard identification and risk analysis reports. Any of these approaches can work, but each also has typical flaws associated with it. Including all instrumentation that is associated with other covered equipment can be a simple approach to setting up the program, but can result in (1) a long list of overdue instrument testing and/or (2) the diversion of resources from more critical AIM tasks. Getting input from the operating department is beneficial, but without guidelines and examples to follow, operations-generated instrument lists are often inconsistent and hard to defend.

Similarly, teams performing hazard identification and risk analyses (HIRAs, also known as process hazard analyses or PHAs) can be an excellent resource. However, using only HIRA reports can be inadequate, such as where an instrument has only a process containment function and not a measurement function. To increase the effectiveness of the HIRA as a resource, HIRA teams can be provided with a specific objective of identifying instrumentation to be included in the AIM program, along with guidelines and examples for instrument selection and for quality and completeness review.

Using the results of studies conducted to satisfy safety instrumented systems standard requirements (Reference 5-3) can help facility personnel establish criteria for selecting instrumentation that provides emergency functions. In addition, when Layer of Protection Analysis (LOPA) has been used to analyze facility safeguards, the LOPA results can identify instrumentation (in addition to other safeguards) that is important to process safety. More information on LOPA is provided in Chapter 15.

Fired Equipment. Equipment such as furnaces and fired heaters often operate at extreme conditions and can involve many possible damage mechanisms to system components including burners, tubes and refractory lining. Not only will these system components need to be monitored and maintained, but also the combustion safeguards associated with the fired equipment as well.

Utilities and Support Systems. Utility and support systems may be critical to process operations and as such ought to be considered for inclusion in the AIM program. For example, loss of nitrogen may lead to ignitable vapor spaces in storage tanks. Loss of containment of nitrogen in an analyzer building could lead to an asphyxiation hazard. Devices such as power generation and distribution, uninterruptible power supplies, emergency communications systems and electrical grounding and bonding systems can serve important safety functions. Specific utility systems for consideration may be found in hazard identification and risk analysis studies.

Mitigation Systems. Another important type of assets for consideration is those assets and systems in place to mitigate or to act as the final means to prevent chemical releases, fires and other catastrophic events. This may include flame arresters, suppression systems, fixed and portable fire protection equipment, sprinkler and deluge systems, emergency inerting, reaction quenching or "kill" systems, emergency dump systems, flammable gas detectors and emergency isolation devices.

AIM program developers at some facilities may be inclined to omit traditional safety, fire protection, emergency response, plant evacuation alarms, building ventilation and/or power distribution equipment from AIM asset lists because this equipment is procured, inspected and tested by personnel from other departments (e.g., safety personnel, fire chief) or by contract personnel supervised by other departments. However, AIM activities for those assets need to also meet or exceed the AIM program requirements, and such AIM activities need to be appropriately managed and documented.

Third-Party Assets. Another area to consider is assets belonging to another company, such as a chemical vendor or bulk gas supplier, which are connected to a process within the host facility. The host facility will ultimately be accountable for any mishaps that lead to safety, environmental, or plant performance issues; however, the company owning the assets is generally responsible for asset maintenance (depending on contractual terms). The vendor's assets might be included in the facility's AIM program in the same manner that the facility is evaluating its own assets. Frequently, the vendor will still perform the AIM activities; however, the host facility can take steps to ensure that the vendor's AIM activities comply with or exceed the AIM program requirements of the host facility.

Transportation Assets. Transportation-related assets, including railcars, ISO containers, tube trailers, trucks, and barges used for on-site storage, are usually considered part of a process facility during the time they are connected to a process and not connected to a motive force (e.g., a truck trailer staged by the process and disconnected from the truck). The AIM activities associated with the transportation asset are frequently the responsibility of the transport company and may be difficult for facilities to track. Facilities need to ensure that transport companies are aware of AIM requirements and make provisions to verify that AIM activities are occurring.

In addition, cross-country pipelines are generally considered transportation assets. AIM activities for such pipelines involve not only the piping system components but also their associated cathodic protection systems. The same is true for other transportation assets such as marine docks.

Temporary Components. Including in the AIM program provisions for addressing temporary components can be integrated with a facility's management of change (MOC) procedure. For instance, any asset that (a) is part of a temporary change, (b) will be in place for more than a defined length of time and (c) would meet the facility's AIM asset selection criteria may be required to have appropriate activities performed as part of managing the temporary change. However, since the change is temporary in nature, the asset may not be included on the AIM asset list. Possible examples are pilot test equipment and temporary repair assets such as leak repair clamps. (Note that some leak repair clamps are more permanent in nature, and as such may be required to have a more formal inclusion in the AIM program.)

Structural and Support Facilities. Structural components, such as foundations and structures that support the weight or movement of other assets (e.g., piping supports/hangers, columns, pipe racks and bolting), are appropriate considerations for inclusion in an AIM program, either separately or as part of associated equipment. Decision criteria may include the age and history of the facility, the apparent condition of the structural components, and geographical issues such as the potential for earthquake or hurricane damage. Even new installations need to be inspected for structural defects.

Some support facilities that may need to be taken into account include floating production and storage facilities with possible buoyancy or stability issues, buildings (in particular, occupied structures providing protection to personnel during incidents), and building components such as ventilation systems and flammable/toxic vapor ingress detection systems.

5.3   LEVEL OF DETAIL

Once the asset selection criteria have been established, facilities can generate an AIM program asset list. The starting point is often a *master asset list* (also termed *master equipment list*). For use with AIM program development, this master asset list may need to be validated such as by a line-by-line piping and instrumentation diagram (P&ID) review, including circuitized piping systems. For process facilities, because consistency between the master asset list and the P&IDs is so important, the process of establishing the master asset list would greatly benefit from a detailed walkdown of the P&IDs for field validation at the same time.

Ideally, the master asset list would consist of a single database organized by asset type. An asset taxonomy may be needed for this purpose. See ISO 55000 (Reference 5-4) for further information on asset taxonomy.

Itemizing the assets in an asset list can be done at different levels of detail. For this reason, facilities may need to establish a consistent approach to find the most appropriate level of detail. Some guidance on level of detail is as follows:

- *Pressure vessels.* Generally, facilities include internal coils, liners, and jackets with the pressure vessel. Determine whether to give special designations to any or all of those components.

- *Storage tanks.* Storage tanks generally have some unique aspects related to tank inspections such as floors, roof seals, internal coils, under-floor heaters and cathodic protection.

- *Rotating equipment packages.* Lube oil systems, seal flush systems, and other support items for large rotating equipment may have drivers, piping, pumps, pressure vessels, and instrumentation to include in the AIM program. Sometimes, these items are installed together on a skid and without individual asset numbers. Some facilities assign numbers and list the items separately. Other facilities group the support items with the rotating equipment items.

- *Utility and support systems.* Some facilities group an entire utility or other support system together. Other facilities list the individual system components. Ensure that entire system listings include a full description of the components in the system.

- *Vendor packages.* Some facilities itemize the separate pieces of vendor assets. Others group vendor assets together system by system. Again, when listing as a system, include a full description of system components.

- *Piping.* In a relatively simple facility, piping can be listed by system description. More often, facilities use a line numbering system. Whichever approach is used, the facility needs to determine how piping system components will be included and identified. Piping appurtenances (e.g., expansion loops, expansion joints, sight glasses, drain lines, nipples, emergency isolation manual block valves, electrical bonding on piping, and cathodic protection) need to be appropriately considered. Within a piping circuit, it is common to have specific components (e.g., critical valves, hoses, expansion joints) called out as unique assets with their own requirements. It is important to correctly document these items so they are clearly defined and managed by the AIM program. Developing isometric circuit diagrams and/or highlighted piping and instrumentation diagrams (P&IDs) to accompany piping system descriptions is beneficial. Not only do P&IDs need to be kept up to date for use in identifying piping system components to be included in the AIM program, but piping isometric drawings also need to be kept up to date in order to identify condition monitoring locations (CMLs).

- *Instrumentation and control loops.* Because AIM activities are performed by different groups and because the frequency of component testing may vary, some facilities prefer to list the components of process control or safety instrumented systems separately (e.g., sensor, controller, transducer, final control element). Other facilities identify and list the safety systems or instrument loops themselves (e.g., jacket cooling flow control loop). Either approach can be used. Ensure that the functionality and associated logic are tested for the entire system even when the asset list identifies the components separately. Section 12.3 includes equipment-specific integrity management considerations for instrumentation and controls, such as whether and how the basic process control system needs to be maintained.

- *Pressure relief / vent systems and devices.* Many relief devices are individually numbered. However, miscellaneous assets such as seal loops and weighted hatches are sometimes included as part of the vessel. Ensure that AIM tasks are in place for these items. Also, relief device discharge piping is frequently inspected visually as part of relief device servicing. This piping is generally catalogued with the devices. Ensure that the integral components of these systems (e.g., flares, relief catch pots, relief dump tanks, emergency scrubbers, burst disk indicators) are recognized.

- *Drainage systems and containments.* If not characterized as piping systems as described above, drainage systems are likely to be inspected and maintained as systems in their own right, with the same being true of dikes, impoundments, and other secondary containment systems that are integral to preventing follow-on effects and minimizing environmental impacts and personnel exposures.

- *Fire and chemical release mitigation systems.* Similar to the instrument loop considerations, these systems can be treated as whole systems or individual components. Also, associated piping systems may be numbered or named. Ensure that functional testing verifies mitigation system performance. Mitigation systems that are used in emergency operations, such as employee alarm systems, may have whole-system functional testing requirements that are already established by codes, standards or regulations.

The considerations listed above are not likely to be all-inclusive. A final review may be needed to detect and correct any inconsistencies found in the application of the AIM program asset selection criteria.

5.4   ASSET CRITICALITY DETERMINATION

Some companies designate certain assets or parts of assets as "critical assets." This criticality may be associated with safety (i.e., "safety-critical equipment"), environmental protection, security and/or business considerations. (These criticality terms are defined in Section 5.4.2.) The implication is that the integrity of critical assets is managed differently than for those assets not considered critical, as considered in this section. For example, a criticality rating for a particular asset may be used to determine how often the equipment is to be inspected, tested or maintained. It may also be used by the maintenance scheduler as a guide to which work orders require immediate attention and which may be rescheduled to a future date when resources become available.

Even so, the integrity of all assets selected for inclusion in the AIM program is still managed with respect to appropriate inspection, testing and preventive maintenance activities as well as quality control and deficiency management, as discussed in the following chapters. Criticality determination is not mandatory in most jurisdictions; some companies may decide to manage all assets the same way with respect to planning and executing inspection, testing and maintenance tasks.

5.4.1   Historical Background

The first approaches to systematically ranking the criticality of equipment were developed by the U.S. military in the 1940s with the publication of a standard (Reference 5-5) for Failure Modes, Effects and Criticality Analysis (FMECA) to evaluate the potential impact of each functional or hardware failure on mission success, personnel and system safety, maintainability and system performance. The technique used failure mode analysis to rank each potential failure by the severity of its effect so that corrective actions may be taken to eliminate or control design risk.

This technique may be applied to any electrical or mechanical equipment or system, so that later in the 1960s NASA contractors (References 5-6, 5-7) used variations before NASA published its own standard (Reference 5-8). Simultaneously, the use of FMECA was expanding into the civil aviation and automotive (Reference 5-9) industries. Meanwhile, in Europe, standards were also published for general use (References 5-10, 5-11).

While initially the main focus of equipment criticality ranking was on production and quality issues, increasingly plants have also used an equipment ranking approach for occupational safety and environmental performance (emissions, ISO 14001, regulatory compliance). More recently, some companies in the process industries have started to apply criticality ranking for process safety purposes, including regulatory compliance (in Europe), and to make decisions regarding which ITPM tasks may be rescheduled and which must be done on the specified date.

Prior to 1988, the UK Department of Energy and certification bodies defined what safety measures were required for offshore oil production installations. This inflexible prescriptive approach did not account for the differences in hazards between each installation. Following the 1988 Piper Alpha disaster and the subsequent report of the Cullen public inquiry (Reference 5-12), a more flexible, goal-setting regulatory regime was introduced that allowed for alternative solutions to hazard management. In particular, operators are required to define and identify their own safety-critical elements and their required performance standards (Reference 5-13) for managing the appropriate hazard(s) in terms of:

- Functionality,

- Availability,

- Reliability,

- Survivability, and

- Interaction/Dependency.

5.4.2 Definitions

The underlying philosophy of managing some assets within an AIM program differently than others, considering them to be "critical assets," is generally a recognition of the realities that resources are limited and the best maintenance planning is thwarted by unforeseen circumstances, resulting in situations that may not allow all inspection, testing, preventive maintenance and deficiency management tasks to be done on time and according to plan. Even so, because some assets are critical to the operation of a process, the priority on the maintenance of critical equipment needs to be elevated such that if not all AIM tasks can be accomplished on time for whatever reason, maintenance of the critical equipment will not be deferred. Although personnel safety, public safety and environmental protection is the primary AIM focus from a process safety perspective, business considerations might also be important when considering equipment criticality.

A suggested definition of the term *critical asset* is as follows:

Critical asset

Asset, the malfunction or failure of which could (or is likely to) cause, contribute to, or fail to prevent or mitigate a major business impact or a major safety, environmental or security incident

---

In the case of a criticality determination for process safety reasons, equipment falling into this prioritized realm has been called *safety-critical equipment (SCE)* or *safety-critical elements (also abbreviated SCE)*. The term "safety-critical" does not fully reflect the objective of identifying and managing critical equipment. The term *major-incident critical* would be more appropriate.

There are not one but two basic ways to define these terms. Suggested definitions are as follows:

Safety-critical equipment/element
(consequence-based definition)

Equipment, the malfunction or failure of which could cause or contribute to a major incident, or the purpose of which is to prevent a major incident or mitigate its effects

---

Safety-critical equipment/element
(risk-based definition)

Equipment, the malfunction or failure of which is likely to cause or contribute to a major incident, or the purpose of which is to prevent a major incident or mitigate its effects

---

A consequence-based definition tends to be less useful for the purposes described in this section, since it tends to capture a large percentage of equipment, particularly in high-hazard operations. For example, in a process handling a toxic and/or flammable fluid under pressure, practically all elements of the primary containment system would be captured by the consequence-based definition, since any loss-of-containment failure could result in a major incident. The use of a risk-based definition is discussed in Section 5.4.5.

Of course, for these definitions to be useful, the term *major incident* also needs to be defined. This term (or the similar term *major accident*) may already be defined in some countries or industries. For example, the definition of *major accident* used by the UK Health and Safety Executive (Reference 5-14) is:

An occurrence such as a major emission, fire, or explosion resulting from uncontrolled developments in the course of the operation of any establishment [to which the COMAH regulations apply] and leading to serious danger to human health or the environment (whether immediate or delayed), inside or outside the establishment, and involving one or more dangerous substances.

It is suggested that, absent a predetermined definition, it be defined as a process safety incident falling into either of the highest two severity categories in the process safety event severity table of ANSI/API RP 754 (Reference 5-15), modified as necessary to meet a particular company's objectives. Table 5-1 shows the severity categories from this source, with the two highest-severity categories identified as "major incident" impacts. Use of a broad definition of the term *major incident* with the Table 5-1 categories (perhaps not limiting economic losses to direct costs) can capture critical equipment for business and environmental reasons as well as safety.

5.4.3 Why Determine Asset Criticality?

Designating asset criticalities will provide opportunities for adapting maintenance strategies to asset criticality, and, if the correct priorities are assigned and followed, it can ensure high plant availability and low business interruption. Several possible reasons exist for designating some assets within an AIM program as critical assets, or assigning a criticality level to all equipment within the program; some of these reasons also have specific potential benefits associated with them:

- To meet specific regulatory requirements that dictate designating and managing critical assets in a certain way.

- To establish priorities for determining where to focus preventive and predictive maintenance activities.

- To set priorities for reliability-centered maintenance (RCM) selection and the application of equipment failure root cause analysis (RCA).

- To define the need for and optimum level of spare parts and other resources for the facility.

- At the design stage, to indicate the need for redundancy and the selection of alternative equipment based on life cycle cost analysis.

TABLE 5-1. Suggested Major Incident Definition

(Levels extracted from Table 4 of Reference 5-15; see Reference 5-15 for details; all values in US$)

Severity Points		Safety / Human Health	Direct Cost from Fire or Explosion	Material Release Within Any 1-Hr Period	Community Impact	Off-Site Environmental Impact
Major Incident severity levels	**27**	• Multiple fatalities of employees, contractors, or subcontractors, or • Multiple hospital admissions of third parties, or • A fatality of a third party	Resulting in ≥ $100,000,000 of Direct Cost Damage	• Release volume ≥ 27x Tier 1 TQ outside of secondary containment	• Officially declared evacuation > 48 hours	• Resulting in ≥ $100,000,000 of Acute Environmental Costs, or • Large-scale injury or death of aquatic or land-based wildlife
	9	• A fatality of an employee, contractor, or subcontractor, or • A hospital admission of a third party	• Resulting in $10,000,000 ≤ Direct Cost Damage < $100,000,000	• Release volume 9x ≤ Tier 1 TQ < 27x outside of secondary containment	• Officially declared evacuation > 24 hours < 48 hours	• Resulting in $10,000,000 ≤ Acute Environmental Cost < $100,000,000, or • Medium-scale injury or death of aquatic or land-based wildlife
Severity levels below Major Incident threshold	**3**	• Days Away From Work injury to an employee, contractor, or subcontractor, or • Injury requiring treatment beyond first aid to a third party	• Resulting in $1,000,000 ≤ Direct Cost Damage < $10,000,000	• Release volume 3x ≤ Tier 1 TQ < 9x outside of secondary containment	• Officially declared shelter-in-place or public protective measures (e.g., road closure) for > 3 hours, or • Officially declared evacuation > 3 hours < 24 hours	• Resulting in $1,000,000 ≤ Acute Environmental Cost < $10,000,000, or • Small-scale injury or death of aquatic or land-based wildlife
	1	• Injury requiring treatment beyond first aid to an employee, contractor, or subcontractor	• Resulting in $100,000 ≤ Direct Cost Damage < $1,000,000	• Release volume 1x ≤ Tier 1 TQ < 3x outside of secondary containment	• Officially declared shelter-in-place or public protective measures (e.g., road closure) for < 3 hours, or • Officially declared evacuation < 3 hours	• Resulting in $100,000 ≤ Acute Environmental Cost < $1,000,000

Assets or parts of assets designated as critical may require personnel maintaining those assets to have special training and qualifications, to follow specific QA requirements and procedures, to conduct all ITPM activities as scheduled with no deferrals, to allow no substitutions without a high level of review and authorization, and to conduct a safety review for preventive maintenance activities on the critical assets.

5.4.4 Criticality Analysis

Criticality analysis identifies how important equipment is to the operation of a facility, taking into account the safety, environmental and economic consequences of the equipment failing or not meeting its intended design function.

Criticality analysis

Quantitative analysis of events and effects and their ranking in order of the seriousness of their consequences

---

A facility needs to define consequence severity categories consistent with overall company business drivers. This takes the guesswork out of ranking and provides uniformity between process units. Criteria often include:

- Plant operations in relation to economic losses; i.e., breakdowns, repair costs, lost production, poor product quality, incidents, regulatory non-compliance, etc.

- Safety; i.e., injury, fatality

- Environment; i.e., air/water/waste emissions.

Other considerations may include maintainability, product quality, food safety or other public health impacts and social media attention if any is above a threshold impact. The levels previously introduced in Table 5-1 are one example of a criticality ranking that can be used in an AIM program.

Rankings are assigned on the basis of the impact of failure in each category. Some companies have employed a risk-based criticality using a risk matrix instead of a purely consequence-based approach. Others use a concept of *overall equipment effectiveness* (*OEE*); see e.g. Reference 5-16:

$$OEE \; = \; A \bullet P \bullet Q$$

where A = Availability (% of scheduled time available for production; uptime)
        P = Performance (% use of the designed production speed/capacity)
        Q = Quality (% of production with good quality)

In addition, a separate performance index for safety and environment may be used.

It should be noted that determining equipment criticalities is not always done by a formal, quantitative criticality analysis. Criticality selection can also be prescriptive, based on other factors such as regulatory requirements, consistency with best practices, and incident histories. For example, a company might prescribe that all emergency relief devices protecting process vessels above a certain volume and/or design pressure, or all emergency relief devices in toxic service, be designated as safety-critical equipment.

Whether done by analysis or otherwise, a clear link needs to be made between identified major incident hazards and the equipment that would make the most significant contribution to their prevention, detection, control and mitigation. When considering process safety, proper management, including timely completion of all ITPM tasks, of the identified safety-critical equipment ensures that the necessary safety barriers to prevent and mitigate major incidents are in place and functioning as intended. Examples of possible safety-critical equipment include pressure safety valves, emergency shutdown valves and deluge systems. Table 5-2 lists other possible safety-critical equipment for oil and gas exploration and production facilities (not intended to be an all-inclusive list, and not all of the items on the list will be determined to be safety critical for every such facility).

TABLE 5-2. Some Typical Safety-Critical Equipment for Exploration and Production Facilities

Typical SCE for E&P Facilities
Downhole safety systems
Wellhead safety systems
Safety instrumented systems including combustion safeguards
Primary containment systems for pressurized processes
Ventilation systems
Pressure relief valves and pressure safety valves
Blast relief panels
Emergency depressuring and blowdown systems
Gas and fire detection systems
Fireproofing
Firefighting equipment; sprinkler, deluge and fire suppression systems
Emergency warning systems; other emergency communications equipment
Structures, havens and refuges for protection of personnel in an emergency
Fixed support structures
Floating support structures
Floating structure mooring/anchoring systems
Other equipment identified as safety critical in process hazard analyses or by other means

5.4.5   Risk-Based Criticality

As described in Section 5.4.1, European regulations for safety-critical elements (SCE) are primarily focused on functionality and availability/reliability while surviving fire/explosion incidents they are intended to prevent and/or mitigate. This consequence-based approach has led to a comprehensive definition for SCE that often includes a large majority of items on the master equipment list. This is hardly surprising, given the origin of the regulations that stem from the Piper Alpha disaster (Reference 5-12), where fires and explosions resulted in:

- Failure of major hydrocarbon inventories (gas risers, piping systems, pressure vessels, diesel storage)

- Destruction of the control room

- Failure of communication systems

- Inaccessibility of fire pumps on manual start

- Impairment of the environment in the temporary refuge (living quarters)

- Inaccessibility of evacuation, escape and rescue systems (egress routes, helideck, lifeboats/rafts).

By contrast, for the purposes of asset integrity management, a narrower definition of safety-critical equipment based upon *risk* rather than consequence may be required. The objective of equipment criticality is to identify the subset of equipment that is absolutely critical to the management of major incident hazards. This is the equipment that ensures the prevention, control and mitigation of the major hazards, and therefore is required to have a high reliability and availability before and during a major incident. Much of this safety-critical equipment will require preventive maintenance and/or calibration to ensure its reliability to function on demand as per specification. Some equipment may become critical or increase in criticality during its life cycle due to deterioration and increased risk of failure, such as a corroded pressure vessel that has been subject to a fitness-for-service assessment and requires more frequent inspections and a greater monitoring effort.

When decisions are taken on the deployment of available resources for ITPM, the designation of safety-critical equipment allows management to optimize resources without deferring tasks associated with safety-critical equipment. As such, the safety-critical equipment inventory in non-regulated operations ideally represents some fraction of the items on the master equipment list (MEL) that covers the majority of the risks. Selecting too large of a fraction of the total MEL will likely result in a lack of focus upon those items that need to receive preferential attention.

Figure 5-1 shows an example flowchart for helping to decide which equipment is safety critical from a risk-based perspective (as defined in Section 5.4.2) and thus is not to be deferred with respect to its scheduled ITPM tasks. *Utilizing Figure 5-1 would require several terms to be defined by the user,* as listed in the figure. Some suggested guidance for these terms is as follows:

- *Stored energy* – Most commonly, a compressed gas or vapor contained in process-related equipment such as a pressure vessel, boiler, piping system or tank with sufficiently high pressure and volume to potentially cause severe injury or loss if suddenly released; the stored energy could also be in a different form such as significant stored thermal, kinetic, hydraulic, positional, electrical or chemical energy.

- *Hazardous material* – The definition of this term, as well as the *stored energy* term, could take into account energy control procedures, prescriptive regulations and process hazards identified in HIRAs/PHAs.

- *Reasonable potential* – This is obviously a key term when using a risk-based determination of equipment criticality. Two examples of how this term might be used are as follows:

  > *Equipment (such as a pressure vessel or piping system component) designed to recognized codes/standards and inspected (in accordance with recognized codes/standards) that does NOT indicate significant degradation (e.g. corrosion, cracking, etc. and most of the corrosion allowance remains) does NOT have a* **reasonable potential** *of failure before the next scheduled inspection.*

  > *Equipment (such as a pressure vessel or piping system component) that DOES indicate significant degradation (e.g. corrosion, cracking, etc. and little or no corrosion allowance remains) DOES have a* **reasonable potential** *of failure. More frequent inspections are probably required and it is essential that these inspections are NOT deferred when due.*

- *Major release* – Sudden, unplanned release of the stored energy or hazardous material having *major incident* potential.

- *Major incident* – See the discussion in Section 5.4.2 above.

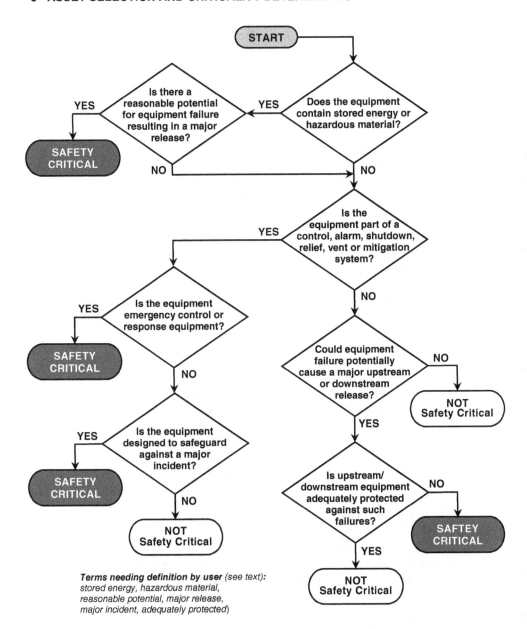

Figure 5-1. Risk-based determination of safety-critical equipment

- *Safeguard* – In this context, equipment in place to decrease the probability *(preventive safeguard)* or mitigate the severity *(mitigative safeguard)* of a cause-consequence scenario. These safeguards might be identified in risk analyses or Layer of Protection Analyses as part of independent protection layers protecting against major-incident consequences.

- *Alarm* – This would typically include safety alarms but not process-aid alarms. It would include alarms credited as *safeguards* against *major incidents*.

- *Mitigation system* – System in place to reduce loss event impacts. Typical mitigation systems include excess-flow valves, secondary containment systems, fire protection systems, emergency ventilation, vent scrubbers, flares and blast protection.

- *Potentially cause a major upstream or downstream release* – The terms *upstream* and *downstream* refer to connected processes. An example of this might be an equipment failure in a utility system providing cooling to an exothermic reactor with runaway reaction potential.

- *Adequately protected* – Having adequate safeguards to reduce risks associated with equipment failure to a tolerable risk level for the facility, taking into account the redundancy built into layers of protection. For example, if failure of equipment XYZ could cause overpressure of upstream and/or downstream equipment, is the upstream and/or downstream equipment safeguarded by a properly designed and maintained safety shutdown system and/or emergency overpressure protection? If YES, then the safety shutdown system and/or emergency relief system is safety critical but equipment XYZ is not.

5.5   DOCUMENTATION

To ensure that the scope of the AIM program is clearly communicated and understood, the assets selected for inclusion in or exclusion from the AIM program, as well as the determination of asset criticalities, need to be documented. This documentation is typically captured in some form of database that indicates, for each asset, the asset name and identifier (e.g., tag number, serial number or loop number), any criticality designation (e.g., whether it is considered safety critical), and perhaps comments related to the AIM program. Note that, in most cases, a facility's computerized maintenance management system (CMMS) will not have the functionality needed to document all necessary AIM information, allow accessibility to all parties needing the information and/or have the required data security to prevent unintended alterations.

If an asset is <u>not</u> to be included in the AIM program, the rationale for exclusion of the asset from the AIM program (i.e., the selection criteria) needs to be documented to help preserve corporate memory. Thorough documentation will also ensure that auditors and regulators can be provided a defensible case for excluding an item. For example, one justification for exclusion may be that the asset is outside of the process boundary with respect to regulatory coverage. Criticality determinations and deferral procedures also need to be documented.

Just as other important documents, the asset list needs to be kept current. Additions, deletions, and significant modifications (such as changes in criticality level) need to be tracked as part of change management, such that the asset list is always kept up to date. In addition, other documents that are based on the asset list (e.g., inspection plans, QA plans) may need to be updated when changes occur.

5.6   ROLES AND RESPONSIBILITIES

The roles and responsibilities for asset selection and criticality determination can be assigned to personnel in various departments. Typically, personnel in the Maintenance, Engineering, EHS (Environmental, Health and Safety) and Operations departments will be involved, perhaps as part of a multidisciplinary team. Example roles and responsibilities for the asset selection and criticality determination process are shown in Table 5-3. While the specific information in Table 5-3 may differ from one facility to the next, what remains consistent is the importance of assigning and communicating roles and responsibilities to the appropriate personnel. This matrix uses the following letter designators:

R   indicates the job position(s) with primary <u>responsibility</u> for the activity

A   indicates <u>approver</u> of the work or decisions made by a responsible party

S   indicates the job position(s) that typically <u>support</u> the responsible party in completing the activity

I   indicates the job positions that are likely to be <u>informed</u> when the activity is completed or delayed.

Note that different assets included in the AIM program may have different accountable parties for actual asset-level management.

TABLE 5-3. Example Roles and Responsibilities Matrix for Asset Selection and Criticality Determination

R = Primary Responsibility A = Approval S = Support I = Informed Activity	Maintenance Department Personnel			Production Department Personnel		Other Personnel		
	Maintenance Manager	*Maintenance Engineers*	*Maintenance Supervisors*	*Area Superintendent*	*Production/Process Engineers*	*Production Supervisors*	*PSM/AIM Coordinator*	*HIRA (PHA) Teams*
Establishing Selection Criteria								
• Review program objectives	A	R	R		R		S	
• Establish rules to follow	A	R	S	I	S	I	S	
• Agree on level of detail	A	R	S		S		S	
• Document selection criteria	I	R	I	I	R	I	S	I
Generating Asset List and Determining Criticalities								
• Apply criteria to asset lists and/or drawings		R			R		S	S
Maintaining Asset List								
• Review equipment additions, deletions, modifications		R	S	I	S	I	S	

CHAPTER 5 REFERENCES

5-1   Center for Chemical Process Safety, *Guidelines for Management of Change for Process Safety*, American Institute of Chemical Engineers, New York, NY, 2008.

5-2   Center for Chemical Process Safety, *Guidelines for Safe Automation of Chemical Processes, Second Edition*, American Institute of Chemical Engineers, New York, NY, 2016.

5-3   IEC 61511, *Functional Safety: Safety Instrumented Systems for the Process Industry Sector - Part 1: Framework, Definitions, System, Hardware and Software Requirements*, International Electrotechnical Commission, Geneva, Switzerland. Also by the same title ANSI/ISA-84.00.01 Part 1 (IEC 61511-1 Mod), International Society for Automation, Research Triangle Park, NC.

5-4   ISO 55000, Asset Management – Overview, Principles and Terminology, International Organization for Standardization, Geneva, Switzerland.

5-5   MIL-STD-1629A, *Procedures for Performing a Failure Mode, Effects and Criticality Analysis*, U.S. Department of Defense, Washington, DC.

5-6   Neal, R.A., *Modes of Failure Analysis Summary for the Nerva B-2 Reactor*, Westinghouse Electric Corporation Astronuclear Laboratory, WANL–TNR–042, 1962.

5-7   Dill, R., et al., *State of the Art Reliability Estimate of Saturn V Propulsion Systems*, General Electric Company, RM 63TMP–22, 1963.

5-8   *Procedure for Failure Mode, Effects and Criticality Analysis (FMECA)*, RA–006–013–1A, National Aeronautics and Space Administration, Washington, DC, 1966. Retrieved 2010-03-13.

5-9   ARP926, *Fault/Failure Analysis Procedure*, SAE International, Warrendale, PA, 1997.

5-10   IEC 60812, *Analysis techniques for system reliability – Procedure for failure mode and effects analysis (FMEA)*, International Electrotechnical Commission, 2006.

5-11   BS 5760–5, *Reliability of Systems, Equipment and Components Part 5: Guide to Failure Modes, Effects and Criticality Analysis (FMEA and FMECA)*, British Standards Institute, 1991.

5-12   Cullen, W.D., *The Public Inquiry into the Piper Alpha Disaster*, HMSO, 1990.

5-13   *The Offshore Installations (Safety Case) Regulations 2005*, UK S.I. 2005/3117, 2005.

5-14   UK Health & Safety Executive, *The Control of Major Accident Hazards Regulations 2015, Guidance on Regulations*, Publication L111 (Third edition), 2015, www.hse.gov.uk/pubns/priced/l111.pdf.

5-15   ANSI/API RP 754, *Process Safety Performance Indicators for the Refining and Petrochemical Industries*, American Petroleum Institute, Washington, DC.

5-16   Muchiri, P. and L. Pintelon, "Performance Measurement Using Overall Equipment Effectiveness (OEE): Literature Review and Practical Application Discussion," *International Journal of Production Research 46*(13), 1 July 2008.

APPENDIX 5A. SAMPLE GUIDELINES FOR SELECTING ASSETS FOR AN AIM PROGRAM

This appendix documents the criteria that a Fictitious Chemicals Company has established for selecting covered assets for its AIM program. Of particular interest are the non-pressure-containing assets (e.g., instruments) included in the program. To start with, all chemical-containing assets (e.g., vessels, tanks, piping, pumps) and relief devices are in the program. Further prioritization (i.e., on a system-by-system basis) will be used to designate the criticality of some chemical-containing equipment, which may result in the exclusion of assets with non-consequential failures.

Emergency shutdown (ESD) systems are all covered by the Fictitious Chemicals AIM program. For the purpose of establishing an AIM asset list, an ESD is limited to shutdown devices that can be actuated manually and remotely to isolate the process. These systems are generally actuated by control room switches. No automatic interlocks are included without meeting the other criteria set below, even if these interlocks result in the shutdown of the process and/or the closing of ESD valves.

In general, the following philosophy is being applied to all other assets at Fictitious Chemicals Company:

- All systems that are used for detecting and/or mitigating releases after they have occurred (e.g., area hydrocarbon detectors, deluge systems, dike areas) are included.

- Any device that can detect a deviation that will directly lead to a release (e.g., high-high pressure alarm in a vessel with a relief valve that discharges to atmosphere), and the failure of that device is unannounced, is included in the program.

- Devices that detect process deviations that may lead to other process deviations (e.g., high temperature alarms or high level alarms on vessels for which those deviations may cause high pressure) are not included if the last process deviation before the release (i.e., high pressure for the example above) can be detected.

- When no detection device is available for the last process deviation before a release, then devices detecting the contributing causes are to be included (e.g., safeguards designed to prevent compressor failures or seal failures on single-seal pumps).

The following specific scenarios are the results of reviews with Fictitious Chemicals personnel:

- *Flare system failures.* AIM activities that exist for ensuring at least one level of defense against flameouts and loss of purge gas are included in the AIM program.

- *Liquid in the flare system.* During upsets from some processes, liquid hydrocarbon can be released from the flare stack. Therefore, instruments that detect deviations that could contribute to those events (e.g., high level alarms for flare system knockout drums) are covered. For the other process areas, liquid hydrocarbons are unlikely to be released from the flare stacks and level alarms are not covered.

- *Relief devices.* All high-pressure alarms/switches on equipment that relieves to the atmosphere are covered. This includes equipment for which the primary relief route is the flare, but a secondary route is to the atmosphere. If no high-pressure detection is available, devices that prevent or detect contributing causes of high pressure are covered.

- *Restriction orifices* intended to limit relief device release requirements or restrict flows for another safety reason (e.g., safe maximum rate of feed to a reactor) are covered. Note that in many services, corrosion and/or erosion of these orifices is unlikely.

- *Pump seal ruptures (single-seal hydrocarbon pumps).* Devices that are intended to prevent deviations contributing to pump seal failures are covered.

- *Pump seal ruptures (double-seal pumps).* Devices that will alert personnel to the loss of the primary seal (e.g., pressure alarms, sealant system level alarms) are covered. If no devices exist (or if only local indicators are installed), these pumps are to be treated as single-seal pumps.

- *Direct releases prevented by process interlocks.* Interlocks that prevent inadvertent releases of process material directly to the atmosphere (e.g., opening of reactor block valves and dump tank drain valves) are covered.

- *Reverse flow of process materials into utility systems or to the atmosphere.* Check valves, backflow preventers, and pressure regulators that provide the primary defense against hazardous reverse flow scenarios are to be included in the AIM program.

- *Dust explosions and/or flammable concentrations of volatile hydrocarbon in finishing areas.* Safeguards protecting against these scenarios include nitrogen blanketing in several plants and air sweeping through the finishing system. Devices that detect the loss of such systems (pressure alarms, flow alarms, analyzers) are to be included in the AIM program. Fictitious Chemicals also relies on administrative controls (e.g., laboratory analyses) to prevent/detect flammable concentration scenarios.

- *Polymer formation in monomer vessels.* Systems and instrument loops (e.g., nitrogen purges) that are used to prevent plugging of nozzles leading to pressure relief devices are included.

- *Releases associated with barge and railcar loading/unloading.* Hose integrity is to be verified on a periodic basis.

- *Overfilling storage tanks (that do not have relief devices).* Instrumentation used to help prevent the overflow of hazardous materials is included in the AIM program.

- *Process motor failures.* In general, motors and other drivers will only be included in the AIM program if the functioning of the driver is critical to process safety (including particular reactor agitator motors, key compressors, pumps, fan motors, etc.). Note: Many motors will be included in inspection programs for reliability reasons.

- *Building ventilation systems* that are used to control the concentration of hazardous vapors and dusts, such as flammable, explosive, or toxic materials, are included in the AIM program. In addition, include as necessary building features such as seals to help ensure positive pressure.

- *Thermal reactions and chemical decompositions leading to vessel temperatures higher than vessel temperature ratings.* Interlocks and alarms in these systems that help detect and/or prevent these events are included in the AIM program.

- *Freezing of water and other chemicals in process equipment.* Generally, heat-tracing systems designed for freeze protection are included in the AIM program.

- *Cooling systems, including tracing systems, that are used to help prevent runaway reactions* (e.g., at monomer storage tanks, in services with heat sensitive materials) are included in the AIM program.

- *Compressor failures.* Systems and instrument loops that are used to protect compressors (e.g., surge protection, thrust and vibration protection, high- and low-pressure instrumentation, high-temperature instrumentation, high liquid level instruments in accumulators, vibration instruments, lube oil systems and related instruments, intercoolers, knockout drums, and liquid removal systems) are included in the AIM program.

- *Releases of process materials to the sewer or liquid to the flare system as a result of low interface levels.* Interface level instrumentation related to these scenarios is to be included in the AIM program.

- *Known scenarios (e.g., polymer formation) for plugging flare lines.* Systems and instrument loops used to prevent blockage of pressure relief device vent headers (e.g., process vessel liquid level instrumentation, nitrogen purge systems, liquid drain systems, vent header pressure monitoring instrumentation) are included in the AIM program.

- *Failures of rupture disks beneath relief valves.* Systems and instrumentation used to detect leaking or burst rupture disks are included in the AIM program.

- *Reactor upsets.* Systems and instrumentation used to prevent runaway reactions (e.g., inhibitor or poison injection systems, dump systems, quench systems, pressure venting systems) are included in the AIM program.

- *High level in vessels (e.g., seal pots) downstream of relief valves.* Systems and instrumentation that are used to maintain or monitor the liquid level in seal pots, as well as systems and instrumentation that are used to prevent overfilling of seal pots, are to be included in the AIM program.

- *Utility and support system failures.* Safeguards on utility systems, including low-pressure alarms, air dryers, emergency lighting panels, and indicators that confirm that backup systems are operating, are included in the AIM program. Emergency utility systems (e.g., electric power supplies, instrument air supplies, emergency cooling systems) are to be included in the AIM program. All of the parts and instrumentation for each emergency system are to be included.

- *Control valves at high pressure/low pressure interface points.* The integrity of the valve, as well as the logical actions governing valve position, are included in the AIM program.

- *Heat exchangers.* When tube leaks will result in adverse consequences, the exchanger tubes are included in the AIM program.

- *Systems that are used to enhance the reliability of covered instrument loops*, such as chemical seals and instrument cabinet gas purges, are included in the AIM program.

- *Miscellaneous other equipment* to consider for the AIM program:
  - Blast walls, bunkers, barricades, and blast-resistant doors for control rooms and other occupied process buildings
  - Barriers used to protect process equipment from vehicles

- Pipe galleries carrying process piping, including spring hangers and other supports
- Excess flow valves
- Structural steel
- Fire protection insulation applied to structural steel and to process vessels
- Instrumentation designed to detect heat, flammable vapors, or toxic vapors
- Vent handling systems such as flares, scrubbers, and thermal oxidizers
- Emergency response equipment, including ambulances, fire trucks, self-contained breathing apparatus (SCBA), radios, and rescue equipment
- Portable oxygen and flammable gas meters
- Testing and inspection test equipment
- Cranes and other lifting equipment
- Tank internals (e.g., heating or cooling coils, agitator baffles, steady bearings, roof supports)
- Process sewers, including liquid seals, purge systems, and sewer ventilation systems
- Valves in pressure relief lines, including three-way valve assemblies and drains
- Weather protection bags or socks placed over vertical vent lines from pressure relief devices
- Liquid expansion bottles, including rupture disks and heat tracing, on lines that do not have pressure relief devices
- Inerting systems and associated instrumentation
- Uninterruptible power supply (UPS) systems for instrumentation, area alarms, public address systems, and communication systems
- High-temperature monitor and interlock instrumentation for solids processing equipment that could have potential hot spots, such as bearings
- Electrical grounding and bonding systems.

6

INSPECTION, TESTING AND PREVENTIVE MAINTENANCE

Once the scope of the asset integrity management (AIM) program has been defined, AIM program efforts often focus next on developing and implementing an inspection, testing and preventive maintenance (ITPM) program. Although they are considered together in this chapter, ITPM program activities might cut across more than one part of an organization. For example, an Inspection Department may perform inspections and tests while the Maintenance Department carries out preventive maintenance tasks. ITPM terminology as used in this document is defined in Table 6-1.

In many respects, the ITPM program is the core of AIM. However, inspections and tests do not solve asset integrity problems; they can only help identify deficiencies and plan maintenance. The objective of the ITPM program is to identify and implement maintenance tasks needed to ensure the ongoing integrity of AIM assets, in order to adequately control process risks.

TABLE 6-1. Definition of ITPM Terms

Term	Definition
Inspection	Assessing the current condition and/or rate of degradation of assets
Testing	Checking the operation/functionality of assets. Includes *proof tests*, which are the exercising of passive (standby) systems.]
Preventive maintenance	Maintenance that seeks to reduce the frequency and severity of unplanned shutdowns by establishing a fixed schedule of routine inspection and repairs *(see note)*
Inspection, testing and preventive maintenance (ITPM)	Scheduled proactive maintenance activities intended to (1) assess the current condition and/or rate of degradation of assets, (2) test the operation/functionality of assets, and/or (3) prevent asset failure by restoring asset condition

Note: This is not intended to imply that every task traditionally viewed as a PM is included in the AIM program; only PM tasks needed to prevent failures of interest, such as loss of containment of hazardous materials, may be included in an AIM ITPM program

In this chapter, ITPM program development and implementation are divided into two major phases:

1. *ITPM task planning.* The activities of this phase include identifying and documenting the ITPM tasks needed to ensure ongoing integrity of AIM program assets, establishing the frequency at which the ITPM tasks are to be performed and translating the tasks into a schedule.

2. *ITPM task execution and monitoring.* The plan becomes reality when the tasks are executed by qualified personnel as scheduled. To help achieve this objective, the ITPM program needs to ensure that processes are established to monitor the schedule, the task results, and overall program performance. In addition, continuous monitoring of the task schedule and results provides a means for optimizing the tasks. ITPM task execution does not stop at performing ITPM tasks. It includes issuing corrective actions and tracking them to completion. Performing inspections and tests by themselves is of little value; they only bring value as a tool to correct deficiencies and to plan future maintenance activities.

> **ITPM task**
> An inspection, testing or preventive maintenance activity that is performed at some interval with the purpose of assessing the current condition of the asset, in order to plan identified corrective activities to ensure ongoing asset integrity

Many organizations find that the ITPM program includes maintenance management activities (e.g., developing and managing the ITPM schedule) and proactive maintenance tasks that are similar to those in reliability programs. An AIM program often focuses on asset integrity issues necessary for ensuring safety and protecting the environment. A reliability program similarly focuses on asset integrity issues needed to ensure business performance (e.g., reducing process downtime). Used together, these programs implement maintenance management activities and proactive maintenance tasks to detect, prevent and manage asset failures before they impact performance. Therefore, concepts and processes used to develop and implement the ITPM program (and the entire AIM program) also can be used to implement and/or improve a reliability program.

6.1   ITPM TASK PLANNING

To develop the ITPM program, facility personnel often identify and document the ITPM tasks and then develop a schedule for each of these recurring tasks. This document refers to the list of recurring tasks and their associated schedule as the *ITPM plan*. The key attributes of the ITPM plan are that it:

> **ITPM plan**
> **List of recurring ITPM tasks and their associated schedule**

- Includes recurring ITPM tasks for all assets within the scope of the AIM program

- Defines the basis for each ITPM task and its corresponding interval

- Provides a link to procedures and other necessary references such as original equipment manufacturer (OEM) manuals

- Defines or references the acceptance criteria location for each ITPM task.

This section outlines the following activities and considerations for developing an ITPM plan that meets those attributes:

6.1.1 – Select ITPM Tasks

6.1.2 – Develop Sampling Criteria

6.1.3 – Other ITPM Task Planning Considerations

6.1.4 – Schedule ITPM Tasks.

6.1.1   Select ITPM Tasks

The starting point for developing the ITPM plan is selecting the recurring ITPM tasks. Each asset within the AIM scope needs to be included in ITPM task selection. A six-step process for selecting ITPM tasks is illustrated in Figure 6-1 and described in detail in the following paragraphs.

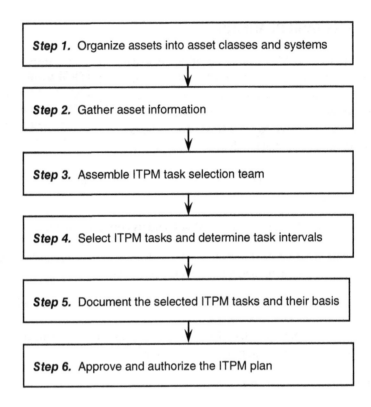

Figure 6-1. ITPM task selection process

Step 1 – Organize Assets into Asset Classes and Systems. While ITPM task selection can be performed for each individual asset, grouping assets into classes (e.g., pressure vessels, centrifugal pumps) and systems (e.g., basic process control systems, safety instrumented systems) can reduce the time needed to select tasks and can help ensure program consistency. The basic concept behind having asset classes and systems is that the ITPM tasks selected are applicable to all assets within the class or system. However, when developing asset classes and systems, unique items and/or items in different service conditions (e.g., different chemicals, higher pressures) need to be identified and separated into sub-classes when different ITPM tasks and intervals are warranted. Furthermore, special cases, such as particularly problematic assets, may merit unique treatment in the ITPM plan.

Organizing assets into classes and systems also provides other advantages:

- *Fewer procedures may be needed.* Some regulations require procedures for every ITPM task. Using the asset class approach, the procedures can typically be developed for each task by asset class or system rather than for each individual item within the asset class or system. Chapter 9 discusses ITPM procedures in detail.

- *Improved identification and communication of responsibilities for ITPM tasks.* ITPM plans based on asset classes and systems typically document task responsibilities for each asset class or system.

- *Consistent selection of ITPM tasks and intervals for added assets.* As processes and assets are added, an ITPM plan based on asset classes and systems makes it easier to define the ITPM tasks and intervals for newly added assets.

- *Efficient ITPM program documentation for reviewers, auditors and regulators.* Early in an AIM program review or audit, information is typically requested on ITPM tasks and their intervals. An ITPM plan based on asset classes and systems provides this documentation concisely.

Step 2 – Gather Asset Information. Compiling the following types of information about each asset and its operation prior to task selection will enable efficient selection of ITPM tasks and their intervals:

- Engineering data, such as design specifications and as-built drawings

- Operational data, such as procedures that contain operating parameters and tables of operating limits

- Maintenance and inspection history, including current ITPM tasks and task schedule, as well as inspection and repair history

- Corrosion Control Documents (CCDs) that identify corrosion loops, degradation mechanisms and/or credible failure modes by process unit

- Safety and reliability analyses, such as process hazards analyses and reliability-centered maintenance (RCM) analyses, which can provide information about types of failures expected and the effects of those failures; such analyses may also identify important protective assets such as critical alarms, emergency shutdown systems and critical utilities, the functionality of which need to be maintained

- Asset criticality determinations

- Codes, standards and practices applicable to assets, especially those that specify ITPM tasks and intervals

- Applicable regulatory/jurisdictional requirements

- Applicable site or corporate environmental, health and safety policies

- OEM manuals

- Risk-based analyses, if the ITPM task selection and/or interval determination is to be risk-based. Examples of risk-based analyses include:
  - o Quantitative risk assessments
  - o Layer of Protection Analyses (LOPAs)
  - o Risk-based inspection (RBI) studies.

  Step 4 of the ITPM task selection process and Chapters 7 and 15 include additional risk-based ITPM task selection processes and applicable risk analysis techniques.

Some of the above information is asset-specific and usually kept in the asset files of the facility. Table 6-2 provides a suggested list of information to include in asset files for selected asset types in order to provide the proper information needed for the AIM program, including the ITPM program. When assembling this information, both completeness and accuracy (data quality) of the information are vitally important.

Asset information can be useful for selecting tasks by:

- Identifying the RAGAGEP and/or risk-based analysis to be considered

- Providing information on potential or known damage mechanisms, including corrosion studies

- Identifying candidate ITPM tasks

- Providing regulatory/jurisdictional and/or company ITPM requirements.

It should be noted that potential damage mechanisms will vary for any given asset depending on several factors, including the specific materials of construction used and how the asset was constructed (post-weld heat treatment, etc.) as well as the particular service involved (process fluids, vibration, thermal cycling, etc.).

TABLE 6-2. Typical Asset File Information for Selected Asset Types

Asset Type	Design and Construction Information	Service History	ITPM History	Vendor-Supplied Information
Pressure Vessels and Atmospheric Storage Tanks	• Design specification • Design code • Materials of construction • Corrosion allowance • Corrosion studies • Welder qualifications • Bolting procedures/ specifications • P&IDs and PFDs • Damage mechanisms • Fabrication QA reports	• Fluids handled • Type of service (e.g., continuous, intermittent, irregular) • Vessel/tank history (e.g., alterations, date of repairs) • Operating parameters • Temperature/pressure excursions • Failures and repair history	• Condition monitoring locations (CMLs) • Inspections performed • Examination techniques used • Inspector qualifications • Inspection reports and results	• Data reports (e.g., U-1 form, API 650 form) • Type of construction • Rubbings/ photocopies of code nameplates • As-built drawings
Piping	• Piping specification • Design code • Corrosion allowance • Corrosion studies • Welder qualifications • Bolting procedures/ specifications • P&IDs and PFDs • Damage mechanisms	• Fluids handled • Type of service (e.g., continuous, intermittent, irregular) • Operating parameters • Temperature/pressure excursions • Failures and repair history	• Circuit definition and CMLs, usually on isometric drawings • Inspections performed • Examination techniques used • Inspector qualifications • Inspection reports and results	• Equipment manuals for system components (e.g., filters, valves)
Relief Devices	• Design specification • Relief design basis and calculations • Materials of construction • P&IDs and PFDs	• Fluids handled • Failures and repair history	• Inspections and tests performed • Inspection and test results • Testing organization certification/ qualifications	• Equipment manuals • Manufacturer's data report • Rubbings/ photocopies of code nameplates
Rotating Equipment	• Asset specification • Materials of construction • Seal configuration and data • Damage mechanisms	• Fluids handled • Type of service (e.g., continuous, intermittent, irregular) • Type of lubricant • Operating parameters • Temperature/pressure excursions • Failures and repair history	• Inspections and PM performed • Inspection and PM results	• Equipment manuals • Manufacturer's data report (e.g., API 610 form) • Performance testing data (e.g., pump curves) • Recommended spare parts list • As-built drawing

TABLE 6-2. *Continued*

Asset Type	Design and Construction Information	Service History	ITPM History	Vendor-Supplied Information
Instrumentation	• Instrument specification • Materials of construction • Line diagrams • Logic diagrams • Required safety integrity level (SIL) and safety requirement specification	• Corrosive or fouling service • Failures and repair history	• Testing performed (e.g., calibration) • Testing results	• Instrument manuals • Factory calibration reports
Electrical Equipment	• Asset specification • Line diagrams • Overcurrent protection information • Logic diagrams (e.g., redundant power supply switching logic)	• Failures and repair history	• Testing performed (e.g., infrared analysis, transformer oil analysis) • Testing results	• Equipment manuals • As-built drawings

Asset information can assist in the following ITPM activities:

- *Defining acceptance criteria,* usually by referencing specific asset file information such as corrosion allowance. (See Step 4 below, as well as Section 11.2, for additional information on acceptance criteria.)

- *Preparing for execution of an ITPM task,* including the preparation of written procedures and the obtaining of information from previous inspections and tests on the same assets.

Step 3 – Assemble ITPM Task Selection Team. With the breadth of knowledge needed to select tasks, employing a multidisciplinary team is highly desirable. Typical personnel included on the team are:

- *Engineering personnel,* who can provide knowledge of the asset design and applicable codes, standards and recommended practices.

- *Operations personnel,* who can provide knowledge of the asset operation and failure history. In addition, having operations personnel involved in the task selection process helps promote buy-in to the ITPM plan, which can pay dividends during task execution.

- *Maintenance personnel,* who can provide knowledge of current maintenance practices and maintenance history. They can likewise help promote buy-in to the ITPM plan.

- *Inspection personnel*, who can provide knowledge of inspection and testing codes, standards, recommended practices, potential damage mechanisms and inspection history.

- *Reliability and maintenance engineers*, who can provide knowledge of inspections, PMs, potential damage mechanisms and asset history.

- *Corrosion engineers*, who can provide knowledge of corrosion and other damage mechanisms (e.g., stress-induced cracking) and corrosion prevention and monitoring techniques.

- *Process engineers,* who can provide knowledge of asset design and operation; asset history; and applicable codes, standards and recommended practices.

- *Inspection and maintenance contractors.* If facility management plans to use outside contractors for inspection and nondestructive testing (NDT) tasks, including contractor representatives on the team can be beneficial.
- *Asset manufacturer and vendor subject matter experts (SMEs).* These individuals are especially helpful when selecting ITPM tasks for licensed processes and new assets because they can provide valuable knowledge about the operation and maintenance of processes and assets when facility personnel lack pertinent experience. .

Step 4 – Select ITPM Tasks and Determine Task Intervals. When selecting the ITPM tasks, the team considers the types of failures to be addressed (e.g., general corrosion, unavailability of an instrumented shutdown system) and the best approach (and most effective tasks) for preventing the failures, including detecting degradation or incipient failures.

Identify Damage and Failure Mechanisms. The team may have many potential failures to consider. However, the task selection process can be simplified by understanding that for most processes that use hazardous materials, the failure modes that most need to be addressed relate to:

- Preventing loss of containment of process materials and/or stored energies

- Preventing or detecting the loss of functionality of critical controls, safety systems (e.g., alarms, shutdowns), emergency response assets and critical utilities

- Preventing unnecessary startup, shutdown and/or cyclic operation of processes in which transient operations introduce a hazardous condition in the process.

In addition, identification of specific failure/damage mechanisms (e.g., localized corrosion, erosion, signal drift) is needed to ensure that proper ITPM tasks and intervals are selected. This is especially true of pressure-boundary equipment such as pressure vessels, storage tanks, process vessels and piping. The following is a partial list of general failure/damage mechanisms (Reference 6-2) for pressure-boundary equipment:

- Equipment loading failures, such as ductile fracture, brittle fracture, fatigue, buckling

- Wear, such as abrasive wear, adhesive wear, fretting

- Internal and/or external corrosion, such as uniform corrosion, localized corrosion, pitting or corrosion under insulation

- Thermal-related failures, such as creep, metallurgical transformation, thermal fatigue

- Cracking, such as stress-corrosion cracking

- Embrittlement.

Note that it is possible for a combination of mechanisms to be present. In particular, corrosion plus erosion can result in an accelerated combination effect. Monitoring of corrosion and erosion as two separate failure mechanisms may not provide early detection of incipient failures unless the possible combination effect is taken into account when present.

API RP 571 and ASME PCC-3 (References 6-3 and 6-4) provide additional information on damage mechanisms for fixed equipment. API RP 583 addresses the specific mechanism of corrosion under insulation and under fireproofing (Reference 6-5).

Determine Approach to Managing Damage/Failure Mechanisms. The team can then determine the best approach or approaches for managing each identified damage or failure mechanism. In general, potential failures can be managed by implementing ITPM tasks in one or both of two broad categories:

- Condition-based tasks
  - o Detect the onset of a failure condition, such as by inspecting for the presence of a crack or measuring for excessive pump vibration.

o Assess the condition of assets, such as by inspecting for plugging or fouling, determining remaining life based on corrosion rate or measuring the accuracy of instrumentation.

o Discover latent (hidden) failures, such as by verifying the operation of emergency shutdown systems, confirming the functionality of interlocks and verifying the operation of backup power supply systems.

- Proactive tasks

o Perform time-based and/or risk-based activities such as lubricating pumps, replacing parts known to wear over time, and in some situations replacing an entire asset on a regular basis.

o Perform restorative activities such as cleaning, reconditioning and/or recalibration to restore assets to an as-new condition.

Assessing the condition of assets is known as *condition monitoring*. Appendix 6D gives details for several condition monitoring (CM) approaches, including temperature measurement, dynamic monitoring, oil analysis, corrosion analysis, nondestructive testing, electrical testing and monitoring, observation and surveillance, and performance monitoring.

> **Condition monitoring**
>
> **Observing, measuring and/or trending of indicators with respect to some independent parameter (usually time or cycles) to indicate the current and future ability of a structure, system or component to function within acceptance criteria.**

Considerations when determining the best approach or approaches to managing potential failures include the following:

- Some failure mechanisms or potential failure locations are more difficult to detect than others. Some, such as holes in tubing or small cracks, may be too small to detect by particular detection methods. For example, tubing inspection methods may be limited to detecting hole sizes of 0.01 to 0.02 inch (0.3 to 0.5 mm) or larger by eddy current or other methods. This may have a bearing, for example, on the selection of the proper NDE method or the degree of coverage.

- Various failure mechanisms have differing lengths of time or number of cycles between a detectable potential failure condition and an actual functional failure. This is known as the *potential-to-functional failure interval*, or *P-F interval*, which can range from days to years or can be highly variable. P-F intervals associated with different condition monitoring techniques are included in Appendix 6D.

> **P-F interval**
>
> **Average amount of time or number of cycles between a detectable potential failure condition and an actual functional failure.**

- Since failure rates may not be constant over time, knowing how different failure mechanisms are likely to vary over the life cycle of an asset may have a bearing on how potential failures are managed. Most asset types are known to have relatively constant failure rates over their useful life, but fatigue or wear-out failures may be anticipated to increase beyond a certain age or service (see Figure 6-2), resulting in a need to possibly decrease ITPM intervals or change inspection techniques.

Determine Types of ITPM Tasks Required. After identifying the desired approach or approaches for managing each failure, the team can then determine the types of tasks needed. In general, the tasks fall into the following three categories:

1. Inspection tasks, which detect the onset of a failure condition (e.g., vessel wall crack) and/or assess the asset condition (e.g., vessel wall thickness).

2. Testing tasks, including predictive maintenance tasks, which assess the condition of the equipment (e.g., drift in an instrument, vibration of a pump) and/or detect hidden failures (e.g., functional test of a shutdown system).

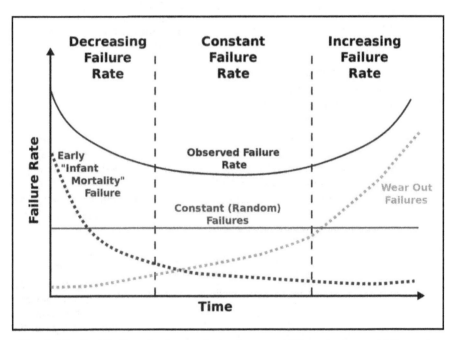

Figure 6-2. Contribution of early, constant and wear-out failures to observed failure rate
(NOTE: Early failures may be minimal; wear-out failures may increase only slightly with time)

3. Preventive maintenance (PM) tasks, which help prevent premature failure of assets by (1) promoting inherent asset reliability (e.g., lubricating a pump) or (2) restoring the asset reliability by replacing the entire item or selected components or parts (e.g., rebuilding of a compressor before its functionality is lost).

Once the desired types of tasks (inspection, testing and/or PM) are identified, the team can employ several resources to assist in selecting specific tasks. In general, the following resources can be used:

- Inspection codes, standards, recommended practices

- Manufacturer's recommendations

- Professional organization or trade group guidance

- Insurance company recommendations

- In-house history of the asset under consideration

- Common industry practices (e.g., vibration analysis of rotating equipment)

- Environmental, health and safety recommendations

- Risk analysis recommendations.

Appendix 6A contains a brief description of some common predictive maintenance and nondestructive testing (NDT) tasks for mechanical equipment. API RP 571 and ASME PCC-3 (References 6-3 and 6-4) contain information on common damage mechanisms and inspection and monitoring techniques for many types of fixed process equipment such as vessels, tanks and piping. A summary table correlating equipment imperfection types and pertinent nondestructive examination techniques is given in Appendix 6B. This information can help the ITPM team determine appropriate ITPM tasks.

Team members need to ensure that each selected task addresses the types of failures targeted for prevention. Furthermore, team members can verify that any ITPM codes, standards and recommended practices selected are consistent with the applicable design codes and standards (e.g., API 510 activities are applied to ASME-coded pressure vessels). For some assets such as specialized equipment and assets not governed by RAGAGEP, OEM manuals may list ITPM recommendations.

The team may need to decide whether inspection or replacement of an asset is preferable for some assets. Section 6.1.3 provides additional information on this issue.

Assets Owned and Maintained by Others. In addition to the assets maintained by the facility, the ITPM plan will need to include any assets owned and maintained by others, but only if the asset (such as a nitrogen storage and supply system for flammable storage tank inerting) is included in the facility's AIM program. To develop this portion of the ITPM plan, a representative from the asset owner's organization can be asked to supply the necessary information. The ITPM plan needs to clearly document which organization is responsible for performing these ITPM tasks (i.e., the facility staff, the asset owner, or each organization responsible for specific tasks). This is normally established as part of contractual requirements, so is determined during contract negotiations. From an AIM perspective, it is important to verify that the ITPM tasks performed are aligned with the facility's AIM requirements and a line of communication is established including auditability.

Determine Acceptance Criteria. Acceptance criteria, consistent with company standards and good engineering practice, provide necessary information for evaluating asset integrity, to help ensure that corrective actions are taken when necessary. Acceptance criteria also define the limits for a specific ITPM activity.

> **Acceptance criterion**
> Technical basis used to determine whether an asset is deficient (e.g., when analyzing ITPM results)

Acceptance criteria can be expressed qualitatively or quantitatively. An example of a qualitative criterion is "no missing or bent pipe supports." An example of a quantitative acceptance criterion is "the wall loss is less than the stated corrosion allowance of 0.125 inch (3.2 mm)." Examples of other quantitative criteria might include speed of response in seconds for an on-demand safety system, or maximum percentage of allowable calibration drift. Additional information on defining and using acceptance criteria is provided in Section 11.2.

Determine Task Intervals. To finalize ITPM task selection, the team can decide whether to change the current ITPM tasks and/or frequencies or to add new ITPM tasks if no tasks are in place to address failures of interest. Presented in Figure 6-3 is a decision matrix that provides guidance for this decision.

> **Task interval**
> Time between successive ITPM task executions

Once a task is selected, the team needs to determine the appropriate task interval. For some tasks, especially new tasks, this task interval can be viewed as the starting point or maximum interval for establishing the ongoing task frequency if continued performance of the task is required. Other factors, such as the operating and maintenance history of the asset, often dictate the ongoing interval. These factors vary somewhat for different types of assets, but in general the factors provide insight into:

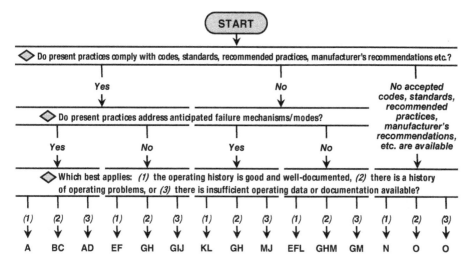

Letter Key:

A Document the present practice in the ITPM plan and reference applicable code(s), standard(s), recommended practice(s), manufacturer's recommendation(s), etc.

B Review the assets and operations to ensure that failure mechanisms/modes are adequately addressed by tasks in the ITPM plan.

C Consider whether more rigorous ITPM would be appropriate.

D Where possible, identify data to be gathered and documented to validate ITPM task selection.

E Review the assets and operations to determine if the unaddressed failure mechanisms/modes are important.

F Consider whether more rigorous ITPM would be appropriate. Otherwise, document the present practice in the ITPM plan and reference applicable code(s), standard(s), RP(s), manufacturer's recommendation(s), etc.

G Review the assets and operations to identify all failure mechanisms/modes <u>not</u> adequately addressed.

H Determine ITPM needed to address all failure mechanisms/modes.

I Determine ITPM needed to address any failure mechanisms/modes. Otherwise, document the present practice in the ITPM plan and reference applicable code(s), standard(s), RP(s), manufacturer's recommendation(s), etc.

J Ensure that steps are taken (e.g., procedures are modified) to improve documentation practices.

K Document present practice in the ITPM plan and provide justification for any variance from applicable codes, standards, recommended practices, manufacturer's recommendations, etc.

L Ensure that documentation supports your decision.

M Consider upgrading the present practices to meet applicable codes, standards, recommended practices, manufacturer's recommendations, etc.

N Document the present practice(s) in the ITPM plan and include a "performance history" notation as the rationale for the practice(s).

O Consider the consequences of the asset failure. If the consequences of failure are hazardous, contact vendors, chemical suppliers, other users, etc., for suggestions.

Figure 6-3. Example ITPM task selection decision tree

- Probability of the asset failing before the next ITPM task
- Probability of the ITPM task detecting the asset failure mode of interest
- Consequences of the asset failure.

For example, the task interval for inspecting pressure vessels, tanks and piping can be influenced by:

- Age of the asset
- Materials handled
- Process conditions (e.g., pressure, temperature)
- Material of construction
- Types of damage mechanisms to which the asset is subject (e.g., general corrosion, localized corrosion, stress-corrosion cracking, high-temperature hydrogen attack)
- Type of inspection or testing technique used, including the effectiveness of the technique to detect and quantify damage
- Measured rate of damage / deterioration (e.g., corrosion rate) and calculated remaining life
- Trending of inspection and testing results
- Inspection/maintenance history
- Current condition of the asset
- Codes, standards and/or jurisdictional requirements.

More detailed information on ITPM tasks and intervals for specific types of assets is provided in Chapter 12.

Many sources of codes, standards and practices provide information and guidance on these and other factors and how they can influence task intervals. Table 6-3 provides some of the factors to consider when establishing task frequencies for relief devices, instrumentation and rotating equipment.

The task interval need not be a single time interval for each asset or asset type. Some companies practice what is sometimes termed "old-age inspections" or similar, where a much more in-depth inspection is performed periodically such as every 10 or 20 years, perhaps involving more condition monitoring locations, 100% coverage or additional condition monitoring techniques.

TABLE 6-3. Factors Affecting ITPM Tasks for Relief Devices, Instrumentation and Rotating Equipment

Relief Devices	Instrumentation	Rotating Equipment
• Type of device • Relief design case • Process conditions • Susceptibility to plugging • Test/maintenance history	• Type of instrument (e.g., pressure, level) • Vintage of the instrumentation (e.g., pneumatic, electronic) • Measurement technology used (e.g., radar level detection) • Process conditions • Environmental conditions (e.g., outside installation) • Type of testing used • Test/maintenance history • Desired performance (e.g., safety integrity level of a safety instrumented system)	• Type of rotating equipment • Age of the asset • Type of seal system (e.g., asset, packing) • Process conditions • Type of ITPM task used • Test/maintenance history

The ongoing task intervals for some assets can be adjusted within allowable limits based on historical test/inspection results after a maintenance history has been established. For example, detection of significant corrosion may result in a need for more frequent inspections, whereas the observation of no significant damage or deterioration over time may allow the extending of task intervals within RAGAGEP limits. This may not be allowable for some types of assets, such as safety instrumented systems that must be functionally tested at a set frequency in order to achieve a reliability goal.

Performing an ITPM task less frequently than recommended by codes, standards, recommended practices, manufacturer's recommendations and common industry practices may not be able to be justified without a formal RBI study. The decision to perform a task less frequently will need to be based on the operating experience and the level of documentation available to justify the variance. Again, the information in Figure 6-3 can help with this decision.

> **Caution**
> Basing task intervals exclusively on operating history may not provide a satisfactory indication of failures with a low likelihood but high consequence severity.

Many companies are using risk-based approaches to determine ITPM tasks and intervals. For example, companies are using one or more of the following:

• RBI assessments to determine the type and frequency of inspection tasks for pressure-containing equipment (e.g., vessels, tanks, piping)

- RCM analyses to determine type and frequency of ITPM tasks for active equipment items (e.g., pumps, controls, compressors)

- Layer of Protection Analysis (LOPA) and similar analysis techniques used to determine required performance such as Safety Integrity Level

- Unavailability analysis (e.g., Fault Tree Analysis, Markov modeling or simplified equations) to determine testing frequency for safety instrumented systems (SISs).

These approaches all focus on identifying potential loss events, understanding the risks of each loss event and defining ITPM tasks to effectively manage the risk. Each of these approaches is discussed in more detail in Chapter 7.

Step 5 – Document the Selected ITPM Tasks and Their Bases. Finally, the selected tasks are documented in the ITPM plan, including the basis (rationale) for selecting each task. In general, the basis will be (1) a specific code, standard or recommended practice; (2) a manufacturer's recommendation; (3) an industry practice; and/or (4) the facility's operating history and/or maintenance practice. Documentation of the plan can be provided in a table format in the facility's AIM program document and/or on its computerized maintenance management system (CMMS). Table 6-4 provides an example tabular format.

Step 6 – Approve and authorize the ITPM plan. The tasks, frequencies, and schedule need to be approved and authorized by the Operations Department, or another person or position if required by code or regulation. This step needs to be included in a documented management system, then the ITPM plan can be entered into the CMMS.

6.1.2 Develop Sampling Criteria

Another element of ITPM task planning is the application of sampling, particularly for inspections, NDT, and condition-monitoring tasks. Sampling can be applied to determine the specific location and number of representative sample points needed for NDT and condition-monitoring tasks. As mentioned earlier, inspection and condition-monitoring tasks focus on detecting failures, detecting incipient failures, and/or assessing the condition of assets. The nature of these tasks and the size of assets for which they are typically used generally require that sampling be employed to assess the integrity of the assets.

TABLE 6-4. Example ITPM Plan in Tabular Format

Asset Class	Required Activity	Interval or Frequency	Basis	Procedure	Assigned Department	Remarks
Pressure vessels	Visual surveillance	Daily	Performance history	AIM-1	Operations	Completed during routine rounds by operators; deficiencies are documented, but surveillance that does not identify a deficiency is not documented
	External inspection	5 years or half of remaining life, whichever is less	API 510	AIM-2	AIM Department	Visual inspection and condition monitoring
	Internal inspection	10 years or half of remaining life, whichever is less	API 510	AIM-2	AIM Department	Visual inspection and condition monitoring
Pressure transmitters, including alarms and interlocks	Visual surveillance	Weekly	Performance history	AIM-1	Operations	Completed during routine rounds by operators; differential pressures are monitored if leak in tube is suspected; deficiencies are documented but surveillance that does not identify a deficiency is not documented
	Calibration of transmitters	Annually	Industry practice	AIM-3	Electrical and instrumentation (E&I) Department	
	Functional test of alarms and interlocks operation	Biannually (every 2 years)	Industry practice	AIM-4	E&I Department	

The location and number of NDT sample points or condition-monitoring locations (Reference 6-7) are determined based upon several factors, such as:

- *The consequences of failure.* Failures with higher potential for safety or environmental consequences warrant more sample points.

- *The expected deterioration rate (e.g., corrosion rate).* In general, cases of high (or higher-than-expected) rates of deterioration require more sample points.

- *The presence of localized deterioration.* Higher potential for localized deterioration requires more sample points.

- *The asset configuration.* For example, piping circuits with more elbows, tees and injection points require more sample points.

- *The potential for unusual damage mechanisms.* In general, assets with the potential for unusual damage mechanisms, such as biological corrosion, require more sample points.

- *The potential for long-latency damage mechanisms*, such as fatigue or cracking, which may not show up for twenty years or more.

Guidance can be found in codes, standards and practices for determining sample number and sample locations. For example, API 570 and API RP 574 provide guidance on location and number of condition monitoring locations (CMLs) for piping inspections (References 6-6 and 6-7). Other pertinent sources of codes and standards are listed in Appendix 6C. In addition, some RBI approaches include statistical techniques for determining, or accounting for (in the data analysis), the number of sample points.

6.1.3   Other ITPM Task Planning Considerations

In addition to task selection and frequency determination, the ITPM task selection team will also have other decisions to make. Specifically, it will need to decide:

1. Which ITPM tasks are best performed by operations personnel instead of maintenance personnel, including consideration of training and qualification requirements, and

2. The value of performing ITPM tasks versus periodic asset replacement.

Operator-Performed Tasks. Some ITPM tasks are best performed by operations personnel. Some of these tasks might be operator activities that were in place before an ITPM planning effort was even begun. Such activities may include:

- Visual surveillance of assets performed during operator rounds

- Routine inspection of pump seals for leakage

- Verification of lubrication fluid levels

- Sensory detection of abnormalities; e.g., listening for unusual sounds

- Functional testing of emergency isolation valves that are also used in normal operation.

In addition, the ITPM plan may contain PM activities that can be performed by operating personnel. Depending on specific facility culture, these activities may include routine lubrication of assets, tightening of valve packings, and changing oil in selected assets. Such activities are important elements of the AIM program; as such, the ITPM plan needs to include these activities and reflect that they are assigned to operations personnel.

Whenever ITPM tasks are to be done by operations personnel, training qualifications need to be specified and formal, documented, verified training completed before the tasks are performed. Some inspection tasks may seem to be straightforward, such as checking for leaks or doing visual inspections of atmospheric storage tanks or secondary containment dikes, but such inspections have many subtleties related to qualitative acceptance criteria such as when to flag external corrosion or how large of a leak requires corrective action.

Inspect or Replace? The ITPM task selection team may also need to decide whether to perform inspections and tests on an asset or to replace the asset on a routine basis. Typically, replacement is considered only for relatively inexpensive assets such as pressure gauges, rupture disks, small relief valves and small chemical dosing pumps. Replacement may also be specified by another source such as codes or standards or by the asset manufacturer; possible examples include unloading hoses and ton container pigtails (flexible tubing).

However, even if assets such as rupture disks or hoses are replaced on a scheduled basis, they still may need to be appropriately inspected when taken out of service and their as-found condition documented, in order to detect anomalies, trends or incipient failures that may prompt corrective action or a change to a replacement frequency. This is particularly true for critical assets, and may also be desirable for other assets based on reliability analysis. Examples of such inspections might include visual examination of check valves for deposits that would restrict proper functioning, or a flow-through pop test on a safety relief valve to confirm the device was functional and that it opened within the acceptable set pressure range.

For some assets, the inspect-or-replace decision requires considering the risk of failure as well as the economics involved. When considering the risk of failure before replacement, the following issues may be important:

- Is the time-to-failure predictable, so that an appropriate replacement time (with little chance of failure) can be determined?

- Is the consequence of asset failure tolerable?

- If the consequence is not tolerable, do cost-effective inspections and/or tests exist that can help detect the onset of failure?

- Where spares are provided, have they been properly inspected and maintained to be considered ready to use?

- Are quality assurance (QA) practices in place to help ensure that the facility is not increasing risks when replacing the asset (e.g., as a result of improper procurement or improper installation)?

Answering these questions does not necessarily require in-depth analysis; however, it is important to consider potential safety and environmental impacts as well as economics.

Run to Failure? It is not expected that any assets included within the scope of an AIM program will be intentionally run to failure (i.e., "breakdown maintenance"). All assets within the AIM program need to be proactively and appropriately inspected, tested and maintained, and replaced as needed prior to such failures occurring.

6.1.4   Schedule ITPM Tasks

Once the ITPM plan is developed, the next step is to translate it into an executable schedule. The objective of the schedule is to logically organize the ITPM tasks into executable work packages by distributing the tasks over time based on several factors, including asset availability, criticality, and resources. The facility staff needs to include plans in the AIM program for managing and maintaining this schedule.

The basic elements of the schedule are obtained by combining information from the asset list and asset criticality determinations (see Chapter 5) with the ITPM plan. For example, if the ITPM plan specifies that external visual inspection of pressure vessels takes place every five years and the asset list contains five pressure vessels, the schedule will include an external visual inspection every five years for each of the five pressure vessels. This information can then be translated into a time-based schedule.

Several factors are important considerations when developing the ITPM task schedule. Five primary factors for most ITPM tasks are:

- *Asset criticality.* Priority is put on performing no-excuse, no-exception ITPM tasks on schedule for critical assets.

- *Asset availability.* For some ITPM tasks, the asset, the process and/or the entire facility will need to be out of service for a specific period of time to execute the task. Often, this requires scheduling out-of-service ITPM tasks during scheduled facility outages, which may need to be adjusted to meet the task frequency in the ITPM plan.

- *Personnel availability.* Sometimes, the schedule is constrained by (1) the number of personnel needed to perform the tasks in the allotted time and/or (2) the availability of personnel and/or contractors with the training and qualifications needed to perform a task.

- *Unique inspection/testing equipment availability.* For example, if a facility has multiple heat exchangers to inspect using advanced techniques such as by eddy current, the schedule may be constrained by the availability of eddy current test equipment.

- *Spare parts and maintenance material availability.* For tasks that involve rebuilding assets, the availability of spare parts and maintenance materials also needs to be considered.

Many scheduling techniques and tools such as computerized maintenance management systems (CMMS) are available to help organizations develop the ITPM task schedule. For most facilities, the tasks are scheduled by (1) entering them as "recurring work orders" in the CMMS and/or (2) organizing the tasks into groups such as lubrication routes. In addition, the tasks assigned to operations will need to be integrated into the operator work routines and procedures.

6.2   ITPM TASK EXECUTION AND MONITORING

After establishing the ITPM plan and schedule, several factors need to be addressed to help ensure the success of the ITPM program. The following ITPM task execution and monitoring activities are discussed in the subsections below:

6.2.1 – Document Asset and ITPM Task Results

6.2.2 – Implement and Execute ITPM Tasks

6.2.3 – Review and Act on ITPM Data and Findings

6.2.4 – Manage the Task Schedule

6.2.5 – Monitor the ITPM Program.

Additional considerations include the implementation of personnel training and procedures to support the ITPM plan. These topics are addressed in Chapters 8 and 9.

6.2.1   Document Asset and ITPM Task Results

Besides putting together an ITPM plan and schedule, other necessary components of the ITPM program include proper documentation and systems to manage this documentation. In general, two types of documentation are needed: (1) asset documentation and (2) ITPM task results documentation.

Asset Documentation. Some of the same asset information used for task selection is needed for ITPM task execution, so that personnel who execute ITPM tasks can review relevant asset file information. Information of interest includes:

1. ITPM history

2. Asset details (e.g., specific components requiring inspection)

3. ITPM task details (e.g., CMLs, inspection technique)

4. Acceptance criteria.

Reviewing asset-specific information helps ensure that any questionable areas are thoroughly examined during the task, specifics related to the inspection such as inspection technique and sampling criteria are known and the inspection is performed consistently (e.g., taking thickness readings at the same location).

ITPM Task Results Documentation. The second type of documentation, ITPM task results, is generated as part of executing the task. For most ITPM tasks, the results are documented every time the task is executed, regardless of the organization or department performing the task. Documenting the task results provides data needed to

1. Determine the integrity of the asset and identify asset deficiencies,

2. Perform evaluations needed to adjust task frequency (e.g., remaining life calculations), and

3. Spot trends that might help predict failures.

In addition, documenting ITPM tasks provides regulators and auditors evidence that the tasks have been performed as scheduled by qualified personnel.

While the content of the results documentation will vary somewhat based on the ITPM task, some minimum documentation requirements are:

- Asset identifier, such as the serial number or asset tag number

- Date the ITPM task was performed

- Name(s) of the person(s) performing the ITPM task

- Description of the ITPM task performed (e.g., magnetic particle testing, ultrasonic testing measurements, radiography, replacement of impeller)

- Results of the ITPM task.

> **U.S. OSHA PSM requirement**
>
> The employer shall document each inspection and test that has been performed on process equipment. The documentation shall identify the date of the inspection or test, the name of the person who performed the inspection or test, the serial number or other identifier of the equipment on which the inspection or test was performed, a description of the inspection or test performed, and the results of the inspection or test.
>
> **29 CFR 1910.119(j)(4)(iv)**

For a more effective AIM program, facilities can document more information. The additional information usually depends on the type of task (i.e., inspection, testing or PM) and the asset type. Some examples of additional information often included in the ITPM results documentation are:

- As-found as well as as-left condition of each asset

- Any asset fitness-for-service evaluation

- Any materials or spare parts used

- Any QA records associated with the task (e.g., verification of replacement part materials of construction)

- The qualification records of personnel performing the task (e.g., API 510 inspector qualification record)

- Any detailed data associated with the task (e.g., ultrasonic thickness readings, vibration spectrums, photographs of tank/vessel internals), especially any data beneficial to be referenced the next time that the task is performed

- Any identified asset deficiencies

- Any recommended corrective actions

- A definitive statement indicating whether or not the asset is fit for return to service or for limited service

- Any remaining life calculations (or similar calculations or assessment)

- The next scheduled task date.

Documenting by Exception. A departure from the usual ITPM task results documentation practices may be appropriate for certain pre-designated ITPM tasks. Often, for these tasks, the results can be better managed by documenting negative results only (i.e., documenting by exception), if used cautiously and if the use of this documentation approach is clearly stated in the ITPM plan (Reference 6-8). This documentation approach is best used:

- For tasks with short intervals (e.g., daily, weekly)

- When trending of data is not needed to determine asset integrity

- When data history is not needed to determine inspection requirements

- When appropriate control or audit mechanisms are in place to ensure that these tasks are performed as intended.

Documentation Methods and Formats. The methods and formats used to document ITPM task results are varied. The results can be documented electronically or on paper. Many times, paper documentation is used when (1) entering the information into a computer system is not cost-effective and/or (2) the available computer system is not capable of managing the results. For example, facilities lacking specialized inspection data management software often find it cost-prohibitive to manually enter thickness measurement / condition monitoring data into a computerized maintenance management system (CMMS). Some inspection software packages (i.e., programs that can better manage data) are becoming more economical and more common. Additional information on use of CMMS and other software packages is provided in Chapter 13.

Facilities also need to decide on the most applicable format for collecting and documenting the task results. In general, format is influenced by the type of task and type of information to be recorded. For example, a lubrication route is usually best documented using a check-sheet format. On the other hand, internal inspections of pressure vessels are usually documented in formal inspection reports that contain all data collected and observations made. Communicating the method and format chosen for each ITPM task to involved personnel is an important part of training as well as managing change. Chapter 9 includes more details on procedure formats.

A policy that defines the retention requirements for ITPM results documentation can help ensure that ITPM results for assets covered by process safety regulations are retained for the life of the process, especially ITPM results needed to show that assets have been maintained in accordance with RAGAGEP. However, retaining results for all ITPM tasks, especially frequently performed tasks (e.g., daily or weekly tasks), is not practical or necessary. Considerations when developing a policy for retaining and discarding documentation are as follows:

- Retain documents needed to demonstrate that assets are maintained, inspected, tested and operated in a safe manner.

- Retain documents needed to demonstrate that the provisions of applicable codes, standards and practices are being met.

- Retain a sufficient number of documents to demonstrate that the ITPM tasks are being performed as required and the results are being appropriately documented.

- Discard documents that are not needed to execute future ITPM tasks or other maintenance activities (e.g., vessel repair).

- Ensure legal, regulatory, company or other requirements that specify record retention requirements are met.

6.2.2   Implement and Execute ITPM Tasks

Once the ITPM plan and schedule are developed, the ITPM tasks are to be implemented and then executed as required by the ITPM plan and schedule. To implement ITPM tasks, facilities need to ensure that (1) procedures are developed, (2) personnel are trained, and (3) sufficient resources are provided for ITPM program startup and ongoing execution of the ITPM tasks. Chapters 8 and 9 provide information on training and procedures. Resource issues are discussed in Chapter 13.

Once the tasks are being performed, facilities need to ensure that the task results and task schedule are appropriately managed. A key part of task management is ensuring that results are reviewed for asset deficiencies. The following subsection of this chapter and Chapter 11 provide additional information on reviewing tasks and managing asset deficiencies. In addition, the task scheduling process needs to (1) track whether tasks are executed when required and that any deferred tasks are properly managed and (2) include activities for reviewing and optimizing ITPM tasks and their frequencies. Subsection 6.2.4 addresses task scheduling issues.

6.2.3   Review and Act on ITPM Data and Findings

Unless facilities are prepared to manage it, the information generated by performing ITPM tasks can easily become overwhelming. Establishing a system for managing this information can provide tremendous benefits, such as:

- Improved asset integrity, resulting in improved process safety, environmental compliance and asset reliability

- Increased employee buy-in of the AIM program, as employees see positive results from their work efforts

- Improved regulatory compliance, since unresolved ITPM task deficiencies are often identified during AIM program audits.

If ITPM task results are not reviewed and acted upon, facilities are likely to miss opportunities to identify and properly manage asset deficiencies and optimize the ITPM schedule. Therefore, successful AIM programs typically include a management system that assigns personnel the responsibility to:

1. Review task results,

2. Perform any needed calculations/evaluations,

3. Update the ITPM schedule and

4. Initiate corrective actions.

A prompt review of ITPM task data and findings helps ensure that *any corrective actions that require assets to be out of service are implemented before they are returned to service.* In addition, prompt review of reports from frequently performed tasks helps ensure that problems noted by personnel are addressed in a timely manner.

Assigning personnel with appropriate skills and knowledge can help ensure proper review of ITPM task results. Specifically, personnel reviewing ITPM task results will need to:

- *Verify that the ITPM activity was completed as required.*

- *Perform a preliminary review of the data and results.* Reviewing the data for anomalies will allow suspicious information to be verified or corrected.

- *Evaluate and analyze the results and recommendations.* This may include:
  - o Entering data into software for analysis

o Performing (or initiating the performance of):

  ♦ Evaluations (e.g., fitness for service)

  ♦ Calculations (e.g., corrosion rates, next inspection date)

  ♦ Statistical analysis (due to potential for variations in data results from inherent NDE method accuracy/repeatability and non-uniform corrosion rates, the use of statistical methods and confidence factors may be employed)

o Reviewing recommendations for practicality and effectiveness

o Adjusting ITPM task frequencies, as discussed in the next section.

- *Identify asset deficiencies or initiate a review to determine deficiencies.* For some ITPM activities, this step may simply involve comparing the task results to established acceptance criteria (e.g., minimum thickness values, no leaks from seals). For other ITPM activities, identifying deficiencies may require personnel with specific expertise such as with pressure vessels.

- *Develop recommendations for resolving any deficiencies.* Have inspectors distinguish between mandatory recommendations and opportunities for continuous improvement. Once a deficiency is identified, the facility's asset deficiency process is used to manage the deficiency. Chapter 11 discusses the asset deficiency process in detail.

- *Track asset deficiencies to resolution.* Usually, a list or log of asset deficiencies is maintained to help ensure they are all resolved. Again, Chapter 11 provides additional information on this topic.

- *Resolve minor issues.* Sometimes, ITPM activities identify asset damage/degradation issues such as minor rust or a slightly higher than normal vibration reading that have not yet resulted in a level of damage/degradation to be considered a deficiency.

- *Resolve anomalous results*, such as measurement data indicating an increasing wall thickness. The resolution process may require retesting, flagging of questionable data and/or an investigation to determine the source of the anomalous results.

- *Identify "bad actors"* such as a standby safety system that fails periodic functional testing more than once or a pump that has an abnormally high failure frequency. Electronic data management can significantly aid in this task.

- *Ensure that recommendations and all necessary corrective actions are implemented.* When no corrective actions are required, ensure this decision and its basis are documented.

- *Implement the facility's quality management process* (as discussed in Chapter 10) to ensure both the completeness and the effectiveness of the ITPMs.

Because of the large number of ITPM reports that can be generated, personnel can be trained to highlight out-of-the-ordinary results. In some facilities, reports of out-of-the-ordinary results are routed differently from routine reports. Similarly, firms contracted to perform ITPM activities can be instructed to highlight areas of concern in a summary report or cover letter that can help keep this important information from being lost or overlooked.

6.2.4 Manage the Task Schedule

All ITPM tasks are ideally performed as scheduled. However, tasks might occasionally be delayed for various reasons. To see that the potential for task delays are properly managed, the AIM program needs to be set up and implemented to ensure that (1) no task delays occur for critical asset ITPMs and (2) a deferral process is followed when the task schedule cannot be met.

Deferral Process. Most facilities attempt to level-load their ITPM task plan, but occasionally peaks in the workload occur due to breakdowns and other unexpected equipment failures. Under these circumstances, the facility will not have resources for both the corrective breakdown maintenance and all the planned ITPM tasks, and therefore some tasks will need to be deferred and re-scheduled for a later date. When this is necessary, it is important that tasks for the less-critical assets are deferred and those considered critical assets are not deferred.

For pressure-containing equipment as well as for other assets, ITPM activities that are not performed at the set frequency are overdue. If suitable reasons for a delay in inspecting, testing or performing preventive maintenance on pressure-containing equipment exist, current codes such as API 510 (Reference 6-9) define a methodology for delaying the inspections:

1. Have a deferral procedure in place for addressing the requirements for the deferral.

2. Include in the deferral procedure a requirement that the determination for deferral be based on a risk assessment.

3. Review, via risk assessment, the equipment history, potential damage mechanisms that may affect the equipment, and an analysis of the potential consequences of the inspection deferral.

4. A qualified inspector must agree with the assessment.

5. Include in the deferral procedure a requirement for concurrence with appropriate pressure equipment personnel and the owner/user management representative.

6. Define in the procedure any additional controls required during the deferral as well as the need for performing alternative inspection / testing.

7. The deferral and alternative safeguards must be documented.

The above illustrates a deferral process for one type of assets. Operating companies need to develop their own internal procedures or standards for dealing with deferral of ITPM tasks for all types of assets.

"Grace Periods." Note that having a selective, authorized deferral process in place is not the same as using "grace periods" to manage an ITPM task schedule. A *grace period* is a site-specific time interval allowed between when an ITPM task is due per the schedule and the actual completion date. The grace period typically varies with the task interval, such as one day for weekly tasks and one week for quarterly tasks. This may occur through a site procedure or by CMMS system default settings.

Grace periods often occur without broad awareness within an organization, giving the possibility of automatically extended time windows within which ITPM tasks are allowed to be completed. RAGAGEP sometimes permit a deferral of ITPMs (as discussed above) with a strict procedural process defined in the RAGAGEP. However, grace periods are not recognized in codes and standards. Regulatory authorities likewise consider any task not performed on schedule to be overdue.

Frequency Adjustments. Another feature of task schedule management is the adjustment of task frequencies based on ITPM results. For example, RAGAGEP for vessels, tanks and piping include formulas for calculating the remaining life of the asset based on inspection results, as well as criteria for adjusting task frequencies (References 6-6, 6-9, 6-10 and 6-11). Similarly, ITPM task results for other asset types can be used to adjust task frequencies. For example, testing frequencies can be increased or decreased based on the results of functional tests, unless a maximum test interval is specified by a code, standard, practice or a reliability requirement such as for a safety instrumented system. Protocols need to be established for frequency adjustment, with management-of-change approval of any deviations from the protocol.

Ensuring that task results are used to adjust task frequencies is important because it provides an opportunity to

- Increase task frequencies to help ensure that failure does not occur before the next execution of the task

- Decrease task frequencies to help reduce ITPM resource requirements

- Possibly eliminate tasks that are unnecessary where this is permitted (e.g., condition monitoring on noncritical assets in processes that are not corrosive to the material of construction).

Most organizations benefit from having a procedure that provides criteria for when ITPM intervals can be adjusted and that defines responsibilities for reviewing test/inspection results and operating histories and optimizing task frequencies. Note that determining whether intervals need to be adjusted may be made more difficult if tests and inspections are outsourced.

Baseline Readings. One common problem encountered with task frequency adjustment is a lack of baseline readings. Facilities have not always obtained baseline readings when assets were initially commissioned. As a result, some nominal value, such as the nominal thickness of the steel plate used to manufacture a pressure vessel wall, would need to be assigned and used to calculate initial deterioration rates (e.g., corrosion rates), and ultimately to determine the remaining life. Problems can arise when the true value is different than the assigned nominal value, leading to overly pessimistic or optimistic calculations of the remaining life. For ITPM tasks with a low frequency (i.e., a long interval between tasks, such as 5 or 10 years), obtaining accurate baseline readings improves a facility's confidence in the remaining life calculations. In addition, providing baseline readings can be a QA check of new installations. Taking and documenting baseline readings may be required or recommended by code, and particularly for re-used vessels when changing ownership or service.

6.2.5   Monitor the ITPM Program

Monitoring ITPM program performance includes establishing appropriate performance measures such as percent of ITPM tasks performed on time, number of assets with ITPM tasks, and/or number of assets with outstanding deficiencies. Chapter 14 provides information on performance measures for AIM programs, including suggested performance measures related to the ITPM program.

Periodic reports regarding ITPM program performance need to be generated and to be reviewed by management. One report to receive management attention is an overdue ITPM task report. Management needs to be made aware of when ITPM tasks are not completed as scheduled, then ensure the causes of the late/deferred ITPM tasks are determined and see that necessary corrective actions are developed and implemented in a timely manner.

6.3   ITPM PROGRAM ROLES AND RESPONSIBILITIES

While the assignment of roles and responsibilities will vary between different organizations, basic roles and responsibilities need to be assigned to appropriate positions within the organization. Many of the roles and responsibilities for the ITPM program are assigned to the inspection and maintenance department(s). Involvement of production personnel is also needed and advisable, as described in Section 6.1. In addition, project and maintenance storeroom personnel may need to be involved.

Example roles and responsibilities for an ITPM program are provided in Tables 6-5 and 6-6. Table 6-5 indicates example roles and responsibilities for the ITPM task planning phase and Table 6-6 indicates example roles and responsibilities for the ITPM task execution and monitoring phase. The matrices use the following letter designators:

R   indicates the job position(s) with primary responsibility for the activity

A   indicates approver of the work or decisions made by a responsible party

S   indicates the job position(s) that typically support the responsible party in completing the activity

I   indicates the job position(s) that are informed when the activity is completed or delayed.

It should be noted that these roles and responsibilities may vary when certain process or protective features or when different aspects of an AIM program are introduced. For example, the inclusion of safety instrumented systems will require more input from instrumentation specialists to determine minimum functional test frequencies to achieve established reliability requirements, as well as additional documentation. The addition of a risk-based inspection protocol will require more analytical support, both on an up-front and a continuing basis.

TABLE 6-5. Example Roles and Responsibilities Matrix for the ITPM Task Planning Phase

R = Primary Responsibility
A = Approval
S = Support
I = Informed

Activity	Inspection and Maintenance Department Personnel							Production Department Personnel					Other Personnel					
	Inspection Manager	Maintenance Manager	Maintenance Engineers	Maintenance Supervisors	Inspectors	Maintenance Technicians	Maintenance Planner/Scheduler	Production Manager	Area Superintendent/Unit Manager	Production/Process Engineers	Production Supervisors	Operators	Plant Manager	Project Managers/Engineers	Instrumentation & Electrical	Process Safety Coordinator	Manufacturers and Outside SMEs	Outside Contractors
Task Selection/Frequency Determination																		
• Inspection tasks	R/A		S		S		S	S	R	S	S		I		S	I	S	
• Testing tasks	R/A	R/A	S	S		S	S	S	S	S	S	S	I		S	I	S	S
• Preventive maintenance tasks	R/A	R/A	S	S	S	S	S	S	S	S	S		I		S	I	S	S
Sampling Criteria	A		S		R											I	S	
Task Scheduling	A	A	S		S		R	S	S				I			I		
Task Assignments (at Department or Craft Level)																		
• Preparation & inspection tasks	R/A	A		R	S	S	S		R		S	S				I		
• Testing tasks		A		R		S	S									I		
• Preventive maintenance tasks		A		R		S	S									I		
• Operations ITPM tasks									R		S	S				I		

TABLE 6-6. Example Roles and Responsibilities Matrix for the ITPM Task Execution and Monitoring Phase

R = Primary Responsibility
A = Approval
S = Support
I = Informed

Activity	Inspection Manager	Maintenance Manager	Maintenance Engineers	Maintenance Supervisors	Inspectors	Maintenance Technicians	Maintenance Planner/Scheduler	Production Manager	Area Superintendent/Unit Manager	Production/Process Engineers	Production Supervisors	Operators	Plant Manager	Project Managers/Engineers	Storeroom Personnel	Process Safety Coordinator	Manufacturers and Outside SMEs	Outside Contractors
Acceptance Criteria Determination																		
• Inspection tasks	A		S		R					S							S	S
• Testing tasks		A	R	S		S				S							S	S
• Preventive maintenance tasks		A	R	S		S				S							S	
Task Execution																		
• Prepare assets for inspection	A		S			R	S		I	S	S	R				I		R
• Inspection tasks	A				R		S				I	I	I			I		R
• Testing tasks		A				R	S				S	R				I	R	R
• Preventive maintenance tasks						R	S				S	I				I	R	
• Operations ITPM tasks											R	R						
Asset Documentation Generation and Maintenance																		
• Design/construction information			R							S				R/A				

TABLE 6-6. *Continued*

R = Primary Responsibility
A = Approval
S = Support
I = Informed

Activity	Inspection and Maintenance Departments							Production Department Personnel					Other Personnel					
	Inspection Manager	Maintenance Manager	Maintenance Engineers	Maintenance Supervisors	Inspectors	Maintenance Technicians	Maintenance Planner/Scheduler	Production Manager	Area Superintendent/Unit Manager	Production/Process Engineers	Production Supervisors	Operators	Plant Manager	Project Managers/Engineers	Storeroom Personnel	Process Safety Coordinator	Manufacturers and Outside SMEs	Outside Contractors
• Service history	A	A	R/A	S	S	S	S			R/A	S	S						
• ITPM history	A	A	S	R	R	S	S								I			
• Maintenance history	A	A		R	R		S								I			
• Vendor-supplied information	A		R											R/A	I			
ITPM Task Results Documentation Management																		
• Documentation requirements	R	R	S	R	S	S			R		R							
• Results documentation			S	R	R	S	S				R	S					R	R
ITPM Task Results Management	I	I	R	R	R	S	S				R	S	I				R	R
ITPM Schedule Management																		
• Delayed or omitted tasks	R/A	R/A	S	S	S	S	S	A	S	S	S		A			I		
• Task frequency adjustment	A	A	R	R	R	S	S	A	S	S	S	I				I		
ITPM Program Monitoring	R/A	R/A	S	S	S		S	I	I		I	I	I			I		

CHAPTER 6 REFERENCES

6-1   *Process Safety Management of Highly Hazardous Chemicals*, 29 CFR 1910.119, U.S. Occupational Safety and Health Administration, Washington, DC, 1992.

6-2   Wulpi, D., *Understanding How Components Fail, 3rd Edition*, ASM International, Materials Park, OH, 2013.

6-3   API RP 571, *Damage Mechanisms Affecting Fixed Equipment in the Refining Industry*, American Petroleum Institute, Washington, DC.

6-4   ASME PCC-3, *Inspection Planning Using Risk-Based Methods*, American Society of Mechanical Engineers, New York, NY.

6-5   API RP 583, *Corrosion Under Insulation and Fireproofing*, American Petroleum Institute, Washington, DC.

6-6   API 570, *Piping Inspection Code: Inspection, Repair, Alteration, and Rerating of In-Service Piping*, American Petroleum Institute, Washington, DC.

6-7   API RP 574, *Inspection Practices for Piping System Components*, American Petroleum Institute, Washington, DC.

6-8   U.S. Occupational Safety and Health Administration, *Clarification on the Documentation of Inspections and Tests Required Under the Asset Integrity Provisions*, Correspondence letter to Mr. Sylvester W. Fretwell, Lever Brothers Company, dated September 16, 1996.

6-9   API 510, *Pressure Vessel Inspection Code: Maintenance Inspection, Rating, Repair and Alteration*, American Petroleum Institute, Washington, DC.

6-10   API 653, *Tank Inspection, Repair, Alteration, and Reconstruction*, American Petroleum Institute, Washington, DC.

6-11   *National Board Inspection Code*, National Board of Boiler and Pressure Vessel Inspectors, Columbus, OH.

Additional Chapter 6 Resources

Center for Chemical Process Safety, *Guidelines for Initiating Events and Independent Protection Layers in Layer of Protection Analysis*, American Institute of Chemical Engineers, New York, NY, 2014.

Center for Chemical Process Safety, *Guidelines for Safe and Reliable Instrumented Protective Systems*, American Institute of Chemical Engineers, New York, NY, 2007.

Center for Chemical Process Safety, *Guidelines for Risk Based Process Safety*, American Institute of Chemical Engineers, New York, NY, 2007.

Center for Chemical Process Safety, *Guidelines for Safe Automation of Chemical Processes, 2nd Ed.*, American Institute of Chemical Engineers, New York, NY, 2016.

IEC 61511, *Functional Safety: Safety Instrumented Systems for the Process Industry Sector - Part 1: Framework, Definitions, System, Hardware and Software Requirements*, International Electrotechnical Commission, Geneva, Switzerland. Also by the same title ANSI/ISA-84.00.01 Part 1 (IEC 61511-1 Mod), International Society for Automation, Research Triangle Park, NC.

ISO 55000, Asset Management – Overview, Principles and Terminology, International Organization for Standardization, Geneva, Switzerland.

Technical Manual TED 1-0.15A, *Pressure Vessel Guidelines*, U.S. Occupational Safety and Health Administration, Washington, DC, 1999.

APPENDIX 6A. COMMON PREDICTIVE MAINTENANCE AND NONDESTRUCTIVE TESTING (NDT) TECHNIQUES FOR MECHANICAL EQUIPMENT

Detecting asset failures before they occur is a primary objective of the AIM program. Because asset failures are many times preceded by an advanced warning period, QA and ITPM programs often include predictive maintenance and NDT techniques aimed at detecting and recognizing these warnings. Both predictive maintenance and NDT are considered to be condition monitoring (CM) activities. Below are brief descriptions of some general CM categories and an overview of different types of CM tasks. Appendix 6D contains examples of the CM techniques listed below.

Temperature Measurement. Temperature measurement techniques help detect potential failures related to a temperature change in assets. Measured temperature changes can indicate problems such as excessive friction (e.g., faulty bearings, inadequate lubrication), degraded heat transfer (e.g., fouling in a heat exchanger), and poor electrical connections (e.g., loose, corroded or oxidized connections).

Dynamic Monitoring. Dynamic monitoring (e.g., vibration monitoring, spectrum analysis, shock pulse analysis) involves measuring and analyzing energy emitted from asset equipment in the form of waves, such as vibrations, pulses, and acoustic effects. Measured changes in the vibration characteristics from equipment can indicate problems such as wear, imbalance, misalignment, and damage.

Oil Analysis. Oil analysis (e.g., ferrography, particle counter testing) can be performed on different types of oils, such as lubrication, hydraulic, or insulation oils. Oil analysis can indicate problems such as machine degradation (e.g., wear), oil contamination, improper oil consistency (e.g., incorrect or improper amount of additives), and oil deterioration.

Corrosion Monitoring. Corrosion monitoring (e.g., coupon testing, electrical resistance testing, linear polarization, electrochemical noise measurement) helps provide an indication of the extent of corrosion, the corrosion rate, and the corrosion state (i.e., active or passive corrosion state) of a material.

Nondestructive Testing. NDT involves performing tests that are noninvasive to the test subject (e.g., x-ray, ultrasonic). Many of these tests can be performed while the equipment is on line.

Electrical Testing and Monitoring. Electrical CM techniques (e.g., high potential testing, power signature analysis) involve measuring changes in system properties such as resistance, conductivity, dielectric strength, and potential. These techniques are helpful in detecting such conditions as electrical insulation deterioration, broken motor rotor bars, and shorted motor stator lamination.

Observation and Surveillance. Observation and surveillance CM techniques (e.g., visual, audio, and touch inspections) are based on human sensory capabilities. Observation and surveillance can supplement other CM techniques in detecting such problems as loose/worn parts, leaking equipment, poor electrical/pipe connections, steam leaks, pressure relief valve (PRV) leaks, and surface roughness changes.

Performance Monitoring. Monitoring equipment performance is a form of CM that predicts problems by monitoring changes in variables such as pressure, temperature, flow rate, electrical power consumption, and/or equipment capacity.

Appendix 6B. Imperfection vs Type of NDE Method

This Appendix contains *Imperfection vs. type of NDE method*, ASME Section V, Article 1, Nonmandatory Appendix A. Reproduced with permission.

ASME SECTION V, ARTICLE 1
NONMANDATORY APPENDIX

IMPERFECTION VS
TYPE OF NDE METHOD

A-110   SCOPE

Table A-110 lists common imperfections and the NDE methods that are generally capable of detecting them.

Caution - Table A-110 should be regarded for general guidance only and not as a basis for requiring or prohibiting a particular type of NDE method for a specific application. For example, material and product form are factors that could result in differences from the degree of effectiveness implied in the Table. For service-induced imperfections, accessibility and other conditions at the examination location are also significant factors that must be considered in selecting a particular NDE method. In addition, Table A-110 must not be considered to be all inclusive; there are several NDE methods/techniques and imperfections not listed in the table. The user must consider all applicable conditions when selecting NDE methods for a specific application.

ASME Section V, Article 1, Appx. A, Table A-110 (reproduced with permission)	Surface[1]		Subsurface[2]		Volumetric[3]				
	VT	PT	MT	ET	RT	UTA	UTS	AE	UTT
Service-Induced Imperfections									
Abrasive Wear (Localized)	▲	■	■		▲	■	■		■
Baffle Wear (Heat Exchangers)	▲			■					
Corrosion-Assisted Fatigue Cracks	♦	■	▲		♦	▲		▲	
Corrosion - Crevice	▲								♦
- General / Uniform				♦	■		■		▲
- Pitting	▲	▲	♦		▲	♦	♦	■	♦
- Selective	▲	▲	♦						♦
Creep (Primary)[4]									
Erosion	▲				▲	♦	■		■
Fatigue Cracks	♦	▲	▲	■	■	▲		▲	
Fretting (Heat Exchanger Tubing)	■			■					■
Hot Cracking		■	■		■	♦		■	
Hydrogen-Induced Cracking		■	■		♦	■		■	
Intergranular Stress-Corrosion Cracks						♦			
Stress-Corrosion Cracks (Transgranular)	♦	■	▲	♦	■	■		■	
Welding Imperfections									
Burn Through	▲				▲	■			♦
Cracks	♦	▲	▲	■	■	▲	♦	▲	
Excessive/Inadequate Reinforcement	▲				▲	■	♦		♦
Inclusions (Slag/Tungsten)			■	■	▲	■	♦	♦	
Incomplete Fusion	■		■	■	■	▲	■	■	
Incomplete Penetration	■	▲	▲	■	▲	▲	■	▲	
Misalignment	▲				▲	■			
Overlap	■	▲	▲	♦		♦			
Porosity	▲	▲	♦		▲	■	♦	♦	
Root Concavity	▲				▲	■	♦	♦	♦
Undercut	▲	■	■	♦	▲	■	♦	♦	
Product Form Imperfections									
Bursts (Forgings)	♦	▲	▲	■	▲	■	■	▲	
Cold Shuts (Castings)	♦	▲	▲	■	▲	■	■	♦	
Cracks (All Product Forms)	♦	▲	▲	■	■	■	♦	▲	
Hot Tear (Castings)	♦	▲	▲	■	■	■	♦	♦	
Inclusions (All Product Forms)			■	■	▲	■	♦	♦	
Lamination (Plate, Pipe)	♦	■	■			♦	▲	♦	▲
Laps (Forgings)	♦	▲	▲	♦	■		♦	♦	
Porosity (Castings)	▲	▲	♦		▲	♦	♦	♦	
Seams (Bar, Pipe)	♦	▲	▲	■	■	♦	■	♦	

Legend: **AE** - Acoustic Emission **ET** - Electromagnetic (Eddy Current) **MT** - Magnetic Particle **PT** - Liquid Penetrant **RT** - Radiography **UTA** - Ultrasonic Angle Beam **UTS** - Ultrasonic Straight Beam **UTT** - Ultrasonic Thickness Measurement **VT** – Visual

▲ - All or most standard techniques will detect this imperfection under all or most conditions.

■ - One or more standard technique(s) will detect this imperfection under certain conditions.

♦ - Special techniques, conditions, and/or personnel qualifications are required to detect this imperfection.

NOTES: Methods capable of detecting imperfections that: [1]are open to the surface only. [2]are either open to the surface or slightly subsurface. [3]may be located anywhere within the examined volume. [4]Various NDE methods are capable of detecting tertiary (3rd stage) creep and some, particularly using special techniques, are capable of detecting secondary (2nd stage) creep. There are various descriptions/definitions for the stages of creep and a particular description/definition will not be applicable to all materials and product forms. GENERAL NOTE: Table A-110 lists imperfections and NDE methods that are capable of detecting them. It must be kept in mind that this table is very general in nature. Many factors influence the detectability of imperfections. This Table assumes that only qualified personnel are performing nondestructive examinations and good conditions exist to permit examination (good access, surface conditions, cleanliness, etc.).

Appendix 6C. Common Codes and Standards

Items	Design or Construction Codes	Inspection, Repair, Alteration, Rerating or Fitness for Service Codes	"Support" or "Referenced" Codes or Publications
Boilers	ASME I ASME IV	NBIC	ASME II, ABCD ASME V ASME VI & VII ASME IX API RP 573 SNT-TC-1A
Pressure vessels	ASME VIII DIV 1&2	NBIC API 510 API 579	ASME II, ABCD ASME V ASME IX API RP 572 API IRE II SNT-TC-1A
Piping	ASME B31.1 ASME B31.3	API 570 API 579	ASME II, ABCD ASME V ASME IX API RP 574 ASME B16.5 SNT-TC-1A
Valves	ASME B16.34 API 600 API 609	API 598 API RP 591	API RP 574 ASME V ASME IX
Aboveground storage tanks	API 12B API 650 API 620	API 653 API 579	API 651, API 652 API 2016, API 2207 API RP 575 ASME V ASME IX SNT-TC-1A
Safety Relief / Safety Valves	ASME I ASME IV ASME VIII API 2000	NBIC API RP 576	ASME PTC-25 API 627 ASME V ASME IX
Pumps	API 610 API 574-676	API RP 683 MFG STDS	MFG STDS AWS D14.5
Instrumentation and Controls	VARIOUS ISA STDS AND RP 551	ISA / MFG STDS	INSTRUMENT ENGINEER'S HDBK MFG STDS
Pipelines (49 CFR 186-199)	ASME B31.4 ASME B31.8 API 1104	ASME B31G	ASME V ASME IX

Appendix 6D. Summary of Selected Condition Monitoring Techniques

This appendix provides brief descriptions of selected condition monitoring (CM) techniques for each of the following general categories of CM techniques:

- Temperature measurement
- Dynamic monitoring
- Oil analysis
- Corrosion analysis
- Nondestructive testing (NDT)
- Electrical testing and monitoring
- Observation and surveillance
- Performance monitoring.

This appendix does not contain a complete listing of condition monitoring techniques. However, the more common techniques have been included. In addition, the technique information provided (e.g., P-F interval, application) is representative of data/experience from a variety of industries. Therefore, this information needs to be carefully evaluated before using it to determine the application and interval of a condition monitoring technique.

A.   General Temperature Measurement CM Techniques

- *Thermography.* This non-contact technique measures infrared radiation emitted from the surface of an object. Infrared radiation is emitted from all objects above the temperature of absolute zero (0 K or -273 °C). This form of condition monitoring results in color or gray-scale images that identify temperature differences in the surface being examined. Test images can be photographed or electronically recorded for future analysis.

- *Thermometry.* The measurement of the surface temperature of an object using thermocouples or temperature indicating paint applied to the surface. This form of condition monitoring results in (1) an electrical current being generated in the thermocouple that is converted to a temperature readout by an instrument or (2) a permanent change in the color of the temperature indicating paint that indicates the highest temperature reached by the surface.

B.   Specific Temperature Measurement Techniques

- *Temperature-Indicating Paint.* This contact measurement technique is used to indicate the surface temperature of objects upon which the special paint has been applied. The paint will change colors as the surface temperature of the monitored object increases, and it retains the color of the highest temperature the surface has encountered. Some applications of the technique include locating hot spots and insulation failures.

  Typical potential-to-functional failure (P-F) interval: Weeks to months

  Skill level: No specific training needed

  Advantages of the technique include:

  – The test is simple and no special training is required to observe the results

  – The paint retains the color of the highest temperature reached providing a permanent record.

  Disadvantages of the technique include:

  – Once the paint color changes, it does not change back to the original color again

  – The effective life of an application of the paint is usually limited to one or two years or until the paint changes colors.

- *Infrared Thermography.* This non-contact technique uses infrared scanners to measure the temperature of heat-radiating surfaces within the line of sight of the scanner. The scanner measures temperature variations on the surface of the object being monitored and converts the temperature data into video or audio signals that can be displayed or recorded in a wide variety of formats. The sensitivity of the technique is affected by the reflectivity of the object being observed. The scanners are available for a wide range of temperature sensitivities and resolutions. Some applications of the technique include scanning elevated, large, distant, or hot surfaces.

  Typical potential-to-functional failure (P-F) interval: Days to months

  Skill level: Trained and experienced technician

  Advantages of the technique include:

  – Scanners can be portable and are generally considered easy to operate

- Provides dramatic images of the object's temperature profile

- Non-contact testing (i.e., safe to measure energized systems, can measure object without disturbing its temperature)

- Temperature of large surface areas can be observed quickly and essentially continuously

- Wide variety of equipment options are available including various lenses and zoom view capabilities.

- Test data can be recorded, printed, logged, or fed to other digital equipment.

Disadvantages of the technique include:

- Equipment costs are considered moderate to expensive

- Interpretation of the results requires training and experience

- Scanners do not measure well through metal or glass housings or barriers.

C.　Dynamic Monitoring CM Techniques

- ***Time Waveform Analysis.*** Time waveform analysis can identify a wide range of mechanical instabilities including problems such as chipped, cracked, or broken teeth; pump cavitation; misalignment; looseness; and/or eccentricity. This technique uses an oscilloscope connected to the output of a vibration analyzer or real-time analyzer. Through manipulation of the analyzer output signal, the oscilloscope can generate a waveform representing vibration in the dynamic system being monitored. Some applications of the technique include monitoring gearboxes, pumps, and roller bearings.

Typical P-F interval: Weeks to months

Skill level: Trained and experienced technician

Advantages of the technique include:

- Analysis is effective when looking for beats, pulses, instabilities, and a multitude of other conditions of interest

- Analysis often provides more information than frequency analysis.

Disadvantages of the technique include:

- Time wave forms can be complex and confusing

- Testing can consume a considerable amount of time

– Personnel need considerable practice and experience to interpret complex wave forms.

• **Spectrum Analysis.** Spectrum analysis transforms data that is in the time domain to the frequency domain using the fast Fourier transform (FFT) algorithm, by either the data collector itself or a host computer. After the data is collected and transformed (i.e., organized by frequency), it is compared to the baseline or expected values. Problems are identified by changes in amplitude at selected frequencies. Some applications of the technique include monitoring shafts, gearboxes, belt drives, compressors, engines, roller bearings, journal bearings, electric motors, pumps, and turbines.

Typical P-F interval: Weeks to months

Skill level: Trained and experienced technician

Advantages of the technique include:

– Equipment is portable and easy to use

– Software is available that makes the mathematical transformation of the data rapid and accurate

– Small performance changes in the equipment being tested can be identified by these tests

– Characteristic frequencies usually allow the user to isolate the problem to a component.

Disadvantages of the technique include:

– Impacts to the equipment, random noise, and vibrations of nearby equipment can interfere with the tests and can look similar in the results.

• **Shock Pulse Analysis.** Shock pulse analysis measures the impact of rollers with the raceway and produces a shock pulse reading that changes as the conditions within the bearing deteriorate. This technique uses a shock pulse analyzer that is set up specifically for the type and size of bearings being tested and is fed a signal from an accelerometer placed on a bearing housing. It can identify issues such as lubricant problems, problems with oil seals and packings, and incorrect bearing installation and/or alignment. Some applications of the technique include monitoring roller bearings, impact tools, and internal combustion engine valves.

Typical P-F interval: Weeks to months

Skill level: Trained and experienced technician

Advantages of the technique include:

- Test equipment is portable and easy to operate

- Test results are essentially immediate

- Sensitivity of the test is generally considered better than conventional vibration analysis.

Disadvantages of the technique include:

- The test is limited to roller-type bearings

- The test is highly dependent on accurate bearing size and speed information.

- **Ultrasonic Analysis.** When used as a dynamic monitoring technique, ultrasonic analysis helps detect changes in sound patterns caused by problems such as wear, fatigue, and deterioration in moving parts. Ultrasound (i.e., high frequency sound waves that are above human perception from 20 Hz to 100 kHz) is detected by an ultrasonic translator and converted to audible or visual output. (Note: See Ultrasonic Testing as a nondestructive condition monitoring technique for its other capabilities.) One application of the technique includes monitoring bearing fatigue or wear.

Typical P-F interval: Highly variable

Skill level: Trained skilled worker

Advantages of the technique include:

- Tests are quick and easy to do

- Location of the noise source can be pinpointed accurately

- Equipment is portable and monitoring can be done from a long range.

D. Oil Analysis CM Techniques

- **Ferrography.** Ferrography is a technique that identifies the density and size ratio of particles in oil or grease caused by problems such as wear, fatigue, and/or corrosion. A representative sample is diluted with a fixer solvent and then passed over an inclined glass slide that is subjected to a magnetic field. The magnetic field provides separation of the ferrous particles (ferrous particles align with the magnetic field lines) and distributes them along the length of the slide (non-magnetic and non-metallic particles are distributed randomly along the slide). The total density of the particles and the ratio of large-to-small particles indicate the type and extent of wear. Analysis of the test slide is done by bichromatic microscopic examination using both reflected and transmitted light sources (which may be used simultaneously). Green, red, and polarized filters are also used to distinguish the size, composition, shape, and texture of both metallic and non-metallic particles. An electron microscope can also be employed in the analysis to determine particle shapes and provide an indication of the cause of failure. Some applications of the technique include analyzing grease and oil used in diesel and gasoline engines, gas turbines, transmissions, gearboxes, compressors, and hydraulic systems.

Typical P-F interval: Months

Skill level: Trained semi-skilled worker to take the sample and experienced technician to perform and interpret the analysis

Advantages of the technique include:

– Ferrography is more sensitive than many other tests at identifying early signs of wear

– The slide provides a permanent record and allows the measurement of particle size and shape.

Disadvantages of the technique include:

– The test is time-consuming and requires expensive equipment

– In depth analysis requires an electron microscope

– Primary target is limited to ferromagnetic particles.

• **Particle Counter.** Particle counter testing monitors particles in both lubricating and hydraulic oils caused by problems such as corrosion, wear, fatigue, and contaminants. There are several types of particle counting tests available. Two in particular are light extinction and light scattering particle counters. In a light extinction particle counter test, an incandescent light shines on an object cell that the oil sample fluid moves through under controlled flow and volume conditions. A particle counter (i.e., photo diode) receives the light passing through the sample and based on the amount of light blocked it indicates the number of particles in a predetermined size range. A direct reading of the ISO cleanliness value can be determined from this test.

In a light-scattering particle counter test, a laser light shines on an object cell that the oil sample fluid moves through under controlled flow and volume conditions. When opaque particles pass through the laser, the scattered light created is measured and translated into a particle count by a photo diode. A direct reading of the ISO cleanliness value can be determined from this test.

Some applications of these techniques include analyzing oil used in engines, compressors, transmissions, gearboxes, and hydraulic systems.

Typical P-F interval: Weeks to months

Skill level: Trained, skilled worker

Advantages of the technique include:

– Test results are quickly available

– The tests are accurate and reproducible

– The tests are more accurate than graded filtration.

Disadvantages of the technique include:

– The tests are dependent on good fluid conditions and are hampered by air bubbles, water contamination, and translucent particles

– The tests provide no information on the chemical nature of the contamination.

• **Sediment Tests (ASTM D-1698).** Sediment testing provides information about sediment (e.g., inorganic sediment from contamination and organic sediment from oil deterioration or contamination) and soluble sludge from electrical insulating oil deterioration. It involves the use of a centrifuge to separate sediment from oil, and the sediment-free portion is subject to further steps (i.e., dilution, precipitation, and filtration), to measure the soluble sludge. The total sediment is weighed, and then baked to remove the organics, which provides an organic/inorganic composition. Some applications of the technique include analyzing petroleum-based insulating oils in transformers, breakers, and cables.

Typical P-F interval: Weeks

Skill level: Electrician to take the sample and trained laboratory technician to perform and interpret the analysis

Advantages of the technique include:

– The test is relatively quick and easy to complete

– Samples can be taken online.

Disadvantages of the technique include:

– Only low-viscosity oil can be sampled

– Testing must be performed in a laboratory.

- ***Atomic Emissions Spectroscopy.***   Atomic emissions spectroscopy identifies problems such as corrosion, wear metals, contaminants, and additives in lubrication and hydraulic oil samples by measuring the characteristic radiation emitted when samples are subjected to high energy and temperature conditions. The test results are in parts per million (ppm) for a wide variety of elements of interest including iron, aluminum, chromium, copper, lead, tin, nickel, silver, and components of oil additives such as boron, zinc, phosphorus, and calcium. Some applications of the technique include analyzing oil used in diesel and gasoline engines, compressors, transmissions, gearboxes, and hydraulic systems.

Typical P-F interval: Weeks to months

Skill level: Trained semi-skilled worker to take the sample and experienced technician to perform and interpret the analysis

Advantages of the technique include:

– The tests are fairly low-cost

– The tests yield rapid and accurate results

– Range of elements identified is large.

Disadvantages of the technique include:

– The tests do not identify the wear process that contaminated the oil

– Large particles in the sample may not be counted in the results.

- ***Infrared Spectroscopy.***   Infrared spectroscopy involves placing a sample in a beam of infrared light and then measuring the absorbent light energy at various specific wavelengths to determine the level of an element in a sample without destroying the sample.   Mathematical manipulations of the absorption data results in a "fingerprint" of the sample oil which can be compared to prior samples or standards by intelligent software. The analysis can provide information about oil deterioration, oxidation, water contamination, or oil additives. Some applications of the technique include nuclear power systems, turbine generators, sulfur hexafluoride or nitrogen sealed systems, transformer oils, and breakers.

Typical P-F interval: Weeks to months

Skill level: Trained, semi-skilled worker to take the sample and trained laboratory technician to perform and interpret the analysis

Advantages of the technique include:

- Data can be used to determine ASTM parameters
- The test is highly repeatable
- Data can be used to generate a total acid number (TAN) and a total base number (TBN).

Disadvantages of the technique include:

- Test equipment manufacturers are not consistent in the processing of data
- Typically, the test is limited to about 1000 ppm water contamination.

- **Potentiometric Titration.** Total Acid Number (TAN) - Potentiometric titration - TAN is used to determine the extent of breakdown in lubrication or hydraulic oil by determining the level of acidity in an oil sample. The test involves mixing the oil sample with solvents and water and then measuring the change in the electrical conductivity as the mixture is titrated with potassium hydroxide (KOH). The more KOH a sample uses is an indication of a higher acid number and more oil deterioration. Some applications of the technique include testing oil used in diesel/gasoline engines, gas turbines, transmissions, gearboxes, compressors, hydraulic systems, and transformers.

Typical P-F interval: Weeks to months

Skill level: Trained, semi-skilled worker to take the sample and trained laboratory technician to perform and interpret the analysis

Advantages of the technique include:

- The test can be performed on any color oil
- The test is considered accurate within 15%.

Disadvantages of the technique include:

- The test is limited to petroleum-based oils
- Some of the chemicals used to complete the tests are hazardous.

- *Karl Fischer Titration Test (ASTM D-1744).* The Karl Fischer titration test measures moisture in a lubrication or hydraulic oil sample, which is an indicator of a degraded oil condition, by measuring electrical current flow between two electrodes immersed in the sample solution. Karl Fischer reagent is metered into the sample until all the entrained water is reacted with the reagent. Results are reported in ppm of water. Some applications of the technique include analyzing oil in enclosed oil systems such as engines, gearboxes, transmissions, compressors, hydraulic systems, turbines, and transformers.

  Typical P-F interval: Days to weeks

  Skill level: Trained lab technician

  Advantages of the technique include:

  - The test is accurate for small quantities of water contamination

  - The test can be completed fairly quickly

  - Results are repeatable.

  Disadvantages of the technique include:

  - Considerable skill is required to interpret the results

  - Automated equipment is relatively expensive and not portable.

- *Kinematic Viscosity.* The kinematic viscosity test provides an indication of oil deterioration over time or contamination of the oil by fuel or other oils. The test measures the fluids resistance to flow under known pressure and temperature conditions and involves forcing a sample to flow through a capillary viscometer. Based on the test results, the dynamic viscosity of the oil sample can be calculated. Some applications of the technique include testing oil used in diesel/gasoline engines, turbines, transmissions, gearboxes, compressors, and hydraulic systems.

  Typical P-F interval: Weeks to months

  Skill level: Trained, semi-skilled worker to take the sample and trained laboratory technician to perform and interpret the analysis

  Advantages of the technique include:

  - The test can be used for most lubricating oils, both transparent and opaque

  - Results are repeatable.

Disadvantages of the technique include:

– The test is not done in the field

– Flammable solvents are used.

- **Dielectric Strength Tests.** Dielectric strength tests are used to measure the insulating quality of electrical insulating oils. Potential quality deterioration is often caused by contamination or oil breakdown. The test is performed by subjecting the sample to an electrical stress at a given temperature by passing voltage through the sample. Some of the applications of the test include testing of insulating oils in transformers, breakers, and cables.

Typical P-F interval: Months

Skill level: Electrician to take the sample and trained laboratory technician to perform and interpret the analysis

Advantages of the technique include:

– The test is rapid and relatively simple

– The equipment does not need to be offline to perform the test.

Disadvantages of the technique include:

– The sampling technique can affect the test results

– The test must be completed in the lab

– The materials and equipment used to complete the test are hazardous.

E.   Corrosion Monitoring CM Techniques

- **Coupon Testing.** Coupon testing involves placing sacrificial coupons, which are usually made from low-carbon steel or from a grade of material that duplicates the material of construction of the equipment being monitored, into the process so that the corrosion from the equipment can be monitored. The coupons are periodically measured and observed to understand the process environment's effect on these test pieces. Measurements include checking for weight loss, dimensional changes, and physical damage such as pitting. Some applications of the technique include testing at petroleum refineries, process plants, underground/undersea structures, cathodic protection monitoring, abrasive slurry transport, water distribution systems, atmospheric corrosion, and electrical generating plants.

Typical P-F interval: Months

Skill level: Trained and experienced technician

Advantages of the technique include:

- Corrosion effects can be accurately predicted when the environment is consistent over the test period
- Testing is relatively inexpensive and yields vivid examples of the corrosion to expect.

Disadvantages of the technique include:

- Testing can take a long period of time to complete
- Determining the corrosion rates can take several weeks or months of testing
- The tests involve working directly with the potentially hazardous corrosive material streams.

- **Corrometer.** Corrometer testing helps measure the corrosion rate of equipment by monitoring the change in the electrical resistance of a sample material. As the sample material's cross section is reduced due to corrosion, the electrical resistance of the sample will increase. The measured resistance change corresponds to the total metal loss and can be converted to corrosion rate. Some applications of the technique include testing at petroleum refineries, process plants, underground/undersea structures, cathodic protection monitoring, abrasive slurry transport, water distribution systems, atmospheric corrosion, paper mills, and electrical generating plants.

Typical P-F interval: Months

Skill level: Trained and experienced technician

Advantages of the technique include:

- Portable equipment is available
- Testing works in many environments
- Testing can be made continuous with an on-line monitor
- Results are easily converted to corrosion rates.

Disadvantages of the technique include:

- Portable equipment does not provide permanent records
- The test does not typically indicate changes in the corrosion rate.

- *Potential monitoring.* Potential monitoring helps understand the corrosion state (i.e., active or passive) of material by monitoring localized corrosion and indicating when active corrosion is in progress. This test takes advantage of the fact that metals in an active corrosion state (i.e., higher corrosion rate) have a different electrical potential than when they are in a passive corrosion state (i.e., lower corrosion rate). A voltmeter is used to measure the potential of the sample area. Some applications of the technique include monitoring in chemical process plants, paper mills, pollution control plants, electrical utilities, and desalination plants. The technique is best suited to stainless steel, nickel-based alloys, and titanium.

  Typical P-F interval: Varies depending on material and rate of corrosion

  Skill level: Trained and experienced technician

  Advantages of the technique include:

  – The test gives a rapid response to change

  – Localized corrosive effects are monitored.

  Disadvantages of the technique include:

  – The test does not provide corrosion rates

  – Testing is influenced by changes in temperature and acidity.

F.   Non-Destructive Testing (NDT) CM Techniques

- *X-ray Radiography.* X-ray radiography helps identify surface and subsurface flaws caused by problems such as stress, corrosion, inclusions, fatigue, poor or incomplete welds, and trapped gases. In addition, it can be used to locate semiconductor faults and loose wires. The technique produces a radiograph by passing x-rays through opaque materials and producing an image of those materials on film or a cathode ray tube. Typically, film exposed to x-rays is darkest where the object is thinnest or absorbs the least radiation. Some applications of the technique include analysis of welds, steel structures, plastic structures, metallic wear components of engines, compressors, gearboxes, pumps, and shafts.

  Typical P-F interval: Months

  Skill level: Trained and experienced technician to take the radiographs and trained and experienced technician or engineer to interpret the radiographs

Advantages of the technique include:

- The technique examines inside the test materials to locate hidden flaws (i.e., areas that cannot be seen externally)

- The technique provides a permanent record of the test.

Disadvantages of the technique include:

- Sometimes, several views are required to locate the flaw

- The test is not very sensitive to crack-type flaws.

- ***Liquid Dye Penetrant.*** The use of liquid dye penetrants can help detect surface discontinuities or cracks due to problems such as fatigue, wear, surface shrinkage, and grinding. The technique involves applying liquid dye penetrant to a test surface and then allowing sufficient time for the dye to penetrate the surface. Next, excess penetrant is removed from the surface, and the surface is retreated with a developer that draws the penetrant to the surface, revealing the location of imperfections. Liquid penetrants are categorized according to the type of dye (e.g., visible dyes, fluorescent penetrant, and dual sensitivity penetrant) and based on the processing (e.g., water washable, post emulsified, or solvent removed) required to remove them from the surface. Some applications of the technique include analyzing ferrous and non-ferrous materials such as welds, machined surfaces, steel structures, shafts, boilers, plastic structures, and compressor receivers.

Typical P-F interval: Days to months

Skill level: Trained and experienced technician

Advantages of the technique include:

- Visible dye penetrant kits are cheap; fluorescent kits are more sensitive but more expensive

- Surface problems on non-ferrous materials can be detected.

Disadvantages of using the liquid dye penetrant technique are:

- Testing will not work on highly porous materials

- Technique is not conducive to on-line testing

- Experienced personnel are required to evaluate the results

- Dark work area is required for fluorescent dye testing.

- **Ultrasonic Analysis.** Ultrasonic analysis helps detect changes in sound patterns caused by problems such as leaks, wear, fatigue, or deterioration. Ultrasound (i.e., high-frequency sound waves that are above human perception from 20 Hz to 100 kHz) is detected by an ultrasonic translator and converted to audible or visual output. (Note: see Ultrasonic Analysis as a dynamic condition monitoring technique for its other capabilities.) Some applications of the technique include detecting leaks in pressure/vacuum systems, leaks in underground pipes or tanks, and static discharge.

  Typical P-F interval: Highly variable

  Skill level: Trained, skilled worker

  Advantages of the technique include:

  – The tests are quick and easy to do

  – Location of the noise source can be pinpointed accurately

  – Equipment is portable and monitoring can be done from a long range.

  Disadvantages of the technique include:

  – Some tests can only be done under vacuum

  – In general, test results do not indicate the size of a leak.

- **Ultrasonic Transmission Technique.** Ultrasonic testing using a transmission technique helps to detect surface and subsurface discontinuities caused by problems such as fatigue, heat treatment, inclusions, and lack of penetration and gas porosity welds. It also provides the ability to measure thickness in test subjects. The test involves using one of the available transmission techniques to apply an ultrasound signal to a test object, and then the signal is received back and analyzed for changes that might indicate the presence of discontinuities in the test object. Some of the ultrasonic techniques include pulse echo, transmission, resonance, and frequency modulation. Some applications of the technique include inspecting ferrous and nonferrous welds, steel structures, boilers, tubes, plastic structures, and vessels/tanks.

  Typical P-F interval: Weeks to months

  Skill level: Trained and experienced technician

One advantage of the technique is that the tests are applicable to a majority of materials.

One disadvantage of the technique is that the test results do not clearly distinguish between types of defects.

- **Magnetic Particle Inspection.** Magnetic particle inspection helps detect the location of surface/near surface cracks and discontinuities caused by problems such as fatigue wear, inclusions, laminations, heat treatment, hydrogen embrittlement, seams, and corrosion. The technique involves magnetizing the test piece and spraying it with a solution containing very fine iron particles. Then discontinuities on the surface of the test piece will cause the iron particles to accumulate and form an indication of the flaw. The results are then interpreted. Some applications of the technique include analyzing ferromagnetic metals such as vessels/tanks, welds, machined surfaces, shafts, steel structures, and boilers.

Typical P-F interval: Days to months

Skill level: Trained and experienced technician

Advantages of the technique include:

- The test is reliable
- The test is sensitive
- The test is widely used.

Disadvantages of the technique include:

- The test is limited to detecting surface imperfections
- The test is time-consuming
- The test is not applicable as an on-line test.

- **Eddy Current Testing.** Eddy current testing helps detect surface and subsurface flaws caused by problems such as wear, fatigue, and stress, and it helps detect dimensional changes that result from problems such as wear, strain, and corrosion. It can also help determine material hardness. The technique involves applying high-frequency alternating current to conductive material test objects and inducing eddy currents around discontinuities. The electrical effects in the test part are amplified and shown on a cathode ray tube or a meter. Some applications of the technique include boilers, heat exchangers, hydraulic tubes, hoist ropes, railroad lines, and overhead conductors.

Typical P-F interval: Weeks

Skill level: Trained and experienced technician

Advantages of the technique include:

– The test can be performed on a wide variety of conductive materials

– Permanent record can be made via data recorders

One disadvantage of the technique is that nonferrous materials respond poorly to this test.

- **Acoustic Emission.** Acoustic emission testing monitors the plastic deformation and crack formation caused by problems such as fatigue, stress, and wear. The technique involves subjecting the test object to loads and listening to the audible stress waves that result. The test results can be displayed on a cathode ray tube or an x-y recorder. Some applications of the technique include testing structures, pressure vessels, pipelines, and mining excavations.

Typical P-F interval: Weeks

Skill level: Trained and experienced technician

Advantages of the technique include:

– The test can be performed remotely in relation to the flaws and can cover the entire structure

– Active flaws can be detected

– Relative loads used in testing can be used to estimate failure loads in some cases.

Disadvantages of the technique include:

– The test object has to be loaded

– Some electrical and mechanical noises can interfere with the results

– Results analysis can be difficult.

- **Hydrostatic Testing.** Hydrostatic testing helps detect breeches in a system's pressure boundaries caused by problems such as fatigue, stress, and wear. The testing involves filling a system to be tested with water or the operating fluid, sealing the system, and increasing the pressure to approximately 1.5 times the system's operating pressure. Then the pressure is held for a defined period while inspections and monitoring are conducted for visible leaks, a system pressure drop, and make-up water/operating fluid additions. The principle of hydrostatic testing can also be used with compressed gases. Some applications include components (i.e., tanks, vessels, pipelines) and completely assembled systems that contain pressurized fluids or gases.

Typical P-F interval: Days to weeks

Skill level: Trained, skilled worker

One advantage of the technique is that the results are easy to interpret.

Disadvantages of the technique include:

– The test has the potential to overpressurize and damage the system

– The test will not identify defects that have not penetrated a pressure boundary

– The test is not applicable as an on-line test.

- **Visual Inspection – Boroscope.** Visual inspections with a boroscope allow internal inspections of the surface of narrow tubes, bores, pipe, chambers of engines, pumps, turbines, compressors, boilers, etc. The inspection helps to locate and orient surface cracks, oxide films, weld defects, corrosion, wear, and fatigue flaws. The boroscope provides a system to channel light from an external light source to illuminate parts not easily visible to the naked eye, and it also provides a means to photograph and/or magnify the illuminated surface of interest.

Typical P-F interval: Weeks

Skill level: Trained and experienced technician

Advantages of the technique include:

– The equipment provides excellent views

– The parts being examined can be photographed and magnified.

Disadvantages of performing a visual inspection with a boroscope include:

– The inspection is limited to surface conditions

– The lens systems are often inflexible

– Technicians can suffer optic eye fatigue during prolonged inspections.

G.   Electrical Testing and Monitoring CM Techniques

- *Megohmmeter Testing.* A megohmmeter can be used to test the insulation resistance of electrical circuits. The technique involves applying a known voltage to electrical circuits of the equipment being tested and measuring the current flow. Based on the leakage current flowing to ground, the resistance of the equipment insulation can be calculated.

  Typical P-F interval: Months to years

  Skill level: Technicians or engineers

  Advantages of using the technique are that it is simple and well-understood.

  A disadvantage of the technique is that on-line testing cannot be conducted.

- *High Potential Testing.* High potential testing helps detect motor winding ground wall insulation deterioration. The test involves applying high direct current voltage to the stator windings in graduated steps to help determine the voltage at which non-linearity in the test current or drop in the insulation resistance occurs. If the insulation withstands a specified voltage, it is considered to be safe, and the motor can be returned to service. Also, trending in the voltage at which the current becomes non-linear or the resistance drops can be used to predict the remaining motor life. Applications of the technique include AC and DC motors.

  Typical P-F interval: Weeks

  Skill level: Experienced electrical technician

  One advantage of the technique is that the test results usually correlate with surge comparison tests.

  Disadvantages of the technique include:

  – Motors must be off-line for testing

  – The test voltage can be destructive to motor parts.

- **_Surge Testing._** Surge testing helps identify insulation faults in induction/synchronous motors, DC armatures, synchronous field poles, and various coils or coil groups. The technique involves using a high frequency transient surge applied to two separate but equal parts of a winding, and then the resulting reflected waveforms are compared on an oscilloscope. Normally, if no problems are detected at twice the operating voltage plus 1000 volts, the winding is considered good.

  Typical P-F interval: Weeks to months

  Skill level: Trained and experienced test operator

  One advantage of the technique is that the test is portable.

  Disadvantages of the technique include:

  – The test is complex and expensive

  – Careful repetition is required to determine the location or severity of a fault.

- **_Power Signature Analysis._** Power signature analysis can be used to detect motor problems such as broken rotor bars, broken/cracked end rings, flow or machine output restrictions, and machinery misalignment. This on-line technique involves monitoring current flow in one of the power leads at the motor control center or starter. The electrical current variations identified in the test indicate changing machine operating conditions and can be trended over time. Also, line frequencies can be compared with motor frequencies to help detect various motor flaws. Some applications of the technique include AC induction motors, synchronous motors, compressors, pumps, and motor operated valves.

  Typical P-F interval: Weeks to months

  Skill level: Experienced electrician to connect the test equipment and experienced technician to perform the analysis and interpret the data

  Advantages of the technique include:

  – Testing is conducted on-line

  – Test readings can be taken remotely for large or high speed machines.

  Disadvantages of the technique include:

  – Equipment is expensive

  – Analysis of results is complex and often subjective.

- ***Power Factor Testing.*** Power factor testing measures power loss through the insulation system caused by a leakage to ground or moisture in the cables. Some causes of a rising power factor include aging, moisture, contamination, insulation shorts, etc. The technique involves applying a known voltage and then measuring the current flow. Based on the test results, the power factor and watts loss values can be calculated and compared to past readings or a baseline reading if available. Some applications of the technique include electrical circuits, transformer windings, high voltage transformer bushings, and high/medium voltage cables.

  Typical P-F interval: Months

  Skill level: Experienced electrical technician to perform the test and experienced engineer to analyze and interpret the data

  One advantage of the technique is that it is highly regarded as the best predictive test for deterioration of windings.

  One disadvantage of the technique is that the test is conducted off-line.

- ***Motor Circuit Analysis.*** A motor circuit analysis helps to yield a complete picture of motor conditions by performing a series of tests. The test applies voltage at the motor control center power bus to measure resistance to ground, circuit resistance, capacitance to ground, inductance, rotor influence, DC bar to bar, and polarization index/dielectric absorption. The results can identify changes in conductor path resistance caused by loose or corroded connections and loss of copper (turns) in the stator; phase-to-phase inductance caused by magnetic interaction between the stator and rotor; stator inductance affected by rotor position, rotor porosity and eccentricity, stator turn, and coil and phase shorting; and winding cleanliness/resistance to ground. Application of the technique includes use on electric motors (e.g., DC, AC induction, synchronous, and wound rotor).

  Typical P-F interval: Weeks to months

  Skill level: Experienced electrical technician to perform the test

  Advantages of the technique include:

  – The test is low voltage and non-destructive

  – Tests can be performed at the motor control center, which does not require motor disassembly.

  One disadvantage of the technique is that the test cannot be performed on-line.

- ***Battery Impedance Testing.*** Battery impedance testing helps to detect battery cell deterioration. The test involves injecting an AC signal between the battery posts and measuring the resulting voltage. The battery impedance is then calculated and compared to (1) the battery's last test and (2) the impedance of other batteries in the same bank. If the comparison results are outside a certain percentage then this could indicate a cell problem or capacity loss.

  Typical P-F interval: Weeks

  Skill level: Experienced electrical technician to perform the test

  One advantage of the technique is that the test can be performed on-line.

  One disadvantage of the technique is that the tests are lengthy for large batteries.

- ***Other Techniques.*** Battery load testing (provides the information needed to fully assess cell strength); other battery inspections/testing depending on battery type; transformer testing.

H.   Observation and Surveillance Condition Monitoring Techniques

- ***Visual Inspection.*** Visual inspection practices are the oldest and most common condition monitoring techniques employed in industry. Human observation helps identify a broad range of potential problems including loose or worn parts; leaks of lubricating oils, hydraulic fluids and process liquids; missing parts; poor electrical or pipe connections; etc. Inspection standards are easy to establish and communicate to assigned personnel. Essentially all machines and equipment in the industrial setting can be monitored with this technique. Also, human sensory based inspections can serve to verify the results from other CM techniques.

  Typical P-F interval: Varies widely

  Skill level: Trained, semi-skilled workers are normally required

  One advantage of the technique is the versatility of human observation combined with experience can identify an extremely wide range of problem types.

  One disadvantage of the technique is unless inspections are scheduled and recorded, observers can become so familiar with their surroundings that changes of interest go unnoticed.

- *Audio Inspections.* Audio inspection practices are common condition monitoring techniques employed in industry. The monitoring of machinery and equipment by listening to it operate helps identify a broad range of potential problems including worn high friction bearings, steam leaks, pressure relief valve leaks or discharges, coupling leaks, excessive loading on pumps, poor mechanical equipment alignment, etc. Humans are particularly sensitive to new or changed sounds and are easily taught to report and investigate unusual sounds. This technique is often a supplemental inspection to visual inspections. Also, human-sensory-based inspections can serve to verify the results from other CM techniques.

  Typical P-F interval: Varies widely

  Skill level: Trained, semi-skilled workers are normally required

  One advantage of the technique is the versatility of human hearing combined with experience can identify an extremely wide range of problem sounds.

  One disadvantage of the technique is that the inspections must be assigned, so that the inspectors gain sufficient experience to detect new or changed noises.

- *Touch Inspections.* Using touch as an inspection technique can be extremely useful. Heat, scaling, and roughness changes can all be detected by touch. Human touch is extremely sensitive and able to differentiate surface finish differences not discernable by the eye. This technique is often a supplemental inspection to visual inspections. Also, human-sensory-based inspections can serve to verify the results from other CM techniques.

  Typical P-F interval: Varies widely

  Skill level: Trained, semi-skilled workers are normally required

  One advantage of the technique is that the hands and fingers are extremely sensitive to surface finish and to heat.

  One disadvantage of the technique is that the inspectors can be burned by touching hot objects and can be injured or shocked by touching operating equipment.

I. Performance CM Techniques

- *Performance Trending.* Performance trending as a technique of condition monitoring involves collecting and analyzing data on pressure, temperatures, flow rates, or electrical power consumption for the process, machinery, and/or equipment of interest. Data are often collected by operations personnel for other reasons (e.g., quality control programs) and may already be available for analysis. Performance trend data is often coupled with other test results to confirm the identification of problems (e.g., equipment degradation, performance deterioration). Monitoring the performance indicators over a long period of time can provide indications of improper maintenance or poor operations practices. Virtually all industrial machines can be monitored in this fashion and targets for data collection include diesel and gasoline engines, pumps, motors, compressors, etc. Data is often already collected for other reasons, and test data can also be used to optimize performance. In addition, most of the computer control equipment (e.g., distributed control systems, programmable logic controller) has data analysis and alarming features that can be used to trend equipment performance.

Typical P-F interval: Varies widely

Skill level: Trained, semi-skilled workers are normally required

One advantage of the technique is that the data are often already collected.

One disadvantage of the technique is that baseline data may not exist, which necessitates longer time periods to develop trends.

7 ESTABLISHED APPROACHES FOR DEVELOPING TEST AND INSPECTION PLANS

This chapter summarizes established approaches for the development of AIM program ITPM test and inspection plans. When test/inspection plans are being developed, it will be helpful to always keep in mind the following basic tenet:

> *An inspection program has little value in and of itself.*
>
> *It brings value to an organization by providing the necessary data to plan any required repairs before a failure event occurs.*

Three approaches are first addressed that tend to be more prescriptive in nature: code/standard, regulatory authority and company-specific requirements. The remainder of the chapter summarizes three additional approaches that are best known by their acronyms: RBI, FMEA and SIS. These approaches are not mutually exclusive; they are often used in combination.

One related approach, known as reliability-centered maintenance (RCM), is addressed in Section 15.3. As discussed in that section, RCM can be used to develop some aspects of test and inspection plans.

The focus of the approaches in this chapter is on equipment at fixed facilities. Transportation aspects such as marine interfaces and pipeline transport are not covered here, although many of the same principles apply.

7.1 CODE/STANDARD APPROACHES

Maintenance inspection protocols, including minimum inspection frequencies, are fully specified in codes or standards for some types of assets. For example, selectively employed requirements incorporated in codes and standards for inspecting fixed equipment are given in API 510, 570, and 653 for pressure vessels, piping, and tanks, respectively (References 7-1, 7-2, and 7-3), making note of the year and revision number of the standard. No further risk-based analysis is needed when employing such approaches, although good judgment is still required in the application of these codes and standards. Similar inspection and testing requirements for other asset types are given in Chapter 12.

7.2   REGULATORY AUTHORITY APPROACHES

Local, state/province, national or regional authorities may adopt and require adherence to more prescriptive approaches for testing and inspection of some equipment types. For example, in the U.S., the *National Board Inspection Code* (Reference 7-4) gives widely used inspection requirements that are often mandated by state or local authorities for inspection of boilers and pressure vessels.

It should be noted that not all regulatory approaches are prescriptive with respect to developing inspection and testing plans. For example, the UK Pressure Systems Safety Regulations give the desired end result but do not prescribe a method of achieving that end result. Regulation 4 states that "The pressure system shall be properly designed and properly constructed from suitable material, so as to prevent danger."

Note that, in most cases, the regulatory inspection and data analysis is solely directed at the question of whether an asset is sufficiently robust so as not to fail catastrophically prior to the next inspection. This approach, basically a "go/no-go" assessment of the equipment, is valid for the regulator to take; however, it does not address the inspection and testing program as a component of the asset maintenance strategy where the goal is the safe operation of the asset while still providing sufficient up-time (rates, quality, availability) to have a profitable operation.

7.3   COMPANY-SPECIFIC APPROACHES

Some companies have their own internal requirements for developing test and inspection plans for particular equipment types. Development of such approaches is generally where code/standard and regulatory approaches do not apply, or a company wants to be even more restrictive or conservative than codes, standards or regulations. An example of such an approach might be for a unique or proprietary type of asset that is critical to safety and/or business, perhaps with which the company has had past failures with severe consequences. Another example might be where regulatory requirements are performance based and a company establishes its own prescriptive requirements for meeting the performance-based regulation or general duty. The company-specific prescriptive approach might dictate precisely what inspections and tests are to be performed by whom, with specified test equipment and procedures and at a specified inspection/test interval.

These company-based approaches usually relate to the safe and efficient operation of the equipment, such as for alignment of critical pump and agitator shafts to prevent material leakage. Another class of equipment typically inspected to company requirements is ancillary components of a piping system such as expansion joints, hangars/supports, bolting, flexible hoses and functionality of valves. Other examples include:

- Glass-lined vessels

- Sight glasses

- Process equipment grounding and bonding for static elimination

- Explosion barricades and walls

- Structural steel

- Dikes and sumps.

7.4   RISK-BASED INSPECTION (RBI)

Risk-Based Inspection (RBI) is an AIM inspection planning approach that has become more common in the process industries (Reference 7-5). Inspection codes such as API 510 and API 570 now recognize RBI approaches and have included many of its concepts along with the traditional time-based methods.

RBI is a risk assessment and risk management tool that assesses the likelihood and consequence of loss of containment in process equipment. It integrates the traditional codes and standards with flexibility to focus and optimize the activities on risk reduction by identifying higher-risk assets. RBI is typically used to develop and optimize the inspection plans for pressure vessels, storage tanks, piping, and relief devices, which together are generally called "fixed equipment."

To assist in the development and implementation of RBI programs, the American Petroleum Institute has published API RP 580, *Risk-Based Inspection* and API RP 581, *Risk-Based Inspection Technology* (References 7-6 and 7-7) as recommended practices for the implementation of RBI. The U.K. Health and Safety Executive makes available an RBI best-practice research report, as well as online guidance for the application of RBI in refineries, chemical process plants and similar establishments onshore and to process plants on offshore installations (References 7-8 and 7-9).

RBI methodologies focus primarily on pressure vessels, storage tanks, piping and other primary containment system components. However, this method can be applied to most other types of process equipment. This approach provides more focus on assets with higher risks through a systematic analysis of damage mechanisms, condition, and the application of the most appropriate inspection techniques.

Conventional inspection programs can be unfocused, applying the same, often less-effective inspections at prescribed intervals. Too many facilities have broadly applied conventional inspection programs with excessive ultrasonic thickness (UT) measurements and have buried their personnel with measurement data, at the risk of losing sight of more critical assets. A successful RBI program provides rule sets or strategies to guide appropriate inspections and to update inspection plans based on information from each round of inspections. Conventional programs sometimes provide little guidance for the inspector regarding when or why to increase or decrease inspection coverage or frequency. RBI programs can be much more responsive in updating inspection plans and can substantially reduce inspection and lost production costs while at the same time reducing risk.

The relationship of risk level to inspection activity and inspection cost is illustrated in Figure 7-1. Focusing inspection activities on higher-risk assets provides more impact for the same effort and cost. As shown in Figure 7-1, some amount of asset risk is not addressable through inspection; this risk may arise from design, operational or maintenance issues. Also note the potential for risk to rise as typical conventional inspection programs become overloaded with data.

Because the implementation of an RBI program often involves extensive cost and time commitments, many facilities use project management procedures to guide the process. Expertise in RBI methodology and implementation will be needed.

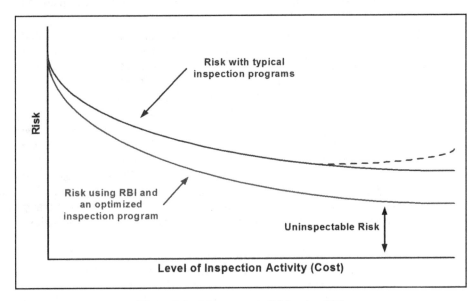

Figure 7-1. Management of risk using RBI

(For illustrative purposes only; not for design)

Whether such expertise is developed in-house or an RBI consultant is used, facilities need to provide work processes, procedures, inspection strategies, training, and software tools.   Facility resources from inspection, process engineering, operations and maintenance groups will also need to be tapped to provide input and expertise on the process and its equipment, with additional involvement of corrosion engineers, equipment engineers, and risk analysts.

7.4.1   Benefit of Software Support for RBI

RBI software has been found to be useful for companies that implement RBI, if personnel with skills in both equipment inspection and computer systems are available to manage the system. Many good RBI software packages exist, with flexibility to adapt to an existing ITPM program. However, many companies have found that a high level of technical computer system management is required, and if this is not adequately addressed in the program implementation it can result in a negative impact on the ITPM system. Also, prior to a facility embarking on an RBI program, it is advisable to estimate the total cost and complexity of managing significant amounts of dynamic data, as well as to discuss and document a budget specific to the RBI program.

7.4.2   Differences from Conventional AIM Programs

Features of RBI that differ from conventional AIM programs are:

- *Equipment and process data.* Data needs are similar to ITPM planning data needs discussed in Section 6.1.3, but are even more fundamental to the analysis. Necessary asset data include design temperatures, design pressures, materials of construction, sizes, details on stress relief, insulation, coatings/linings, current condition and number and type of prior inspections. Also, the following process information is needed: process stream compositions, fluid properties, operating temperatures, operating pressures, flammability, toxicity and inventories. The analysis will focus on the normal/routine process stream compositions; however, non-routine operating modes (e.g., startup, shutdown) need to be considered, as do abnormal conditions if they can initiate or accelerate damage to the equipment.

- *Risk modeling.* Personnel need to identify the underlying methodology for determining the likelihood and consequences of a release. Facilities typically use computer software to support the methodology. To properly interpret the results, personnel need to understand the methods and assumptions used.

- *Inspection strategies.* RBI programs use a set of rules or guidelines for determining appropriate inspection methods, levels of inspection and maximum intervals. These guidelines provide the methodology for creating inspection plans for each asset, based on its risk ranking, equipment type and deterioration mechanisms. It is important that a facility using a particular software implementation of RBI agrees with the underlying assumptions and inspection strategies that accompany it.

- *Inspection planning.* RBI takes current inspection results, enters data into the risk model, recalculates the risk ranking and modifies the inspection plan accordingly, based on the inspection strategy and the expertise of the inspector. Although many commercially available RBI software packages offer ITPM scheduling capabilities, it is advised that this not be done automatically. Instead, the program recommends an inspection plan that is then reviewed and approved by a designated person with experience in equipment inspection and analysis. Many factors outside of the RBI need to be merged for a proper ITPM plan to be generated, such as turnaround schedules, equipment redundancy and availability of parts and manpower.

- *Management systems and tools.* RBI generally uses work processes and computer tools to collect, interpret, integrate and report the inspection data, as well as to plan and schedule inspection tasks. Management of the RBI program also involves reporting on activities, status, exceptions and trends.

7.4.3   RBI Flowchart

Figure 7-2 is a flowchart of an RBI program. Note that, at the front end of the work process and at the update cycle, RBI requires activities that are distinctly different from those of traditional AIM approaches. The shaded boxes, and the arrows entering and exiting those boxes, represent these activities. Details of the steps required to establish an RBI program are given in API 580 (Reference 7-6).

7.5   FAILURE MODES, EFFECTS AND CRITICALITY ANALYSIS APPROACHES

Industry has used the FMEA technique to support ITPM development, in order to identify potential equipment failures in complex systems, understand the effects of each failure on system performance such as safety and mission impacts, determine if sufficient safeguards are provided and identify asset and system improvements, including those involving ITPM tasks. For AIM programs, FMEA can aid in developing elements of the inspection and testing plan (ITP). It does not directly generate the ITP itself.

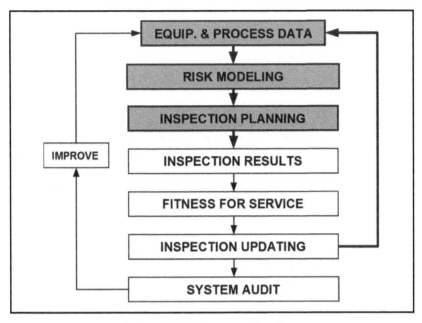

Figure 7-2. RBI program flowchart

FMEAs have been used to evaluate many different types of systems, especially mechanical and electrical systems. FMEA is most commonly used in AIM to analyze critical equipment and complex systems for which failure causes and/or impacts (effects) on process/system performance are not known or not well-understood.

FMEA concepts were introduced in Section 4.7 of this document. This section gives additional details of the FMEA approaches, particularly as they may be used in the development of ITPM test and inspection plans.

One extension of FMEA is known as *FMECA*, which stands for Failure Modes, Effects and Criticality Analysis. FMECA is an FMEA that assesses the criticality (severity of consequences) of the failure modes and resulting effects using qualitative or quantitative criticality categories. This enables the user to focus attention on failure modes that could lead to the most severe impacts.

FMEA and/or FMECA can be used in the ITPM task planning process to better understand the cause-effect relationship between asset failures and resulting effects, usually in terms of loss events such as those resulting in safety impacts, environmental releases or production downtime. This attribute makes FMEA and/or FMECA a useful tool for the ITPM planning process. Benefits include:

- Identification of specific asset failure modes/causes on which to focus ITPM tasks.

- Documentation of the rationale for ITPM task decisions, especially run-to-failure decisions if any.

- Risk ranking, which can be used to prioritize ITPM resources.

Determining the appropriate level of analysis detail as early in the process as possible is one key to successfully using FMEA and/or FMECA to develop an ITPM task plan. An analysis performed with too little detail might not provide the desired information. On the other hand, too much detail will increase analysis time and effort with little added benefit. Typically, the FMEA can be performed at the level of detail at which the ITPM tasks will be defined (e.g., pump, tank, vessel, piping circuit).

FMEA is often used during the design stage to identify and evaluate potential failures of mechanical and electrical systems. However, an FMEA can also be used on existing systems to provide a better understanding of potential failures, the impact of those failures and the existing safeguards. These failure descriptions provide facility personnel with a basis for identifying design improvements and asset failures requiring ITPM tasks. In addition, FMEAs can identify:

- *Consequences* of the asset failures, allowing facility personnel to determine which asset failures are most important to asset integrity.

- *Causes* of the asset failures, so that ITPM tasks can be selected to address underlying causes. (FMEA teams can also assess whether other means such as operator training or better operating procedures are more appropriate.)

- *Risk* of the asset failures, so that asset failures can be prioritized and the appropriate level of resources can be allocated.

An FMEA generally involves the following steps:

Step 1. Define the assets and process of interest.

Step 2. Define the effects of interest to consider during the analysis.

Step 3. Subdivide the process into subsystems or asset items for analysis.

Step 4. Identify potential failure modes of the subsystems/asset items.

Step 5. Evaluate the immediate and ultimate effects of each failure mode and (for FMECA) the criticality of the system effects.

Step 6. For failure modes capable of producing effects of interest, identify existing safeguards capable of detecting the failure mode and interrupting the sequence of events leading up to the effects of concern.

Step 7. Evaluate the adequacy of the safeguards, taking into account the likelihood of the failure mode occurring, the criticality of the effects, and the existing safeguards; recommend changes if weaknesses are found or the scenario risk is judged to be too high.

Step 8. If needed or desired, perform a quantitative evaluation.

Table 7-1 provides an example qualitative FMEA worksheet, with action items highlighted that are related to the development of ITPM plans.   Additional information on FMEAs can be found in CCPS' *Guidelines for Hazard Evaluation Procedures* (Reference 7-10) and in IEC 60812 (Reference 7-11).

7.6   SAFETY INSTRUMENTED SYSTEMS

As described in IEC 61511, *Functional Safety: Safety Instrumented Systems for the Process Industry Sector - Part 1: Framework, Definitions, System, Hardware and Software Requirements* (Reference 7-12), a *safety instrumented system* or *SIS* is a system composed of sensors, logic solvers and final control elements for the purpose of taking the process to a safe state when predetermined conditions are violated.   Other terms used for SIS include automatic emergency shutdown system, safety shutdown system and safety interlock system.

The functional integrity of safety instrumented systems is addressed by IEC 61511 and its U.S. equivalent ANSI/ISA-84.00.01 (IEC 61511 Mod). This standard is focused on the process industries including chemicals, oil and gas production and refining, pulp and paper, and non-nuclear power production. It is considered to be a recognized and generally accepted practice for the specification, design, and functional testing of safety instrumented systems.   Guidance for implementing the IEC 61511 standard, including the aspects related to developing functional testing and inspection requirements, is given in IEC 61511-2 (Reference 7-13).

The IEC 61511 standard addresses all safety life cycle phases, from initial concept, design, implementation, operation, and maintenance through to decommissioning. Although the focus in this section is on the maintenance and functional testing aspects, the earlier life cycle phases will also be summarized to show where the development of test and inspection plans fits into the overall SIS life cycle.

TABLE 7-1. Sample FMEA Worksheet *(test/inspection plan actions in italics)*

Item Number	Description	Failure Mode	Immediate to Ultimate Effects	Safeguards	Comments, Action
TCV-201	Temperature control valve for steam heat to HCl column	Fails open	Increased heating of the HCl column; potential column overpressure and HCl release to surroundings	• Operator response to increasing column temperature (multiple TIs on column/ TAH-201); adequate time to respond and manually isolate TCV-201 • High-high temperature safety shutdown system closes remotely actuated block valve XV-100 on signal from independent temperature sensor TE-100 (SIL 1 SIS) • Excess overhead condensing capacity (spare condenser)	All process indications and alarms are currently only on temperature 1. Add a high-pressure alarm to HCl column 2. Develop an emergency checklist for operators to follow on high temperature and add to SOP, training *3. Include in ITPM plan the functional testing of TCV-201 operation and a positioner calibration check*
		Fails closed	HCl column cools down; loss of production; no further consequences anticipated	—	
		Fails to respond (stuck in last position)	Loss of HCl column temperature control; quality impact downstream; no further consequences anticipated	—	If last position is open too far, effects and safeguards will be the same as leaking through or failing open
		Leaks externally	Less steam to HCl column, compensated by TCV-201 opening further; loss of steam; potential for personnel exposure to leaking steam	• Flange guards	TCV-201 is located near walkway *4. Include in ITPM plan a periodic inspection for presence and integrity of flange guards on steam lines to HCl column*
		Leaks internally	Increased heating of the HCl column; however, at a slow rate; potential for eventual column overpressure and HCl release to surroundings	• Operator response to increasing column temperature (multiple TIs on column/ TAH-201); ample time to respond and manually isolate TCV-201 • High-high temperature safety shutdown system closes remotely actuated block valve XV-100 on signal from independent temperature sensor TE-100 (SIL 1 SIS) • Excess overhead condensing capacity (spare condenser)	

The general steps involved in fulfilling the IEC 61511 requirements are shown in Figure 7-3, with an emphasis on ITPM test and inspection interfaces. It should be noted that this figure represents one suggested approach to meeting the IEC 61511 requirements, but not all of the steps are required to be performed as shown. For example, it is possible to assign a safety integrity level (SIL, as described in the paragraphs that follow) to a safety instrumented system without conducting a Layer of Protection Analysis, and all of the steps through SIL assignment can be performed as part of the process hazard analysis. Nevertheless, the steps in Figure 7-3 will be used as the basis for the following discussion.

1. A *process hazard analysis* (*PHA*, also known as a *Hazard Identification and Risk Analysis* or *HIRA)* is conducted to identify initiating causes in a defined process operation (asset failures, operational errors and external events that put the process in an abnormal situation), the potential consequences of each initiating cause, and safeguards that are included in an existing facility or a new-facility design. Some of these safeguards may be safety instrumented systems. Guidelines for using various PHA methodologies are given in Reference 7-10.

2. A risk analysis is done, such as a *Layer of Protection Analysis* (*LOPA*), to identify available and necessary independent protection layers to reduce the likelihood of hazard scenarios to a tolerable level. This may require better definition or ranking of the consequences being addressed. Note that it is possible to perform a qualitative or order-of-magnitude risk analysis within the PHA or HIRA rather than as a separate analysis. A summary of the LOPA methodology is given in Section 15.4 of this document.

3. Where it is determined that existing or designed safeguards are inadequate to achieve a tolerable risk level, risk-reduction measures including possible *inherently safer design options* (Reference 7-14) are evaluated. It is not necessary to reduce risks using only safety instrumented systems. It is often possible to achieve the tolerable risk level by means of a non-instrumented safeguard such as emergency relief protection. However, it should be noted that both safety instrumented systems and non-instrumented safeguards such as emergency relief protection will have testing and inspection requirements as well as the potential for false trips or premature activation.

4. When risk reduction is still needed, *Safety Instrumented Functions* (*SIFs*) are identified to provide the protection required for all identified scenarios having potential safety impacts.

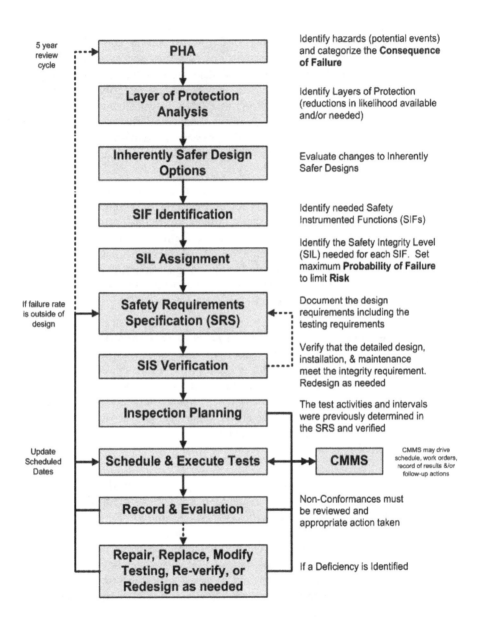

Figure 7-3. SIS development and implementation steps showing test
and inspection interfaces

5. A *Safety Integrity Level* (*SIL*) of SIL 1, SIL 2 or SIL 3 as specified in IEC 61511 is assigned to each SIF to achieve the required risk reductions. In essence, SILs are functional integrity/reliability requirements that are specified in terms of the probability of failure on demand for the SIF as a whole system. Note that it may not be necessary to perform a formal SIL assignment if only one order of magnitude risk reduction is required and the safety function is to be implemented in the basic process control system (BPCS).

6. A *Safety Requirements Specification* (*SRS*) is compiled for each SIS that takes the highest SIL rating of the associated SIFs as its SIL specification. The SRS details all the requirements that will need to be met as the SIS is designed, implemented, operated, and maintained, including inspections and functional testing. It includes such items as the anticipated demand frequency, the maximum allowable spurious trip rate and the required response time. This SRS is then used to design the SIS and specify the functional test interval in such a way as to meet all of the target requirements.

7. A formal *verification step* is required to assure that the system as designed, implemented, operated, and maintained meets the integrity level required.

8. *Inspection planning* is establishing the ITPM plan for the SIS; i.e., setting up for the execution of the inspections and test activities at the intervals required to meet the reliability specifications.

9. *Inspection and testing* is then scheduled and performed per the ITPM plan.

10. *Non-conformances* are evaluated *and deficiencies* addressed, with appropriate follow-up of issues for review and correction, such as redesigning, modifying inspection and test intervals and/or repairing system components.

All of the above steps have documentation requirements associated with them. IEC 61511 also includes requirements for installation, commissioning, pre-startup acceptance testing, operations, management of change, and decommissioning. These requirements essentially outline a quality management system for SISs. Conforming with all of the requirements of IEC 61511, including functional testing and inspection requirements, obviously requires significant management commitment and resource allocation.

The designer of safety instrumented systems selects instrumentation (sensors, final elements, etc.) and logic solvers, decides on redundancy requirements, sets the testing activities, and, along with the information in the SRS, verifies that the design achieves the integrity requirement of the SIL. The methods used for these verification calculations can require sophisticated probabilistic analysis. Some specific considerations include the following:

- Caution is warranted when considering the confidence limits on the data used for SIL determinations (Reference 7-15) and out-of-service time as it affects probabilities of failure on demand (Reference 7-16).

- Instruments can be subject to functional safety assessments using Failure Modes, Effects and Diagnostic Analysis (FMEDA), an extension of traditional FMEA for the development of failure modes and rates (including safe versus dangerous and detected versus undetected).

- Additional analysis may include evaluation of useful life and evaluation of the effectiveness of proof tests.

- ITPM activities still need to be defined for the components of the SIS; for example, calibration of sensors and stroke-testing of valves.

- The demand mode or interval (how frequently the safety function is expected to be needed) adds limits on the SIS architecture that can be used and the effectiveness of automatic diagnostics.

- Fault Tree Analysis and Markov models may be used. These analyses require specific knowledge and skills to perform them.

Many other considerations are also involved throughout the life cycle of a safety instrumented system. Training and certification is available for the Certified Functional Safety Professional and the Certified Functional Safety Expert (CFSP, CFSE).

CHAPTER 7 REFERENCES

7-1   API 510, *Pressure Vessel Inspection Code: Maintenance Inspection, Rating, Repair and Alteration*, American Petroleum Institute, Washington, DC.

7-2   API 570, *Piping Inspection Code: Inspection, Repair, Alteration, and Re-rating of In-service Piping*, American Petroleum Institute, Washington, DC.

7-3   API 653, *Tank Inspection, Repair, Alteration, and Reconstruction*, American Petroleum Institute, Washington, DC.

7-4   *National Board Inspection Code*, National Board of Boiler and Pressure Vessel Inspectors, Columbus, OH.

7-5   Folk, T., "Risk Based Approach to Asset Integrity Success in Implementation," *Process Plant Safety Symposium*, New Orleans, LA, American Institute of Chemical Engineers, 2003.

7-6   API RP 580, *Risk-Based Inspection*, American Petroleum Institute, Washington, DC.

7-7   API RP 581, *Risk-Based Inspection Technology*, American Petroleum Institute, Washington, DC.

7-8   Wintle, J.B. et al., *Best Practice for Risk Based Inspection as a Part of Plant Integrity Management,* CRR 363/2001, UK Health and Safety Executive.

7-9   SPC/Technical/General/46, *Risk Based Inspection (RBI) – A Risk Based Approach to Planned Plant Inspection,* UK Health and Safety Executive, www.hse.gov.uk/foi/internalops/hid_circs/technical_general/spc_tech_gen_46.htm.

7-10   Center for Chemical Process Safety, *Guidelines for Hazard Evaluation Procedures, Third Edition*, American Institute of Chemical Engineers, New York, 2008.

7-11   IEC 60812, *Analysis Techniques for System Reliability – Procedure for Failure Mode and Effects Analysis (FMEA)*, International Electrotechnical Commission, Geneva, Switzerland.

7-12   IEC 61511-1, *Functional Safety: Safety Instrumented Systems for the Process Industry Sector - Part 1: Framework, Definitions, System, Hardware and Software Requirements*, International Electrotechnical Commission, Geneva, Switzerland. Also by the same title ANSI/ISA-84.00.01 Part 1 (IEC 61511-1 Mod), International Society for Automation, Research Triangle Park, NC.

7-13   IEC 61511-2, *Functional safety – Safety Instrumented Systems for the Process Industry Sector – Part 2: Guidelines for the Application of IEC 61511-1*, International Electrotechnical Commission, Geneva, Switzerland. Also by the same title ANSI/ISA-84.00.01 Part 2 (IEC 61511-2 Mod), International Society for Automation, Research Triangle Park, NC.

7-14   American Institute of Chemical Engineers, *Inherently Safer Chemical Processes: A Life Cycle Approach, Second Edition*, Center for Chemical Process Safety, New York, NY, 2008.

7-15   Freeman, R., "Quantifying LOPA Uncertainty," *Process Safety Progress 31*(3), September 2012.

7-16   Broadribb, M. and M. Currie, "HAZOP/LOPA/SIL – Be Careful What You Ask For!" *6th Global Congress on Process Safety,* San Antonio, TX, American Institute of Chemical Engineers, 2010.

Additional Chapter 7 Resources

Geary, W., *Risk Based Inspection - A Case Study Evaluation of Onshore Process Plant,* HSL/2002/20, Health & Safety Laboratory, Sheffield, UK, 2002.

8

AIM TRAINING AND PERFORMANCE ASSURANCE

An important ingredient of an effective asset integrity management program is personnel competency, achieved in part by training and performance assurance. Proper training helps to ensure that only qualified personnel develop and perform AIM tasks and that AIM tasks are performed appropriately and consistently; i.e., with fewer opportunities for human errors. Reducing human errors can greatly reduce the overall rate of asset failures. Figure 8-1 provides a flowchart for the development and implementation of an AIM training program.

This chapter addresses training and qualification of a facility's AIM workforce as well as the selection of qualified contractor personnel. It does not cover general safety-related training such as for safe work practices. The following topics are included:

- Management awareness training
- Skills/knowledge assessment
- Training for new and current workers
- Verification and documentation of training effectiveness
- Certification, where applicable
- Ongoing and refresher training
- Training for maintenance technicians and for operators performing maintenance tasks
- Training for technical personnel
- Contractor issues
- Roles and responsibilities.

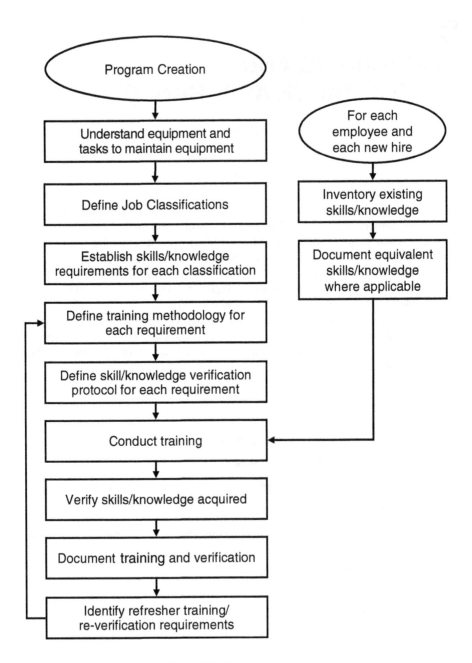

Figure 8-1. Training flowchart

Larger facilities might have a multi-departmental group specifically called together to establish an AIM training program and curriculum for the facility. Such a group may involve Human Resources, Production, AIM/Maintenance, Process Safety, and Quality Management, and may also have input from contractor companies. The group may be tasked to determine what AIM training is to be required for different personnel, who will conduct the training, and how it is to be conducted and verified.

Asset-specific training and/or certification requirements are discussed in more detail in Chapter 12 and in the asset-specific matrices in Appendix 12A. Further information on training is available in ANSI Z490.1, *Criteria for Accepted Practices in Safety, Health and Environmental Training* (Reference 8-1).

8.1   SKILLS AND KNOWLEDGE ASSESSMENT

Formal skills and knowledge assessments can be used to identify gaps in knowledge of the current workforce and to evaluate new hires. Knowing where weaknesses exist allows a facility to develop training or seek outside assistance to address the greatest training and qualification needs. The following describes a skills and knowledge assessment technique that has been successfully applied:

1. Review the AIM program components, such as inspection and testing plans, preventive maintenance activities, quality assurance plans, and common repair activities, to identify and list AIM tasks to be performed.

2. Develop a list of the job classifications for the personnel who will perform the tasks. Note that some tasks will be performed by personnel outside the traditional maintenance workforce, such as storeroom workers, buyers, and process operators; other tasks may be designated for contractors. When possible, identify job classifications and sub-classifications that are independent of pay classifications. For example, a facility may have one pay grade for the instrument and electrical workforce, but for training purposes, classifications (e.g., asset maintenance technician, I&E technician) and sub-classifications (e.g., millwright, DCS technician) may be different or more specific.

3. Solicit input from experienced, knowledgeable personnel when assembling lists of the skills and knowledge required within each job classification and sub-classification. Certification requirements (particularly for welders and fixed-equipment inspectors) and other skills required by codes, standards and/or jurisdictional requirements may also be included. Add these skills/knowledge items to the list of AIM tasks identified in Step 1 above.

4. Determine how best to integrate the training function with any additional training that is required for other process safety programs, process overviews and hazards, safe work practices and any other required training such as management of change, emergency response or hazardous material (HAZMAT) training.

5. Survey the workforce, including supervisory personnel, to determine which skills and knowledge areas need improvement. Consider using a survey format that asks personnel to evaluate both their personal capabilities and those of their co-workers. Note: Ensure that any surveys of co-worker abilities are done in a non-personal manner. See Appendix 8A for a sample survey format.

6. Aggregate the information from the first steps into a baseline training matrix by listing all pertinent roles, training requirements for each role, and refresher training frequencies. The roles will be a mixture of operations, maintenance, technical, and supervisory positions. The resulting matrix can then be used as a training requirements grid for the site.

7. At this point, the site needs to decide whether or not "grandfathering" of tasks is allowed and, if so, how it will be documented for the task. In most cases, this will require comparable or equivalent knowledge assessment including skills demonstration. For example, an electrician having a journeyman status who is hired into the facility may be exempt from some of the general electrical training that would be required of a less-skilled new hire. Similarly, a facility operator who transfers to the maintenance department may have already received facility-specific or process-specific training that would be required of a new hire from outside of the facility.

Often, the existing facility workforce has been performing AIM tasks for many years, so personnel may show some resistance to the concept of a skills and knowledge assessment. Some of the following suggestions may help overcome such resistance:

- Present the skills and knowledge assessment in a positive light. Most workers will welcome the chance to upgrade their skills and knowledge.

- Involve experienced workers when identifying necessary skills and knowledge requirements for each job classification.

- Evaluate general gaps in the workforce skills and knowledge in addition to worker personal needs.

- Develop a draft AIM training program outline.

- Present the outline for the entire training program. Demonstrate that the skills and knowledge assessment is just one of the steps that management is taking to enhance the training program.

- Do not use the skills and knowledge assessment (or other aspects of the training) as a tool to discipline or dismiss ineffective workers.

- Update and finalize (approve) the AIM training program document and implement it based on the feedback from the skills and knowledge assessment.

8.2   TRAINING FOR NEW AND CURRENT WORKERS

For each of the skill and knowledge needs identified, the next step is to determine whether new skills, skill enhancement, or skill remediation can best be accomplished by developing or enhancing procedures, training, or both. AIM procedures are addressed in Chapter 9.

Several training approaches are available to most facilities. Training can occur in a variety of settings: in classrooms, in front of a computer, on shop floors, in the field, at community colleges, at a vendor school and/or through a correspondence school. Facilities need to determine which training approaches would be most effective for meeting the AIM objectives, then ensure that resources are allocated to support the training. Often, facilities use a combination of training approaches. For example, safe work practices may be taught in a classroom, whereas training on compressor maintenance may require field work in addition to (or instead of) classroom education. Table 8-1 compares some of the typical strengths and potential weaknesses inherent in some common training approaches.

Training can be provided by a variety of instructors: experienced personnel, engineers, vendors, contractors or technical colleges. In all cases, the basis for declaring someone qualified to provide the training needs to be documented and included with the training records. Selection and training of the trainers are very important. Facilities need to ensure that trainers know and follow accepted work procedures. Trainers can be selected on the basis of their knowledge of the training topic and on their ability to effectively transfer knowledge to others. A highly skilled mechanic does not necessarily have the talent and ability to be a good teacher. Many facilities implement a training program for the trainers. Some facilities send designated trainers to manufacturers' or suppliers' training classes, then rely on those trainers, in turn, to pass on the knowledge to the rest of the workforce.

Many facilities create standard guides to be used to facilitate each requirement of a training program. Appendix 8B includes a sample training guide for repairing a mechanically sealed pump.

TABLE 8-1. Training Approach Considerations

Approach	Typical Strengths	Potential Weaknesses
Self-paced training *(e.g., computer-based training packages)*	• Consistent information delivered • Verification that training is understood for each individual • Training schedules can be catered to individuals • May minimize "trainer" expenses (in the long term)	• Modules may not be relevant to actual tasks and may not be consistent with company practices (may require review and customization of modules to ensure relevance) • No interaction with a live instructor or classmates • No field skills verification • Potentially high startup expenses
Classroom sessions	• Consistent information delivered • Curriculum will be relevant and can be adjusted to the needs of the workforce • Interaction with classroom instructor and with classmates • Acquired knowledge can be tested readily (by trainer/student interaction as well as by written tests)	• Training material development expenses • Trainer and student schedules need to be coordinated • Requires a system to ensure that employees who miss training sessions are trained in make-up sessions • Sessions may emphasize group needs, but will not accommodate individual needs • No field skills verification
On-the-job training	• Training can be readily adjusted to meet individual needs • Curriculum will be relevant to day-to-day activities • Experienced (knowledgeable) resource person is readily available to coach and answer questions • Field skills can be verified by the trainer	• Inconsistent information -- a trainee may work with multiple people and become confused if different personnel approach tasks differently (Consider designating a limited number of personnel to be trainers) • Incomplete information -- unstructured day-by-day training may never encounter some tasks (Consider establishing checklists and other tools to ensure that all training topics are covered consistently for each employee) • Improper practices and unconventional "wisdom" may be carried from employee to employee, leading to a workforce-wide deficiency (Consider supplementing on-the-job training with other methods to address workforce-wide deficiencies) • Training verification may appear subjective (Establish specific criteria for skills verification) • Temporary loss or less productive use of skilled workers involved in training the new employees

8.3   VERIFICATION AND DOCUMENTATION OF PERFORMANCE ASSURANCE

After selecting appropriate approaches to conduct training, the next step is to establish criteria for judging whether training has successfully enhanced a person's skills and/or knowledge. A variety of means can be used for training verification, including written tests, computerized exams, verbal questioning, field demonstration, or a combination approach. It should be noted that local or national regulations may have specific requirements for defining, demonstrating and/or certifying competency. Those specific requirements are not included in this document.

Knowledge can be verified by testing or questioning. Pre-established criteria such as percent correct are used to decide whether the training has been successfully completed. Many companies have a requirement to go over all missed questions with the trainee to help fill in knowledge gaps, with retraining as necessary until knowledge proficiency is attained. (Reviewing what questions were missed by trainees can also be used as feedback to the trainer to improve future training effectiveness.) For some knowledge areas such as rebuilding of relief valves, the success criterion might be to perform the task unaided and be 100% correct, with retraining and formal retesting until proficiency is demonstrated.

For skills verification, including proficiency at performing certain stepwise procedures, field/shop demonstration is often more appropriate. Field training in particular needs to have specific objective criteria for judging proficiency. In both of these cases, a checklist or other means needs to be used to document that all procedural steps were taken and properly executed.

Training requirements and training completed need to be documented for each individual. Many facilities establish a database that lists each employee and the employee's job classification(s) in a matrix versus the skills/knowledge and/or training classes required. Table 8-2 shows a format and part of a training matrix for general electricians. In the supplemental files accompanying this document are more complete samples of skills/knowledge lists for a general electrician and for a general mechanic.

Backup documentation (e.g., completed written tests, completed checklists) need to be retained for each entry in the employee skills/knowledge matrix to demonstrate that the employee has met the established criteria. Documentation of successfully completed training includes the name of the individual and, if applicable, the trainer; date(s) of the training; how the skills/knowledge were verified (e.g., test, observation); and the results of the verification activity (e.g., test score, pass/fail). Including a copy of the training course material or outline along with the training records can be helpful in verifying content of the training.

TABLE 8-2. General Electrician Training Matrix Format Example

Enter date performance test was completed:	Requirement								
Name	Introduction to Basic Electricity	Fundamentals of DC Circuits	Fundamentals of AC Circuits	Safety in Electrical Maintenance	Electrical Drawings	Electrical Test Equipment	DC Troubleshooting Techniques	AC Troubleshooting Techniques	Electronic Process Transmitters

A management system needs to be in place to ensure personnel training is kept current, especially when periodic refresher training is required. Some facilities incorporate training reminders into the work order system or into another computer program that flags "training required" dates. Numerous training software databases are commercially available that incorporate these features.

8.4   CERTIFICATIONS

Although many AIM skills have no third-party certification requirements, facilities need to be aware of, and make use of, widely recognized certification requirements when they do apply. In particular, certifications are available for AIM activities such as fixed-equipment inspection and welding. In some cases, jurisdictional rules such as state laws also include certification requirements. For example, in the state of New Jersey, an API 510 inspector needs to have a New

Jersey supplemental certification to be a valid inspector. Many jurisdictions have similar requirements for NBIC-authorized inspectors.

Although codes and laws are subject to change, the trend has been to require more certification for maintenance and engineering work. In addition, for some codes such as for welding and equipment examination, systems need to be established to track performance in order to keep certifications current and/or to achieve recertification as required. Table 8-3 contains a list of some of the widely accepted certifications available for AIM work. For those skills that have no external certification requirements, many companies establish qualification or internal certification requirements as part of their training verification and documentation process.

8.5   ONGOING AND REFRESHER TRAINING

There are normally differences in scope between initial training and refresher training, with initial training being comprehensive (all-inclusive) and refresher training being more selective. The materials used for training are typically the same. The following gives some additional considerations for ongoing/refresher training as compared to initial training:

- Initial training must cover all the training required for the work done by the individual's role. In other words, all the training that applies to a person's role is covered in initial training, to ensure that they can safely and successfully perform the work for the role. This includes basic skills and job task training.

- For refresher training, an appropriate evaluation method can be applied to select the refresher training needed for a role, rather than refresher training on 100% of the pertinent tasks and skills. Note that even if a particular task is not selected for refresher training for a role in general, an individual may still receive refresher training on that task if they have an individual need.

- Basic skills do not normally receive refresher training, since refresher training on general and equipment-specific tasks demonstrates the use of the basic skills to an acceptable level.

- The maximum interval for required refresher training on job tasks is typically three years, but a site may elect to have shorter refresher training frequencies based on the criticality of the job tasks, employee input, etc.

- A site can establish refresher training schedules and frequencies such as to maintain consistent cycles in the refresher training program.

TABLE 8-3. Widely Accepted AIM Certifications

Skill	Organization	Certificate or Standard Used
Quality management	API	API Source Inspector
Atmospheric storage tank inspection	API	API 653
	Steel Tank Institute	STI-qualified inspector
Boiler, pressure vessel, and piping inspection	National Board of Boiler and Pressure Vessel Inspectors (NBBPVI)	National Board (NB)-23
	API	API 510, API 570
Corrosion prevention and control	NACE International	• Cathodic Protection Specialist • Corrosion Specialist • Chemical Treatment Specialist • Materials Selection / Design Specialist • Protective Coatings Specialist
	API	API 571
Inspection supervision, technique selection, procedure preparation and approval, code interpretation	American Society of Nondestructive Testing (ASNT)	ASNT Central Certification Program (ACCP) Level III
NDT techniques	ASNT	SNT-TC-1-A, for various individual types of inspections and tests
NDT techniques, equipment calibration, interpretation of results	ASNT	• ACCP Level II • Magnetic particle testing • Liquid penetrant testing • Radiographic testing • Ultrasonic testing • Visual testing
Welding	ASME	Boiler and Pressure Vessel Code (BPVC) Section IX, Welding (for qualifications specific to metals, rods and weld positions)
	American Welding Society (AWS)	Certified Welder
Weld inspection	AWS	• Certified Welding Inspector • Senior Certified Welding Inspector
Control systems	International Society for Automation (ISA)	Certified Control Systems Technician® (CCST)
Safety instrumented systems	ISA	• Certified Functional Safety Expert (CFSE), Certified Functional Safety Professional (CFSP) • ISA 84 SIS Expert

The task and skill lists developed for the skills and knowledge assessment will need to be updated as new assets and/or new maintenance techniques are introduced to the facility, as well as on an ongoing basis as refresher training. Ideally, updates will occur as part of the facility's management of change program. Many facilities have traditionally provided training on new assets with introductory classes taught by vendors and/or project engineers associated with the assets. In other cases, companies have sent employees to factory schools. Such courses may be appropriate, but the training program also needs to consider how the next person will acquire these new skills; i.e., how the facility will train personnel who did not attend the class.

For some topics, refresher training makes sense (e.g., based on employee needs or desires) and, in some cases, is required such as by jurisdictional laws, codes or standards. As an example, a periodic review of process hazards is generally appropriate. For workers who perform tasks or work in specific areas infrequently (e.g., a technician who spends eleven months a year in the shop but works in the process area during annual shutdowns), just-in-time training may be beneficial to review necessary tasks before they are performed. This may be especially critical for maintenance turnarounds at continuously operating facilities such as refineries.

For tasks that are rarely performed, such as rebuilding a compressor once every six years, just-in-time training can be more effective than periodic refresher training. When this approach is used, it is essential that the time required for the just-in-time training is built into the turnaround or project schedule.

8.6   TRAINING FOR MAINTENANCE TECHNICIANS AND OPERATORS PERFORMING MAINTENANCE TASKS

Trained and qualified personnel to perform maintenance tasks are essential to maintaining process equipment in a safe and reliable condition so that significant unplanned events are avoided. Procedures and training are together used to manage the quality of work performed by site personnel on safety-critical equipment, to help ensure defects are not introduced that could lead to loss events.

Training requirements stem from company policies and applicable regulations. Ideally, value is gained from the training, in addition to meeting policy and regulatory requirements, and individuals continue to improve their capabilities through ongoing learning and cumulative experience. The following paragraphs give a stepwise process for training an effective maintenance workforce.

Step 1 - Understand the Equipment and the Tasks to Maintain It. The work to maintain equipment includes routine maintenance, repairs, inspections and tests, and preventive maintenance – all the activities required to keep the equipment in proper working condition. The goal of this activity is to gain an understanding of how the equipment works and what maintenance work needs to be performed.

Step 2 - Strategize to Execute Maintenance Tasks. Factors impacting a site strategy for executing maintenance tasks can include, but are not limited to, the size of the site, the structure of the maintenance organization, the work roles, and the tasks associated with the work roles. One of the more significant strategic decisions for a site is what maintenance work the site intends to contract. Often sites think of maintenance work in three categories:

- Core – The site intends to maintain the skills and perform the work.

- Variable (or non-core) – The site does not intend to maintain the skills, and will contract the work as needed.

- Strategic Mix – Work will be done using a combination of both categories. For example, a site may have sufficient personnel to perform the work normally, but may supplement by contracting work during shutdowns or other peak work periods.

In some cases, resident maintenance contractors will perform work using equipment-specific procedures that are provided by the site to the contractor for site-owned equipment. For that case, specific systems and requirements need to be met. These may include:

- A system developed and managed by the contractor to help ensure the contractor's employees are trained and qualified on the procedures.

- Periodic evaluation and auditing of the contractor training/qualifications.

- A system to help ensure that any revisions or updates to procedures are coordinated between the site and the contractor company.

- A process to manage personnel changes, administered by the contractor for newly assigned contractor supervisory personnel who oversee work on safety-critical equipment, as well as training and qualifications of contractor personnel and related functions.

Step 3 - Organize to Perform Maintenance Work. Having understood the work and maintenance strategy, a site can then develop (or re-evaluate) the organization used to perform maintenance work.

One of the most basic considerations is whether personnel outside the maintenance organization will perform maintenance tasks. For example, some autonomous maintenance might be performed by operating personnel. Those performing such tasks need appropriate training and must be appropriately included in the maintenance training program.

Within the maintenance organization, a site needs to consider what roles there are (types of maintenance technicians) and how work is divided between them. In some situations, this may be dictated by regional or local training and certification. For sites with unionized maintenance organizations, the site Human Resources organization is usually involved in this process.

The outcome of this process is an overview of what roles in the organization are expected to perform specific tasks. Consideration can be given to work that logically fits together, either based on how it is done in the field or the underlying skills required to perform it. Note that it is often appropriate that some work be performed by more than one role.

Step 4 - Identify Required Knowledge, Skills and Abilities to Do the Work. Once the organization has (a) determined the work to be done to maintain the equipment, (b) decided what work will be done by site personnel, and (c) determined what roles do what work, then training requirements can be established by role for each category of training. Essentially, this applies the job analysis to roles in the organization.

Categories of Training and Requirements. Table 8-4 is a table of categories of suggested training groups, with a description of the content of that training (i.e., what the category includes), how it is addressed in initial and refresher training, and some examples for illustration. For the first three categories, additional explanation is given as follows.

- Basic skills training is the prerequisite for general task or equipment-specific task training. In other words, there are certain basic skills a person must possess in order to learn the general or equipment-specific tasks. Examples of basic skills could be measurement techniques or use of hand tools. In some countries, this training is regulated and/or coordinated by government organizations. Typically, apprenticeship programs will cover basic skills, possibly along with additional training such as some general tasks and a certain amount of work in the field.

- General task training and equipment-specific task training together are referred to as Job Task Training.

- General task training and equipment-specific task training are similar in that individuals receive initial training for the tasks their role will perform. Refresher training is provided as required on appropriate general and equipment-specific tasks, based on applicable process hazards, relevant safe work practices, safety-critical equipment and operating experience.

TABLE 8-4. Training Categories and Example Requirements

Category	Content	Initial Training	Refresher Training	Examples
Basic skills	Knowledge and skill training that provides the foundation to achieve successful job task training performance (general and equipment-specific)	Sites may hire to this skill level (verifying content and completeness of training), rather than provide training. Training may also be provided through a third party	Not required. Site may elect to provide. May be provided individually as needed (for cause)	• Basic measurements • Use of hand tools • Reading drawings • Basic metallurgy
General tasks	Skills/tasks that apply to multiple types of equipment, or broadly within a category of equipment	Must receive training on all general skills required to perform the work for the assigned role	Select appropriate refresher training content based on applicable process hazards, relevant safe work practices, PSM-critical equipment, and operating experience. Does not require 100% refresher training	• Work management processes applicable to the work role • Flanged joint makeup • Standard transmitter calibration
Equipment-specific tasks	Skills/tasks that apply to a unique piece of equipment, or several similar pieces of equipment within a category	Must receive training on all equipment specific skills required to perform the work for the assigned role	Select appropriate refresher training content based on applicable process hazards, relevant safe work practices, PSM-critical equipment, and operating experience. Does not require 100% refresher training	• Maintenance on unique/ custom equipment

8.7   TRAINING FOR TECHNICAL PERSONNEL

Sometimes, training discussions do not extend beyond the hourly workforce. However, because managers, supervisors and engineers are also involved in the AIM program, they need to be competent and knowledgeable. Examples of the technical knowledge that may be necessary for supervisory and engineering personnel include training in codes and standards and training in the information needed to assess the suitability for degraded assets to stay in service (e.g., corrosion mechanisms, fracture mechanics, stress analysis).

Table 8-5 is a listing of possible training requirements for a position of Asset Integrity Engineer supporting a risk-based inspection (RBI) program for pressure vessels, storage tanks, and piping. If this person will also be involved in RBI as a risk analyst, additional training may be needed for that role (Reference 8-10). Other technical positions may have much less extensive training requirements. For example, AIM training for a process control engineer may include elements of Process Safety Management, Quality Management, IEC 61511 (Safety Instrumented Systems), ANSI B11.19 (Criteria for Safeguarding Machinery), and NFPA 70 (National Electric Code), as well as Project Cost Estimating.

TABLE 8-5. Example Asset Engineer Potential Training Requirements

Requirement
NDE Limited Level II, ASNT SNT-TC-1A, Visual Testing, waived if API 510 certified
NDE Limited Level II, ASNT SNT-TC-1A, Liquid Penetrant, waived if API 510 certified
NDE Limited Level II, ASNT SNT-TC-1A, Radiographic Interpretation, waived if API 510 certified
NDE Limited Level II, ASNT SNT-TC-1A, Ultrasonic Straight Wave, waived if API 510 certified
Process Safety and Risk Management
Mechanical (Asset) Integrity Management
Quality Management
Fitness for Service – API 579
API 510 Pressure Vessel Inspection (certification optional)
API 570 Piping Inspection (certification optional)
API 653 Tank Inspection (certification optional)
ASME Section VIII, Division 1, Design, Repair and Alteration of Pressure Vessels
ASME B31.3 Process Piping
ASME Section V, Nondestructive Testing
ASME Section IX, Welding
ASME/TEMA Heat Exchangers Design/Fabrication
API 580/581 – Risk-Based Inspection
Root Cause Failure Analysis
Chemical Engineering for Non-Chemical Engineers
Project Cost Estimating
Basic Corrosion/Corrosion Concepts
Welding Inspection
NFPA 70 Overview of the National Electric Code
Motor Protection (electrical)
Pressure Vessel Design Software
RBI Methodology and Software Training

8.8   CONTRACTOR ISSUES

Facilities hire contractors for a variety of activities: shutdowns and turnarounds, new construction, specialty services, and supplemental facility workforce. All maintenance work is contracted at some facilities. In all cases, the facility is responsible for verifying that contractors have a training program that ensures contractor employees are trained in the skills and knowledge necessary to perform their tasks and perform them safely. Meeting this responsibility can involve directly training personnel; more often, it involves a review of the contractors' training programs and verification that each contract employee is trained. Facilities may have jurisdictionally required contractor safety programs with similar requirements for safety training such as in safe work practices and evacuation plans. A contractor has primary responsibility for complying with all safety requirements, ensuring the safety of its own employees, and training its employees on safe work practices. However, coordinating contractor oversight with the facility personnel who are administering contractor safety is important.

Any time that contract workers supplement a facility's workforce by performing tasks comparable to those performed by facility personnel, the facility staff needs to ensure that training is comparable, including the definition of what specific AIM-related competencies and certifications are required. Measures to promote comparable training often include some combination of the following:

- Establishing and implementing a contractor prequalification program

- Training contract workers directly

- Providing contract employers with procedures and/or training materials

- Reviewing contract employers' training programs, then verifying that training is documented for each contract worker

- Requiring third-party certifications such as API, ASNT or EPRI for specific maintenance activities.

Facilities at which all maintenance is contracted (often with subcontractors as well) face similar choices for verifying that the training program is adequate and that all personnel are trained. References 8-11 and 8-12 discuss important process safety issues in the contractor-client relationship.

Often, contractors are hired to perform tasks that the facility cannot, or chooses not to, perform with facility personnel. In such cases, the facility may have no internal procedures to compare to the contractor training program; however, contractor training and its documentation still need to be reviewed.

Certification or licensing is often required for specialty contractors (e.g., welders, inspectors, heavy equipment operators). Facility personnel need to have procedures implemented to verify contractors hold all required certifications and licenses before work is conducted. In-house or third-party verification of NDE and inspection services is vitally important.

Contracting for AIM-related services needs to be done with due diligence and care. Issues needing to be addressed that may have a bearing on contractor training include:

- Interfacing with key safety procedures such as lockout/tagout, confined space entry, line breaking, hot work, excavating, and heavy lifting over or near process equipment

- Interfacing with plant operations, such as when working on utilities

- Controlling access of contractors to facilities and process areas

- Ensuring contractors are kept up to date with facility changes, including changes in emergency response procedures

- Potential co-employment issues (contracts should specify that contractor has primary responsibility for training its employees).

8.9   ROLES AND RESPONSIBILITIES

The roles and responsibilities for the training program are generally assigned to the Training, Engineering and/or Maintenance departments. However, production personnel provide guidance for topics such as process overviews and hazards. Example roles and responsibilities for an AIM training program are shown in Table 8-6, presented in a matrix format that can be useful for many facility types. The matrices use the following letter designators:

R   indicates the job position(s) with primary <u>responsibility</u> for the activity

A   indicates <u>approver</u> of the work or decisions made by a responsible party

S   indicates the job position(s) that typically <u>support</u> the responsible party in completing the activity

I   indicates the job position(s) that are <u>informed</u> when the activity is completed or delayed.

TABLE 8-6. Example Roles and Responsibilities Matrix for the AIM Training Program

R = Primary Responsibility
A = Approval
S = Support
I = Informed

Activity	Maintenance and Engineering Department Personnel								Other Personnel				
	Maintenance Manager	Maintenance Supervisor	Maintenance Engineer	Reliability Engineer	Project Managers/Engineers	Maintenance Technicians	Maintenance Planner/Scheduler	Production/Process Engineers	Process Safety Coordinator	Training Manager	Safety Manager	Plant Manager	Human Resource Personnel
Skills and Knowledge Assessment													
• Identify job classifications	R	S	S	S	I	I		I	I	I	I	I	I
• Identify skills/knowledge requirements	A	R	R	R	S	S		I		I	S	I	I
• Survey the workforce	A	R	I	I	I	S	I		I	I	I	I	S
• Evaluate survey results and identify needs	A	R	S	I	S	I	I	I	I	I	S	I	I
Training Program Development													
• Evaluate different approaches	A	R	S	S	I	S	I		S	S	S	I	I
• Obtain/develop training materials	S	S	S	S	I	I		I	I	R	S	I	
• Establish documentation and tracking system		I	I	I	I	I			S	R	I	I	S
• Identify refresher training topics	A	S	R	S	S	R	I		I	I	I	I	
• Establish certification programs	I	S	R	R	S	I	I		I	S		S	S

TABLE 8-6. *Continued*

R = Primary Responsibility
A = Approval
S = Support
I = Informed

Activity	Maintenance and Engineering Department Personnel								Other Personnel				
	Maintenance Manager	Maintenance Supervisor	Maintenance Engineer	Reliability Engineer	Project Managers/Engineers	Maintenance Technicians	Maintenance Planner/Scheduler	Production/Process Engineers	Process Safety Coordinator	Training Manager	Safety Manager	Plant Manager	Human Resource Personnel
Training Program Development *Continued*													
• Schedule training	I	S	I	I		I	R	I	I	I	I	I	
• Establish ongoing training program	I	S	R	S	S	I	I	I	I	I	I	I	
• Develop/implement technical training program	A	I	R	S	R			S	I	I	I		
• Maintain training records		S	S	S	S	I	S		I	R	I		
Develop Contractor Skills Evaluation Program	A	S	S	S	S			I	R		A	I	

CHAPTER 8 REFERENCES

8-1   ANSI Z490.1, *Criteria for Accepted Practices in Safety, Health, and Environmental Training*, American National Standards Institute, Washington, DC.

8-2   API 510, *Pressure Vessel Inspection Code: Maintenance Inspection, Rating, Repair and Alteration*, American Petroleum Institute, Washington, DC.

8-3   API 570, *Piping Inspection Code: Inspection, Repair, Alteration, and Rerating of In-Service Piping*, American Petroleum Institute, Washington, DC.

8-4   API 653, *Tank Inspection, Repair, Alteration, and Reconstruction*, American Petroleum Institute, Washington, DC.

8-5   American Society of Mechanical Engineers, *International Boiler and Pressure Vessel Code*, New York, NY.

8-6   ASNT Document CP-1, *ASNT Central Certification Program*, American Society of Nondestructive Testing (ASNT), Columbus, OH.

8-7   ANSI/AWS B2.1, *Specification for Welding Procedure and Performance Qualification*, American Welding Society, Miami, FL.

8-8   AWS B5.2, *Specification for the Qualification of Welding Inspector Specialists and Welding Inspector Assistants*, American Welding Society, Miami, FL.

8-9   *National Board Inspection Code*, National Board of Boiler and Pressure Vessel Inspectors, Columbus, OH.

8-10   API Recommended Practice 580, *Risk-Based Inspection*, American Petroleum Institute, Washington, DC.

8-11   Center for Chemical Process Safety, *Contractor and Client Relations to Assure Process Safety*, American Institute of Chemical Engineers, New York, NY, 1996.

8-12   Center for Chemical Process Safety, *Guidelines for Risk Based Process Safety*, American Institute of Chemical Engineers, New York, NY, 2007.

APPENDIX 8A. SAMPLE TRAINING SURVEY

Please complete the following survey to help management in its planned enhancement of our maintenance training program. The results of this survey will be combined with results of a similar survey of the maintenance foremen to establish priorities for the program.

This survey asks for an honest evaluation of your own knowledge and abilities. It also requests an honest general evaluation of your co-workers on the same topics. Thank you for helping us to improve the training program.

AIM Training Survey	Personal capabilities			Co-worker capabilities		
Topic	Very capable	Some knowledge	Basic training necessary	Very capable	Some knowledge	Basic training necessary
Pipefitting						
Knowledge of piping specs						
Knowledge of gasket selection						
Bolt torquing						
Crane operation						
Forklift operation						
Knowledge of emergency response procedures						
Knowledge of the Alpha process						
Knowledge of the Beta process						
Knowledge of chemical hazards						

APPENDIX 8B. SAMPLE TRAINING GUIDE

Included in this appendix is an example of a training guide that might be used for repairing a mechanically sealed pump.

Training Guide **Repair / Replacement of a Mechanically Sealed Pump**	
Trainee's Name (print): _____ *Trainer's Name* (print): _____	
Job Area	Maintenance
Purpose	Provide the on-the-job portion of the qualification criteria for maintenance employees responsible for repairs of mechanically sealed pumps.
References	1. Procedure # ACU 12615 2. Procedure # SAF 161 3. Procedure # SAF 245 4. Procedure # MEC 296 5. Procedure # MEC 322 6. _____ Vendor Manual 7. Safety Data Sheets for chemicals in work area
Trainer Responsibilities	• Demonstrate proper/safe task performance of all elements of the checklist • Perform consistent training on the job tasks for all trainees • Advise and monitor trainees during training on task performance • Evaluate trainees upon completion of the training phase
Trainee Responsibilities	• Acquire skills required to perform tasks associated with this checklist • Notify trainer when they are ready for evaluation

Knowledge Prerequisites				Page 2 of 3
Step	**Process / Training**	**Support Graphics**	**Trainee Initials / Date**	**Trainer Initials / Date**
1	Obtain related documents for performance of procedure. *Supervisor / Trainer:* Identify location and review applicable area procedures, reference data and discuss with Supervisor / Trainer.			
2	Identify equipment in the work area. *Supervisor / Trainer:* During a work area tour, correctly identify all major equipment components and explain their purpose.			
3	Locate paperwork required for these tasks. *Supervisor / Trainer:* Identify paperwork requirements of work to be performed.			
4	Secure proper PPE for these tasks. *Supervisor / Trainer:* Identify proper PPE for job and demonstrate safe and proper use.			
5	Showing a pinch point hazard of pump. *Supervisor / Trainer:* Identify work area hazards (pinch points, chemical hazards).			

	Performance Requirements			Page 3 of 3
Step	Process / Training	Support Graphics	Trainee Initials / Date	Trainer Initials / Date
6	**Field repair of pump:** Proper use of reference data during pump repair. *Supervisor / Trainer:* Demonstrate proper use of reference data.			
7	**Field repair of pump:** Showing one lockout / tagout activity required00 for performance of this procedure. *Supervisor / Trainer:* Perform / simulate lockout / tagout activities.			
8	Showing a de-magnetizing bearing heater. *Supervisor / Trainer:* Identify tools required for job completion and demonstrate proper tool usage.			
9	Performing a pre-shutdown assessment. *Supervisor / Trainer:* Demonstrate, then have Trainee demonstrate performing a pre-shutdown assessment of pump, driver and associated components.			

Trainee has successfully completed all requirements listed above in accordance with the applicable procedures and is qualified to perform the tasks covered by this Performance Checklist.

_____   _____

Trainer   **Date**

Trainee comments:_____

Trainee comments:_____

9
ASSET INTEGRITY PROCEDURES

An effective asset integrity management (AIM) program needs written procedures for program activities and specific tasks such as inspections, testing and preventive maintenance. Written procedures help ensure that AIM program activities and tasks are performed adequately, safely and consistently. To develop and implement written procedures effectively, it is helpful for facility staff to recognize that procedures offer value beyond fulfilling regulatory compliance requirements. Specifically, procedures can add value by:

- Serving as an integral part of the personnel training program

- Reducing human error

- Helping management ensure that activities are performed as intended

- Helping to provide continuity of the AIM program during and after organizational and personnel changes.

An important feature of successful AIM programs is conformity among the written procedures, the corresponding training and the actual field practice for a task or program activity. Differences among procedures, training and practice can result in personnel confusion and mistakes. Therefore, to help ensure that consistent direction is provided, the information in the written procedures needs to be reflected in the training materials. In addition, management personnel can be directly involved in the AIM procedure process by (1) providing resources needed to develop quality procedures, (2) ensuring that training resources are provided, (3) monitoring procedure conformance, and (4) providing review and approval.

Studies of major industrial incidents have shown that many major incidents occur during or after maintenance activities (Reference 9-1). Human errors during the performance of a task or activity figure prominently in these events. Examples include the 1989 explosion and fire in Pasadena, Texas when isolation valve actuator air hoses were reversed when reconnected after maintenance, and the 1988 loss of the Piper Alpha offshore oil platform by a series of events that included lack of communication that a large pump was not ready for service because a relief device was not in place.

Human error can occur for many reasons, but procedure deficiencies are often contributing factors. These deficiencies can include:

- Lack of a current written procedure for a task or activity

- Discrepancies between written procedural steps and actual training

- Failure to follow written procedures, usually because of

  o Inadequate, missing, confusing, or incorrect information in procedures;

  o The existence of multiple conflicting procedures;

  o A facility culture that does not support procedures as the mandatory way to perform tasks or activities (i.e., personnel are permitted to perform tasks their own way); and/or

  o Insufficient management systems such as procedure audits to ensure that procedures are followed

- Lack of personnel training on the procedures.

An effective program for developing and managing AIM program procedures helps reduce human errors by ensuring that accurate procedures are established for AIM tasks and activities. When developing written procedures and keeping them up to date for every AIM task and program activity is impractical, management personnel may need to evaluate and prioritize tasks and activities to help ensure that all truly needed procedures are developed and kept up to date. For example, a written procedure may not be considered necessary for a simple, low-risk, skill-based maintenance task such as testing and changing alarm panel lights or tightening packing of valves in low-pressure, nonhazardous service. Items such as this may be aggregated into a basic skills training package.

Task and activity performance can also be improved with well-written procedures that are accurate and user-friendly. Studies related to human errors have found that employing simple writing techniques can effectively reduce human errors during maintenance tasks and activities. These writing techniques involve the use of easy-to-follow procedure formats, the correct placement and use of warnings and cautions, and the practice of writing instructions using clear, imperative statements. Procedures can be improved by involving the personnel who will be actually performing the AIM tasks and activities. They can help identify tasks and activities needing procedures and can help develop, validate and maintain the written procedures.

The remainder of this chapter briefly discusses features of successful AIM procedure programs. The CCPS book *Guidelines for Writing Effective Operating and Maintenance Procedures* (Reference 9-2) provides more detail on developing and writing effective procedures. The following are discussed in this chapter:

- Types of procedures supporting the AIM program
- Identification of AIM procedure needs
- Procedure development process
- AIM procedure format and content
- Other sources of AIM procedures
- Implementing and maintaining AIM procedures
- Procedure program roles and responsibilities.

9.1   TYPES OF PROCEDURES SUPPORTING THE AIM PROGRAM

AIM programs will often include different types of procedures. The following types of procedures are usually found in an AIM program:

AIM Program Policy. This is generally a brief policy statement giving a strategic direction to the AIM program.

Corporate Guidance/Standard. This is a statement of mandatory performance requirements and regulatory constraints, including such items as requirements for equipment to be maintained by qualified personnel and definition of safety-critical equipment.

AIM Program Procedures. These procedures, which are usually site-specific, outline the activities and the roles and responsibilities for the different elements of the AIM program. In addition, documents that provide guidance or establish standards for AIM activities, such as ITPM plans, inspection standards, and rationale for determining the assets to include in the AIM program, are part of AIM program procedures. Procedures for related programs that affect or are affected by the AIM activities (e.g., MOC, process safety information) will be referenced as part of the program procedure.

Administrative Procedures. These procedures provide instructions for performing administrative tasks associated with the AIM program.

Quality Assurance (QA) Procedures. These procedures define the QA tasks to be performed and/or provide detailed instructions for performing QA tasks.

Maintenance Procedures. These procedures provide instructions for performing core maintenance activities, job-specific or unique repair/replacement tasks, and asset troubleshooting tasks.

ITPM Procedures. These procedures provide instructions for preparing for and performing ITPM tasks as well as recording and reacting to ITPM results.

In addition to the types of procedures listed above, safety procedures (lockout/tagout, confined space entry, line breaking and clearing, hot work permit, scaffold erection, hydroblasting, etc.) are an important part of the AIM program. These procedures are often referenced in various AIM procedures (e.g., maintenance repair/replacement procedures, preparation for inspection procedures, ITPM procedures) and are vital to helping ensure that AIM tasks are performed safely.

How each facility categorizes the different types of AIM procedures will vary. Most facilities will have one or more AIM procedures for each of the above-mentioned categories. The number of procedures needed for each category will also vary from facility to facility. Table 9-1 provides examples of the procedures that are typically developed by category. An example task procedure is given in Appendix 9A.

In addition to the types of procedures mentioned above, procedures established by other departments may cover activities that are part of the AIM program and need to be recognized as AIM procedures. Usually, such procedures include ITPM tasks or QA tasks that are performed by personnel outside the maintenance or inspection departments. For example, operating procedures (or portions of operating procedures) may provide instructions for such ITPM tasks as daily visual inspections of assets, lubrication of pumps, maintenance of safety assets in the process areas, and testing of alarms. Other departments that may have procedures related to the AIM program are the Project Engineering, Environmental, Emergency Response, and Purchasing departments.

The developers of the AIM program procedures may consider the different types of procedures needed, as well as the variety of users within a facility. The format, content, level of detail, and number of procedures will likely vary between different categories. For example, the format and content for a program procedure is usually quite different from that for an ITPM task because the frequency of use, the users' needs, and the detail required are different. For most facilities, more maintenance procedures are required than program or administrative procedures. Figure 9-1 groups the procedures into a common ISO 9000 document tier structure and summarizes the general differences among the procedure types.

The following two sections provide additional information on (1) determining what procedures are needed and (2) determining procedure content and selecting a good procedure format.

TABLE 9-1. Example AIM Procedures

Type of Procedure	Example Procedures	
AIM Program	• AIM program description • Asset selection and AIM program applicability • ITPM program development	• Asset deficiency resolution • Determining repair/replacement procedure needs • ITPM plan/guidelines/standards
Administrative	• Operation of the computerized maintenance management system (CMMS) • Managing work orders • Maintenance work planning • Task schedule management	• Maintaining asset files • Reporting and coding asset failures • Program reviews and auditing
Quality Management	• Selecting and auditing contractors • Selecting and auditing suppliers • Project engineering technical and administrative activities • Project engineering installation/construction standards • Technical and administrative activities for asset repair or replacement	• Ordering, receiving, stocking and issuing of spare parts and maintenance materials • Decommissioning of assets • Assets receipt inspection • Shop-fabricated assets inspection • Positive material identification (PMI) • Maintenance task oversight/review
Maintenance, Repair/ Replacement Tasks	• Rupture disk replacement • Hydrogen compressor asset seal replacement • Large horsepower motor replacement	• Centrifugal pump disassembly and assembly • Flammable solvent piping repairs • Motor starter replacement • Circuit breaker racking
ITPM Procedures *(Note: Regulations may require written procedures for all ITPM tasks)*	• Inspection preparation • Pressure vessel external and internal inspection • Atmospheric storage tank periodic surveillance • Atmospheric storage tank external and internal inspection • Chemical cleaning of heat exchanger tubes • Centrifugal pump vibration analysis • Pump seal visual inspection • Pump lubrication	• Relief valve removal and pop testing • Conservation vent inspection and testing • Safety instrumented system testing • Transmitter calibration • Instrument loop check • On/off valve and position switch testing • Fire sprinkler inspection • Hose house inventory • Tank farm dike visual inspection

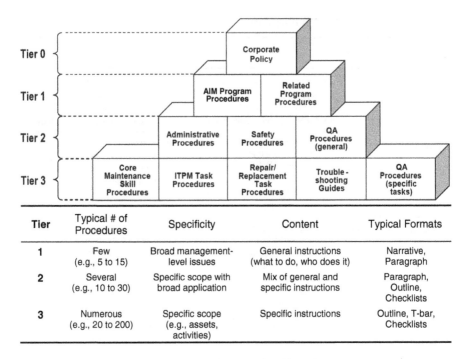

Tier	Typical # of Procedures	Specificity	Content	Typical Formats
1	Few (e.g., 5 to 15)	Broad management-level issues	General instructions (what to do, who does it)	Narrative, Paragraph
2	Several (e.g., 10 to 30)	Specific scope with broad application	Mix of general and specific instructions	Paragraph, Outline, Checklists
3	Numerous (e.g., 20 to 200)	Specific scope (e.g., assets, activities)	Specific instructions	Outline, T-bar, Checklists

Figure 9-1. AIM procedure hierarchy

9.2 IDENTIFICATION OF NEEDS

Determining the objectives of the procedure program is a good starting point for identifying procedure needs. The objectives can be determined by answering the following questions:

- Are there regulatory requirements that must be met (i.e., for what tasks/ activities do the regulations require procedures)?

- What procedures are required by corporate standards/requirements?

- Do applicable codes or standards require a procedure for the activity?

- What system-related tasks are necessary for successful completion of the work (reading P&IDs, design specifications and data sheets; ordering materials and procuring/kitting/retrieving parts; etc.)?

- What benefits (other than meeting regulatory requirements) are expected from the procedures? To train employees? To reduce human errors? To ensure program continuity as personnel change?

- What scope is to be included in the procedure program? Specific types of maintenance tasks? Specific units/processes/buildings?

- Who are the intended users for the different types of procedures?

- How are the different types of procedures to be used? For reference only? For training? In the field each time the task is performed? A combination?

Assessing these issues helps a facility begin to understand the types of procedures needed, as well as some specific issues to be considered during procedure development.

The following may be considered when developing a corporate standard for written maintenance procedures:

- Is a written procedure necessary to ensure that the task/activity is executed correctly?

- Is the risk of incorrectly performing or not performing the task high enough to warrant a written procedure?

- Do the intended procedure users believe that a written procedure is needed for the task?

- Is the task more than a simple sequence of normal craft skills?

- Will a checklist be useful for documenting completion of the task?

Facility personnel can use a variety of information sources to compile a list of tasks and activities for each procedure type (e.g., AIM program, ITPM procedures, QA procedures). Typically, the information can be found in the following:

- The survey used to assess training needs (see Section 8.1); alternatively, consider conducting a survey to assess procedure needs

- The applicable portions of the safety and environmental training manuals

- Any existing maintenance craft and job training protocols

- The ITPM plan (see Section 6.1 in Chapter 6)

- The QA plan (see Chapter 10).

Personnel can use these information sources as the primary tools for identifying the procedures needed for all of the procedure types except for repair/ replacement and troubleshooting task procedures. The procedure lists for repair/ replacement and troubleshooting tasks can often be developed by reviewing the work order history and/or interviewing the maintenance craftsmen and inspectors who perform these tasks. Original equipment manufacturer (OEM) manuals may also contain troubleshooting information, as well as repair or replacement recommendations in the specific service and/or operating context.

Once compiled, the task lists need to be evaluated to determine which activities warrant written procedures. To help improve employee buy-in, a team of personnel that includes one or more of the intended procedure users can participate in this evaluation.

Evaluating most of these factors is usually straightforward and can be accomplished quickly. However, identifying higher-risk tasks and determining if a procedure is warranted can be more involved. For these tasks, expert judgment or the application of a simple risk-ranking tool can be used. Table 9-2 provides an example result from applying a simple risk-ranking tool. The following questions need to be considered when assessing risk:

- Who can be assigned to perform the task?

- What is the consequence if the task is not performed correctly?

- How often is the task performed?

- How likely is it that the task will be performed incorrectly?

- What safeguards are in place, such as safe work practices, error detection, interlocks or independent checks?

- Is the risk of the task best managed by a procedure?

- How much training is required to manage the risk?

- How effective is the training?

TABLE 9-2. Example Qualitative Risk Ranking Results for Procedure Determination

Activity	Task Number	Activity Frequency	Qualitative Risk Ranking			Procedure Required?	Comments
			Severity (Note 1)	Likelihood (Note 2)	Risk Ranking		
Requesting materials and parts from stores	1.1	Daily	Significant	Occasional	Moderate	Yes	
Repairing and rebuilding pumps	1.2	Weekly	Major	Probable	High	Yes	
Repairing pump leaks	1.3	Weekly	Significant	Probable	Moderate	Yes	
Aligning pumps	1.4	Weekly	Significant	Probable	Moderate	Yes	
Repairing and rebuilding agitators	1.5	Once per year	Significant	Possible	Low	No	Team believes a procedure is needed
Repairing and overhauling compressor valves	1.6	2 times per year	Catastrophic	Possible	High	No	Manufacturers' manuals are sufficient
Repairing and overhauling compressors	1.7	2 times per year	Major	Possible	Moderate	Yes	

Note 1. Severity ranking represents credible worst-case incident scenario if the task is incorrectly performed.

Note 2. Likelihood ranking considers both the frequency that the task is performed and the probability of the task being performed incorrectly, resulting in the credible worst-case incident scenario.

A risk assessment often results in a list of procedures to locate or develop for each procedure type (e.g., AIM program, ITPM or maintenance procedure). In addition, Chapter 3 and Appendix C of the CCPS book *Guidelines for Writing Effective Operating and Maintenance Procedures* (Reference 9-2) provide more information on determining which activities require written procedures. The following section outlines a process for developing effective procedures.

9.3   PROCEDURE DEVELOPMENT PROCESS

An effective AIM procedure program uses a structured procedure development process. A good procedure development process will help ensure that:

- The intended procedure users are involved, thus improving employee acceptance and ultimately improving the use of and compliance with procedures.

- Necessary information is incorporated by gathering information from a variety of sources.

- Procedure information is presented in a logical and easy-to-use format by selecting an appropriate layout and following simple procedure-writing guidelines.

- Procedure content is accurate by using a multi-staged review and validation process.

Figure 9-2 provides the basic steps of a good procedure-development process.

The following paragraphs provide details and offer suggestions for improving the effectiveness of each step. Additional information on these steps can be found in Reference 9-2. Note that careful document control (Section 9.6) needs to be maintained throughout the development process as well as during use of the procedures.

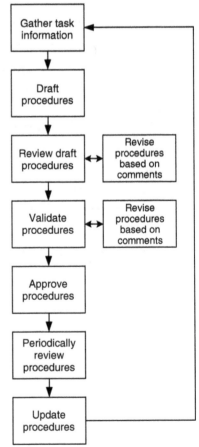

Figure 9-2.
Basic procedure
development process

Gather Task Information.   The objective of this step is to gather all of the information needed to draft the procedure. Necessary information includes (1) the main steps of the task/activity; (2) the hands-on steps (i.e., the discrete, physical actions needed to complete each main step); (3) the specific details related to the task (how, why and who); (4) any special hazards involved in the procedure; (5) special materials and tools, PPE and necessary precautions; and, for ITPM tasks, (6) acceptance criteria and actions to take if criteria are not met. The procedure developer can obtain needed information by (1) interviewing those who actually perform the procedure as well as other relevant subject matter experts (SMEs); (2) reviewing existing documents such as procedures, manuals and drawings; (3) observing the task being performed; and/or (4) obtaining the relevant information from the original equipment manufacturer, which may be in a format other than a written document (e.g., an online video). During the information-gathering step, the procedure developer (who may be an SME as well) needs to be careful to not filter or screen out information as unimportant and to use open-ended questions when interviewing SMEs to avoid biasing the responses.

In addition to the above considerations, it may also be necessary to take into account when equipment was designed and fabricated. For example, the routine inspection date and methodology for inspecting a heat exchanger built 70 years ago may be significantly different than one built recently. The two heat exchangers may be in similar or even identical service, but design and fabrication standards and upgrades in quality control may require them to have significantly different maintenance schedules.

Draft Procedures. During this step, the information gathered in the previous step is converted into a user-friendly procedure. In doing this, the procedure developer determines which information to include in the procedure and where the information is to be incorporated. Certain information such as hazards and special tools is best placed in the introductory section of the procedure; some information belongs in the warning, caution, and note statements (e.g., hazard warnings, safety issues related to incorrectly performing a step); while other types of information is placed directly in the procedure instruction steps (e.g., major steps, hands-on steps). Other procedures or documents might selectively be referenced or linked for some standard practices or supplemental information. Pictures, drawings and even video can be effective visual aids to supplement the procedures. Some procedures may be simple one-point lessons. Once these decisions have been made, the procedure developer incorporates the information into the procedure following an approved format and some simple procedure-writing guidelines; see Section 9.4 for additional information on procedure format and content. Many facilities have found that operations and maintenance personnel can be very effective in drafting both core maintenance procedures as well as those involving operator preparation activities such as line clearing and energy isolation.

Review Draft Procedures. The objective of the review step is to (1) verify that the approved procedure format has been followed, (2) confirm that the procedure meets regulatory (or other similar) requirements, (3) determine if the information provided by the SMEs and other sources has been appropriately incorporated into the procedure, and (4) assess the technical accuracy and readability of the procedure. Because of its multiple objectives, this step usually involves numerous people in the organization. For example, safety department personnel may need to review certain procedures such as ITPM procedures that involve confined space entry. A good practice is to include at least one of the SMEs interviewed in each procedure review. Establishing a formal review process helps ensure that procedures are reviewed by the appropriate personnel. In addition, performing a proofreading review prior to submitting procedures for the full review process allows subsequent reviewers to focus on the technical aspects of the procedure rather than typographical and grammatical errors.

Validate Procedures. This step verifies that the task can be executed as described in the procedure. The SME(s) who provided the information and the SME(s) involved in the procedure development to this point often perform the validation step. In addition, SMEs with different experience levels may also participate in this review. The validation can be accomplished by performing a mock walk-through of the task as described or by following the procedure while actually performing the task. (Note: If the procedure is validated by actually performing the task, personnel performing the task must be made aware that the procedure has not yet been validated and may need to be altered.)

Revise Procedures. Procedures can be revised after the review step, after the validation step and at any time it becomes necessary. The objective of this step is to address comments concerning the procedure. The procedure developer can resolve the comments by modifying the procedure to address the issue, or by convincing the commenter that the issue is already addressed in the procedure. If modified, the procedure needs to receive an appropriate review before progressing to the next step.

Approve Procedures. The objective of this step is to obtain formal approval of the procedure from the appropriate person. If the review process is reliable and the approver is available, the approval step can usually be accomplished without delay.

Periodically Review Procedures. Effective procedures contain current and accurate information. To help maintain this accuracy, the procedure development process needs to contain a step requiring periodic review of procedures. Typically, this is accomplished by periodically resubmitting the procedure to a review process and/or by periodically observing the task while checking the procedure. Including the procedure users in the periodic review process can help ensure that the procedure reflects actual task steps and provides all of the necessary information.

Update Procedures. If deemed necessary (e.g., based on the periodic review), the procedures are updated. The extent of the update will dictate which of the earlier procedure development steps to employ. For example, a simple update that involves expanding or correcting information for a couple of steps may not require a draft procedure review. On the other hand, a procedure that requires a complete rewrite may require all of the review steps. In addition, facilities need to have a process (often part of a management of change procedure) to ensure that procedure updates are communicated to all affected personnel. When the procedure change is substantial, formal training on the changed procedure may be necessary.

9.4   FORMAT AND CONTENT

To create effective AIM procedures, developers need to use a format appropriate for the subject task and ensure that the content is appropriate. For the procedures to add value, they need to both satisfy regulatory requirements (and other objectives, such as corporate mandates) and communicate task information in a manner that is clear, accurate and easy to use.

Several factors influence the selection of the procedure format, including word processing capabilities and company requirements. However, the primary factor is the type of task covered in the procedure. For example, AIM program procedures are typically written in a narrative (paragraph) format or an outline format. On the other hand, ITPM task procedures may be written using an outline format with short sections or a two-column procedure format with one column containing the main steps and the other providing the step details. They may also be developed and implemented in an electronic format, with the approved version residing on a central server accessible through a portable electronic field device. Chapter 4 and Appendix F in the CCPS book *Guidelines for Writing Effective Operating and Maintenance Procedures* (Reference 9-2) provide additional information on the importance of procedure format and contain some sample procedure formats. The format selected should:

- Be user-friendly
- Help simplify the procedure
- Help procedure developers input information in a consistent way
- Be appropriate for the type of task it is explaining.

Another format consideration is the effective use of white space (on the page) and borders, and the selection of an appropriate type style and size. Benefits of using white space and borders effectively (especially for detailed, step-by-step maintenance task procedures) include:

- Uncluttered procedure presentation
- Logical grouping and/or separation of information
- Improved comprehension
- Improved readability.

Ways to effectively create white space on a page include (1) increasing line spacing (e.g., use 1.5 or double spacing versus single spacing), (2) using indentation (e.g., indent sub-steps) and (3) presenting information in columns, tables, and/or bulleted lists. Any of these approaches can increase the readability of the procedure and help the user keep track of steps being performed.

Similarly, the type style and size can enhance or detract from the procedure. Obviously, the type size needs to be large enough to read easily. The procedure developer can use uppercase words to emphasize important items. For example, if the standard use of the uppercase and lowercase is used, then items in all caps (e.g., WARNINGS, IF/THEN statements) will stand out and draw attention to information that is especially important. Some procedure writers make the active verb uppercase in all procedural steps (e.g., VERIFY the vent valve is closed).

Content is the most essential part of a procedure. Obviously, the information needs to be accurate and complete. Level of detail and ease of use are also important. Procedures that do not have these attributes are less likely to be used. This is especially true of detailed maintenance task procedures.

The required level of procedure detail is dictated by the user. The procedure needs to provide enough detail so that the least-experienced worker is provided sufficient information to execute the task. On the other hand, veteran craftsmen may also use these procedures. For these individuals, too much or inappropriate detail is unnecessary and can make a procedure too long and difficult to use.

These are manageable issues. For every procedure, define and use an adequate task information gathering process (see Section 9.3) and employ simple, proven procedure-writing guidelines, such as:

- Each instruction is written as a command
- The proper level of detail is used throughout the procedure
- On average, only one action is implied per instruction step
- The procedure indicates when instruction step sequence is important

- The procedure is written using only words in common usage

- Only acronyms/abbreviations/jargon that aid in readers' understanding of the procedure are used

- Each step is specific (i.e., there is no room left to guess/interpret)

- The procedure is free of steps that require in-your-head calculations

- Any graphics used are for the users' benefit

- The references have been used to the readers' advantage

- Each page header contains the document revision number, procedure issue date and page number

- The procedure includes an indicator at the end to confirm that the user is at the end of the procedure.

Figure 9-3 contains a detailed procedure-writing checklist that outlines some of these simple guidelines. Note that the guidelines in the Figure 9-3 checklist were developed specifically for detailed, step-by-step instructions for a maintenance task procedure; however, some of these guidelines can be applied to procedures covering other types of work. Section 5.3 of the CCPS book *Guidelines for Writing Effective Operating and Maintenance Procedures* (Reference 9-2) provides a more extensive list and explanation of these guidelines.

These writing techniques presented in this section can help improve understanding by:

- Presenting the information in clear and concise statements, starting each step with a command, having only one implied action per instruction step or sub-step and ensuring appropriate use of referenced material such as other procedures or OEM manuals

- Enhancing clarity by using common words and terminology

- Minimizing usage interruptions by providing sufficient details, being specific and using graphics when helpful.

Procedures for many tasks also need to guard against potential deviations that might occur during the task, such as (1) failing to obtain proper permits, (2) performing critical steps out of order and (3) mistakenly replacing a part with a part of different materials of construction. Potential deviations are often addressed by including appropriate precaution, caution and/or warning statements in the procedure.

PROCEDURE WRITING CHECKLIST

1. **Is each instruction written as a <u>command</u>?**

2. **Is the proper <u>level of detail</u> used throughout?** This is judged based on...
   - *How/when* the procedures will be used
   - *Who* will use the procedures
   - Whether substeps are *critical* to safety, environmental, or quality goals
   - Whether the *sequence* of substeps is critical
   - Level of detail used in similar procedures

3. **On average, is there only <u>one implied action per instruction</u>?**
   - Single sentences instead of paragraphs
   - Multiple, concise action statements instead of long sentences
   - Bullet lists instead of commas

4. **Does the procedure indicate when <u>sequence</u> is important?**
   - If sequence matters, each step should be numbered (with an integer or letter)
   - If sequence does not matter, bullet-lists should be used
   - If sequence is safety-critical, a warning/caution should be included before the steps

5. **Has the writer used only <u>common words</u> (no 50-cent words)?**
   - Apply the three-syllable test

6. **Do all <u>acronyms, abbreviations</u>, and <u>jargon</u> aid understanding?**
   - Terms used should be those that are best understood by the workers
   - Should not feel the need to define or initialize them
   - They should be quicker to say than the terms they represent

7. **Is each step <u>specific</u> enough?** No room left to guess/interpret:
   - The meaning of a word or phrase
   - The intent of a step or series of steps
   - A desired quantity or value
   - To what equipment the step applies

8. **Is the procedure free of steps that require in-your-head <u>calculations</u>?**
   - Values expressed as ranges rather than targets with error bands
   - Conversion tables, worksheets, or graphs provided where needed

9. **Have <u>graphics</u> been used to the writer's advantage?**
   - No explanatory paragraphs or lengthy instructions that could be replaced by a picture
   - No "impressive" graphics that provide no real advantage

10. **Have <u>references</u> been used to the writer's advantage?**
    - No lengthy explanations or instructions that could be replaced by a reference
    - Referenced documents should contain details that apply to several procedures
    - No references to a procedure that references still another
    - No gaps or overlaps between this procedure and a referenced document

Figure 9-3. Procedure-writing checklist

9.5   OTHER SOURCES OF AIM PROCEDURES

Facilities often use outside sources of procedures in addition to those developed by the facility. While many regulations require procedures for AIM activities, the requirements rarely state that the facility must develop its own procedures for every task. Many companies have also determined that contractors performing a task or vendors supplying equipment are better able to develop procedures for some AIM tasks.

Some common reasons for using outside procedures are:

- Facility personnel lack the expertise or experience needed to develop the procedure

- A contractor or vendor already has the procedure available

- OEM manuals may provide sufficient information to function as the procedure for some tasks. (If an OEM manual is employed, then it needs to be onsite or otherwise available for use.) Note that manufacturers' manuals often lack the necessary permit and safety information needed to safely perform tasks in the field.

Any outside procedures that are used or referenced need to be reviewed by the facility. The following are some questions to answer during such a review:

- Does the procedure include steps for the tasks to be performed?

- Is sufficient detail presented to ensure safe and consistent execution of the task(s)?

- Do the procedure details match the intended users' skill/knowledge base?

- Does the procedure contain necessary safety information?

- How well does the procedure interface with the facility's safety procedures such as safe work practices?

- Does the procedure contradict any facility safety procedures or practices?

- Does the procedure address facility quality or environmental concerns (e.g., product contamination, proper disposal of wastes)?

Facilities often find that they need to supplement contractor or supplier procedures to address site-specific issues or company policies.

9.6   IMPLEMENTING AND MAINTAINING AIM PROCEDURES

The previous sections of this chapter have outlined many of the activities needed for an effective AIM procedure program. However, two further phases of the procedure program are needed to ensure success of the program; namely, implementation of the AIM procedures and maintenance of the procedures.

While implementation may seem as simple as approving the procedures and then assembling them in manuals, successful implementation requires that the following items be addressed:

- *Document control.* Facilities with a document control system can include procedures in this system. Steps need to be taken to dispose of old and out-of-date procedures when issuing new documents.

- *Access.* Procedures need to be readily accessible to employees. In addition, employees are more likely to use procedures that are kept in a location that does not intimidate them. For example, procedures kept in a supervisor's office may be less likely to be used by employees if they need to go to the supervisor to get a copy. Note that many companies use computer-based systems for procedures, or have procedures incorporated into PM documentation.

- *Training.* Procedure users need to be trained on all new procedures to ensure that they fully understand the procedures. Training also provides an opportunity to verify that the procedures reflect how the work is actually performed. AIM training is addressed in Chapter 8.

Procedures have a tendency to become inaccurate or otherwise degrade over time. The following are some key methods for maintaining procedures:

- *Management of Change.* Changes to tasks covered by procedures need to be managed for the procedures to remain accurate. This includes managing changes that result from improved ways to perform the task/activity, physical changes in the field such as addition of new equipment, and/or organizational changes. Facilities can use existing MOC practices or develop a stand-alone change control process for the AIM procedure program. One challenging aspect of keeping procedures up to date is when the original equipment manufacturer (OEM) changes its procedure or, more likely, only part of its procedure. This may require keeping up with OEM technical bulletins or requesting the latest version of the OEM maintenance procedure, even when updating procedures or when performing repairs or replacements in kind.

- *Periodic Review.* Periodic reviews of AIM procedures provide an effective means to ensure that the procedures are current and accurate and, therefore, more likely to be used. In some cases, periodic procedure reviews can be incorporated into refresher training activities.

- *User Feedback.* Having a means to capture feedback as AIM procedures are actually used in the field can be a very effective means of keeping the procedures up to date. Users may find that asset identifiers are not correct, steps cannot be performed in the stated sequence (requiring a variance to be authorized), or a potentially better or more flexible way of performing the procedure is conceived. Employees can be encouraged to somehow capture this feedback and put it into the MOC process to have the procedure change reviewed and approved. For in-hand procedures, users can make notes and comments right on the printed procedure, then submit them to a supervisor, reviewer or administrator upon completion.

Maintaining procedures requires management commitment. Specifically, this requires that management provide leadership to communicate expectations, then provide the resources needed to update and revise the AIM procedures. Additional information on the topics discussed in this section can be found in the CCPS book *Guidelines for Writing Effective Operating and Maintenance Procedures* (Reference 9-2).

9.7   AIM PROCEDURE PROGRAM ROLES AND RESPONSIBILITIES

The roles and responsibilities for the AIM procedure program can be assigned to personnel in various departments. Typically, the personnel in the Maintenance, Engineering and EHS departments will primarily be involved. However, production personnel are likely to be involved in procedures for AIM activities performed by production personnel. Outside personnel such as contractors may also be involved, and procedures obtained from or developed by contractors and other third parties (such as equipment manufacturers) need to be included in the review and approval process.

Example roles and responsibilities for an AIM procedure program are shown in Table 9-3. The matrix uses the following letter designators:

R indicates the job position(s) with primary <u>responsibility</u> for the activity

A indicates <u>approver</u> of the work or decisions made by a responsible party

S indicates the job position(s) that typically <u>support</u> the responsible party in completing the activity

I indicates the job position(s) that are <u>informed</u> when the activity is completed or delayed.

CHAPTER 9 REFERENCES

9-1   Marsh & McLennan, *Large Property Damage Losses in the Hydrocarbon Chemical Industries – A Thirty Year Review*, 20th Edition, New York, NY, 2003.

9-2   Center for Chemical Process Safety, *Guidelines for Writing Effective Operating and Maintenance Procedures*, American Institute of Chemical Engineers, New York, NY, 1996.

Table 9-3. Example Roles and Responsibilities Matrix for the AIM Procedure Program

R = Primary Responsibility
A = Approval
S = Support
I = Informed

Activity	Inspection Manager	Maintenance Manager	Maintenance Engineers	Maintenance Supervisors	Inspectors	Maintenance Technicians	Maintenance Planner/Scheduler	Engineering Manager	Project Managers	Project Engineers	Area Superintendent/Unit Manager	Production/Process Engineers	Production Supervisors	Operators	Plant manager	EHS Manager	Process Safety Coordinator	Contractor	Equipment Vendor
	Inspection and Maintenance Departments							Engineering Department Personnel			Operations Department Personnel				Other Personnel				
Reviewing and Validating Procedures (including procedures from third parties)																			
• AIM program & related procedures	R	R	S	S	S										I		S		
• Administrative procedures	I	I	R	S	R	S	S								I		S		
• ITPM procedures	I	I	R	R	R	S	S	R			I	R	R	S			I	S	S
• QA procedures			R	S	R	S		R	S	S							I	S	S
• Maintenance procedures			S	R		S	S										I		
Approving Procedures (including procedures from third parties)																			
• AIM program & related procedures	R/A	R/A													R/A		R/A		
• Administrative procedures	R/A	R/A													I		I		
• ITPM procedures	R/A	R/A									R/A						I		
• QA procedures	R/A	R/A						R/A									I		
• Maintenance procedures		R/A															I		

Table 9-3. *Continued*

R = Primary Responsibility
A = Approval
S = Support
I = Informed

Activity	Inspection and Maintenance Departments							Engineering Department Personnel			Operations Department Personnel				Other Personnel				
	Inspection Manager	Maintenance Manager	Maintenance Engineers	Maintenance Supervisors	Inspectors	Maintenance Technicians	Maintenance Planner / Scheduler	Engineering Manager	Project Managers	Project Engineers	Area Superintendent / Unit Manager	Production / Process Engineers	Production Supervisors	Operators	Plant manager	EHS Manager	Process Safety Coordinator	Contractor	Equipment Vendor
Implementing Procedures																			
• AIM program & related procedures	R	R	S	S	S										A		S		
• Administrative procedures	R	R	SI	S	S	S	S								A		I	R	
• ITPM procedures	A	A	S	R	R	S	S	R	S	S	R	S	S				I	R	R
• QA procedures	A	A	S	R	R	S	S	R	S	S							I	R	R
• Maintenance procedures	A	A	S	R		S	S										I	R	R
Maintaining Procedures																			
• AIM program & related procedures	R	R	S	S	S			R	S	S	R				A		S		
• Administrative procedures	A	A	R	S	R	S	S										I	R	
• ITPM procedures	A	A	S	R	R	S	S				R	S	S				I	R	R
• QA procedures	A	A	S	R	R	S	S	A	R	S							I	R	R

APPENDIX 9A. EXAMPLE AIM PROCEDURE

Included in this appendix is only one example of many procedures in a typical AIM program. The actual format and content will of course differ from facility to facility and for different types of procedures and processing operations.

ASSET INTEGRITY PROGRAM MANUAL	Document #:	**AIM-001**
Piping Circuit and Vessel System Leak/Pressure Test	Page:	1 of 6
	Revision #:	
	Revision Date::	

Approved By:

_____

Production Manager   Date

_____

Health and Safety Manager   Date

_____

Plant Manager   Date

_____

A. PURPOSE

The purpose of this procedure is to establish consistent and safe practices for pressure testing piping and vessel systems as part of the Maintenance Program at the _____ Plant.

B. SCOPE

This procedure applies to all piping systems, pressure vessels, storage tanks, and components, including flexible hoses, covered by PSM-10, Maintenance.

C. TOOLS

The following tools may be required to properly perform this procedure.
- Pipefitting tools
- Nitrogen supply and regulator

ASSET INTEGRITY PROGRAM MANUAL	Document #:	**AIM-001**
Piping Circuit and Vessel System Leak/Pressure Test	Page:	2 of 6
	Revision #:	
	Revision Date::	

D. SAFETY

1. Hazards

The following hazards may be present when performing this procedure:

- Potential exposure to process chemicals having toxicity, flammability and/or reactivity hazards (refer to SDSs for specific hazards)
- Nitrogen (asphyxiant)
- Pressurized piping / compressed gas.

2. Personal Protective Equipment

The following personal protective equipment may be required to safely perform this procedure and should be referenced by the work permit:

- Chemical splash goggles
- Safety glasses
- Hard hat
- Safety shoes
- Hearing protection (in high-noise areas)
- Respirator (must conform to HSP-202, Respiratory Protection Program)
- Chemically impervious gloves.

3. Safe Work Permits

The following safe work permits or practices may be required in order to safely perform this procedure:

- HSG-006, Fall Protection
- HSP-204, Lockout/Tagout.

E. PROCEDURE

WARNING:

MAKE SURE THAT THE EQUIPMENT IS SAFE FOR OPENING PRIOR TO PERFORMING THIS PROCEDURE. HAZARDOUS CHEMICALS, PROCESS ENERGY, ELECTRICAL ENERGY AND OTHER ENERGY SOURCES NOT PROPERLY ISOLATED COULD CAUSE SERIOUS INJURY OR DEATH.

ASSET INTEGRITY PROGRAM MANUAL	Document #: **AIM-001**
Piping Circuit and Vessel System Leak/Pressure Test	Page: 3 of 6
	Revision #:
	Revision Date::

1. <u>Operational Personnel</u> shut down, purge, and isolate the equipment to be tested, and place Production Tags per applicable operating procedures and HSP-204, Lockout/Tagout.

2. PUT ON required PPE.

3. OBTAIN the proper test pressure for the equipment to be tested, per ASME B31.3 or other applicable codes and standards.

4. VERIFY that only essential personnel are present, and that there are no other activities in the area that may interfere with the test.

5. PERFORM line breaking per practices in HSP-204, Lockout/Tagout, as required to complete the physical isolation and installation of blinds, and tag those points per HSP-204, Lockout/Tagout.

6. CONNECT the nitrogen regulator to the nitrogen supply hose.

7. CONNECT the nitrogen regulator to the process equipment.

WARNING:

DO NOT EXCEED THE INTENDED TEST PRESSURE TO AVOID DAMAGING OR RUPTURING THE EQUIPMENT AND CAUSING INJURY AND LOSS.

Note:

Typical hose test pressure is 60 psig.

---

8. SET the nitrogen regulator to the test pressure.

ASSET INTEGRITY PROGRAM MANUAL	Document #: **AIM-001**
Piping Circuit and Vessel System Leak/Pressure Test	Page: 4 of 6
	Revision #:
	Revision Date::

9.   Using the HTU pump, APPLY the proper test pressure to the equipment.

10.   Making use of local pressure gauges, MONITOR the pressure until the desired pressure is obtained.

11.   CLOSE the nitrogen regulator and nitrogen supply valves.

12.   MONITOR the pressure for the holding period as specified in ASME B31.3 (typically, a two-hour test is sufficient).

13.   For hard piping and vessel systems, IF the pressure loss is greater than 2 psi, INSPECT the equipment for leaks using soap bubble or equivalent method. For flexible hoses, IF the hose withstands the test pressure, the test is successful.

14.   IF a leak is found:

        i.   INFORM Operational Personnel.

WARNING:

ATTEMPTING REPAIRS ON PRESSURIZED PIPING COULD CAUSE IT TO RUPTURE, WHICH COULD RESULT IN SERIOUS INJURY OR DEATH.

---

        ii.   REPAIR the leak as directed by Operational Personnel and/or the Production Manager in accordance with applicable safety and procedural requirements.

        iii.   REPEAT the pressure testing of the piping system starting with Procedure Step 2 above.

ASSET INTEGRITY PROGRAM MANUAL	Document #:	**AIM-001**
Piping Circuit and Vessel System Leak/Pressure Test	Page:	5 of 6
	Revision #:	
	Revision Date::	

CAUTION:
Exercise caution when relieving pressurized gas from equipment.

15. When Operational Personnel are satisfied with the results of the pressure test, OPEN bleed valves on the equipment being tested.

16. Slowly BLEED the pressure from the equipment using the vent valve.

17. REMOVE blinds from the equipment per line breaking practices in HSP-204, Lockout/Tagout.

18. CLOSE the equipment per MP-2004, Line Closing Procedure.

19. When the equipment has been properly closed, RETEST for leaks at the proper test pressure.

20. REMOVE all personal locks and tags from equipment isolation points.

21. DISCONNECT the nitrogen line and nitrogen regulator from the system.

22. CLOSE OUT safe work permits per applicable plant standards.

23. SUBMIT the work order for closeout per PSM-10, Maintenance.

End of Procedure.

ASSET INTEGRITY PROGRAM MANUAL	Document #:	**AIM-001**
Piping Circuit and Vessel System Leak/Pressure Test	Page:	6 of 6
	Revision #:	
	Revision Date::	

F. RECORDKEEPING

The work order is stored in the Maintenance Management System per PSM-10, Maintenance.

G. ACCOUNTABILITIES

- Operational Personnel or Maintenance Personnel are responsible for shutting down, purging and isolating the piping system being inspected per applicable operating procedures and HSP-204, Lockout/Tagout.

- Maintenance Personnel or Contractor Personnel are responsible for performing the test per Section E of this procedure, signing and submitting safe work permits to appropriate personnel, and submitting the work order for closeout.

H. TRAINING AND CERTIFICATION

In addition to site-wide training requirements (refer to PSM-11, Training), Maintenance Personnel or Contractor Personnel are trained in safe work practices, as specified in applicable site HSPs and HSGs (see Section D3).

I. REFERENCES

- HSG-006, Fall Protection
- HSP-204, Lockout/Tagout
- PSM-10, Maintenance
- PSM-11, Training
- MP-2004, Line Closing Procedure
- ASME B31.3, Chemical Plant and Petroleum Refinery Piping
- API RP 574, Inspection of Piping, Tubing, Valves, and Fittings.

10
QUALITY MANAGEMENT

A life cycle approach to managing asset quality considers quality from the time the asset is designed until the time it is taken out of service for retirement or re-use. Effective quality management can be a powerful tool for upgrading a facility's management of asset integrity.

Presented in Table 10-1 are some AIM-related quality definitions. In an AIM program, quality control (QC) and quality assurance (QA) work together to help ensure that appropriate tools, materials and workmanship combine to provide assets that perform to meet their design intentions. Throughout this document, the primary focus will be on quality assurance, although AIM QA may also include QC activities.

It should be noted that the terms QC and QA can carry different connotations in different organizations. They are sometimes used interchangeably; however, they are not the same. For example, a company might have a contractor performing procurement, installation and/or ITPM. Quality control (following QC procedures, looking at the nuts and bolts, etc.) throughout procurement, installation and/or ITPM would likely be conducted by the contractor or possibly by a third-party inspector. Quality assurance would then be accomplished by some sort of oversight (sampled and/or random) being performed by the company itself and/or by a third party on its behalf. QA can be thought of as being at a higher level than QC. QA may require some sort of certification for those who perform this function.

This chapter presents several suggestions for QA activities applicable to different phases of an asset life cycle. Most facilities are not likely to implement every suggestion, and some QA activities may be more or less rigorous depending on the importance of particular assets. For example, some facilities use positive material identification (PMI) only for specific processes or unique metallurgy.

TABLE 10-1. Definition of Quality-Related Terms for Asset Integrity Management

Term	Definition
Quality management	All the activities that an organization uses to direct, control and coordinate quality. These activities include formulating a quality policy; setting quality objectives; and executing quality planning, quality assurance, quality control and quality improvement
Quality control (QC)	Execution of a procedure or set of procedures intended to ensure that a design or manufactured product or performed service/activity adheres to a defined set of quality criteria or meets the requirements of the client or customer
Quality assurance (QA)	Activities performed to ensure that equipment is designed appropriately and to ensure that the design intent is not compromised, providing confidence throughout that a product or service will continually fulfill a defined need the equipment's entire life cycle

Examining existing practices at each stage of asset life makes it possible to determine whether QA deficiencies exist and, if so, to develop a quality improvement plan to upgrade vulnerable areas and to be the basis of procedures and training for those activities. This chapter discusses QA activities for the following life cycle stages:

- Design/engineering

- Procurement

- Fabrication

- Receiving

- Storage and retrieval

- Construction and installation

- In-service repairs, alterations, and rerating

- Temporary installations and temporary repairs

- Decommissioning / re-use

- Used assets.

In addition, this chapter discusses QA practices for spare parts and for contractor-supplied assets and materials. QA-related incidents highlighted in this chapter are summarized from CCPS Process Safety Beacons and from References 10-1 and 10-2.

The Need for Quality Management.   Much of the reason for having a quality management program is to counter the multitude of possible human errors when managing asset integrity throughout the life cycle of a facility. Such human errors can take many forms. A small sample is as follow:

- Making an error in judgment, having inadequate technical background, or using wrong information when selecting a material of construction

- Executing a poor weld when fabricating pressure equipment

- Switching identical-looking parts during equipment assembly/reassembly

- Using a wrong or outdated procedure when maintaining equipment

- Omitting a step when doing an inspection/testing/maintenance procedure

- Using the wrong NDT procedure for a particular degradation mechanism

- Not selecting a replacement in kind when ordering a spare part, equipment item or a maintenance material

- Selecting the wrong part when retrieving a spare part from the storeroom

- Omitting preventive maintenance tasks for a particular item of rotating equipment.

Each of these errors, as well as many others, can be caught and corrected by well-designed and well-executed quality control or quality assurance activities.

10.1   DESIGN

Design is usually the only opportunity a facility has to "build in" quality to its assets.   The remainder of quality management activities generally contribute toward preserving initial asset integrity.

Quality designs start with competent and creative engineering.   When possible, designs employ features that have been tested and proven, so designs benefit from lessons learned in order that mistakes are not repeated.   Many of these proven designs then become the foundation for codes and standards and, more specifically, for company asset specifications.

All facilities can benefit from having asset or equipment specifications. (Note that some companies refer to their own specifications as their *design criteria* or *design standards*.) Facilities without such documentation can add "Develop asset specifications/design criteria" to their quality improvement plans.

Equipment specifications can begin with a reference to the codes and standards applicable to various equipment types.   Table 10-2 contains a few widely used design codes. Another source of specifications for many processes is the design manuals or other information in the original engineering and purchasing records.   In creating its own specifications, a facility using the original manuals and/or codes and standards may consider supplementing that information with (1) lessons learned that are specific to that facility (e.g., those documented during failure analysis investigations) and (2) updated or new information that may have become available after original construction.

Asset specifications are not static. Updates may be made necessary by lessons learned from experience and investigations, changes in the underlying codes and standards or technology advances.

The challenge of keeping specifications current is compounded following company mergers. Merged companies can be hampered by the use of "legacy specifications," leaving engineers with multiple versions of similar specifications. A similar problem occurs for companies that rely on vendor specifications when different vendors were used to build different process units. A person or team can be assigned to address these issues and to be responsible for producing and maintaining a definitive set of specifications.

TABLE 10-2. Typical Design Code Applications

Application	Design Code or Standard
Boilers (power)	ASME Boiler and Pressure Vessel Code, Section I, Power Boilers
Electrical Systems	NFPA 70, National Electric Code
Instrumentation	Various standards including IEC 61511 for safety instrumented systems
Piping (process)	ASME B31.3, Process Piping
Pressure Vessels	ASME Boiler and Pressure Vessel Code, Section VIII, Pressure Vessels
Pressure Equipment	Pressure Equipment Directive 97/23/EC (European Union)
Pumps	Numerous standards, including:
	API 610/ISO 13709, Centrifugal Pumps for Petroleum, Petrochemical and Natural Gas Industries
	ASME B73.1, Specification for Horizontal End Suction Centrifugal Pumps for Chemical Process
	ASME B73.2, Specifications for Vertical In-line Centrifugal Pumps for Chemical Process
Storage Tanks	API 620, Design and Construction of Large, Welded, Low-pressure Storage Tanks
	API 650, Welded Steel Tanks for Oil Storage
	UL 142, Steel Aboveground Tanks for Flammable and Combustible Liquids

Maintaining proper design specifications is important. Documenting the design specifications and developing the supporting drawings and data sheets (i.e., process safety knowledge) are often required by regulations, such as the Process Safety Information requirements of the U.S. OSHA PSM Standard, 20 CFR 1910.119(d). Applying the specifications correctly and having appropriate safeguards in place help produce a quality design. Companies use a variety of methods, including safety and design reviews (e.g., piping and instrumentation diagram reviews, relief system reviews and various hazard reviews) to help ensure the quality of designs. The CCPS publication *Guidelines for Hazard Evaluation Procedures* (Reference 10-3) includes descriptions of many hazard review techniques. Appendix B of this reference also includes an extensive hazard evaluation checklist that companies can use to aid their design efforts.

During the design stage, facility personnel need to establish and implement a quality management plan for evaluating and checking the design at the different design phases, as well as establishing a QA plan to be used during fabrication and construction. The quality management plan needs to start all the way back at the verification of the basic data upon which the design is based. It will need to include quality and accuracy checks throughout the design process. For example, most companies employ engineering contractors to do significant design work. Relief calculations done by a young engineer would be verified by a senior engineer who checks them and signs off on them. QA plans would be a deliverable item as part of the engineering design from the contractor.

10.2 PROCUREMENT

Quality management for procurement helps ensure that purchases adhere to a specified design, that change guidelines (i.e., knowing when substitution is acceptable and having appropriate approval for substitutions) are understood, and that qualified vendors are used. Often, QA is simpler if fewer people are involved. If parts and asset purchasing can be carried out centrally (rather than having multiple departments involved) and only approved vendors are used, errors are less likely. All personnel involved in procurement need to understand the specifications well enough to recognize if inappropriate parts or materials are being ordered and to know when authorization is required for a material or part substitution.

> **Procurement incident**
>
> Use of a liquid chlorine transfer hose made of similar-looking stainless steel braid instead of Hastelloy C led to corrosion and hose failure resulting in a 48,000 lb (22,000 kg) chlorine release during railcar unloading. Both user and hose vendor quality management systems were found to be inadequate.

A few words of caution are warranted with respect to procurement. These items may possibly warrant QA checks.

- Those involved in procurement are influenced by cost and may not understand the detailed requirements and/or technical specifications for equipment items being procured. For example, a regular flashlight may be able to be purchased at a significantly lower cost than an intrinsically safe flashlight; however, it would not provide the same ignition source control.

- Sometimes, procurement is contracted out as part of an engineering, procurement and construction (EPC) contract. Technical specifications need to be carried forward through the procurement process, and those involved in procurement need to be informed that less-expensive alternatives may not meet technical specifications.

Facilities may consider establishing a vendor approval procedure and limiting purchases to qualified vendors. Pre-qualifying vendors helps to eliminate sources of improper parts and materials. A good vendor can benefit a facility QA program by maintaining internal and external quality controls of its own. An example vendor QA plan is presented in Appendix 10A.

10.3   FABRICATION

Quality management for fabrication includes verification that specifications are followed and that shop practices do not compromise quality. The quality of fabricated equipment is an integral part of what the fabricator is expected to deliver, with QA oversight by the owner as required by the asset. Depending upon the importance of the equipment involved, facilities may use shop inspection and shop approval processes.

Shop and field fabrication site inspections are common QA tools. For pressure vessels and other critical assets, the QA process often identifies hold points in the fabrication process. For example, before fabrication can continue, a company inspector or a third-party inspector may be required to verify the quality of root weld passes. Shop/field site inspections may also be used more generally to oversee the fabrication procedures, shop/site conditions, and/or recordkeeping. To help make these inspections more effective, experienced inspectors and engineers may consider providing training, checklists, and/or procedures

Fabrication incident

A loading arm used for crude oil export collapsed at a UK oil terminal as the arm was being maneuvered onto the jetty deck. Examination found one sheared pivot pin was made from inferior material, having been formed locally in a terminal workshop, most likely due to the logistics of supplying genuine spares to the remote location of the terminal.

for others to follow. Some companies are using a shop qualification process to verify the procedures and QA practices of shops that may potentially fabricate their assets. Such preapproval practices help ensure quality asset fabrication, particularly if less-experienced personnel may perform project-specific inspections.   The American Petroleum Institute (www.api.org) has Source Inspector Certifications for both fixed and rotating equipment inspectors acting on behalf of purchasers to help ensure that materials and equipment being purchased will meet contractual requirements.

Many jurisdictions require using a code-approved shop for fabrication of some assets such as relief valves and pressure vessels. These shops have previously undergone inspection and may continue to be inspected regularly by third parties such as jurisdictionally authorized personnel. Some facilities have reported fabrication errors by these code shops, despite the authorizations they hold. Therefore, many facilities inspect and approve shops even when the shops already hold jurisdictional approval.

As part of quality management, some companies have also implemented some form of positive material identification (PMI) as a quality control function to check for proper materials of construction. PMI has gained popularity with the realization that substandard, incorrectly shipped, or accidentally switched materials (e.g., stainless steel with lower-than-specified chromium content) have been sold to unsuspecting (and non-inspecting) facilities. PMI is a process that facility procurement personnel can use to verify that the construction materials of a part or component fall within certain specifications.   Companies and facilities implementing PMI perform material testing, material tracking, and/or documentation tracking. In addition, the point at which companies execute PMI varies.   Some companies start with the steel mill run, others begin during fabrication, while still others begin PMI during receiving. A facility may hire a third party to provide independent testing services, or company personnel may perform the testing themselves. In most programs, after the material composition has been positively identified, the results are documented and the component is tracked until installation. One approach for companies interested in PMI is to identify opportunities such as projects with critical materials concerns and include PMI steps (material tests, tracking, documentation) in the project QA plan. More information on PMI can be found in Appendix 10B (Reference 10-4).

> **PMI incident**
>
> A major fire at the world's largest delayed coker unit killed two operators and a contractor. It was caused by failure of a carbon steel 45° elbow in a feed circuit specified to be 5% chrome steel alloy. Thickness tests at statistically selected inspection points, as well as PMI spot checks, did not include the failed elbow. An updated inspection program found other incorrect metallurgy (less than 1%).

Other quality control checks during fabrication may include such items as QC of welds and hydraulic pressure tests. These may be performed at the fabricating shop, at the construction site before installation, or both. The purchasing company or a third party may provide some sort of oversight of these QC checks as a QA activity.

Another practice that may start at the fabrication step is to document baseline information on newly fabricated assets such as pressure vessels. For example, thickness measurements of as-built equipment can provide baseline data for subsequent in-service inspections.

Final verification tests for operability and functionality, to demonstrate that more complex equipment items meet design specifications, may take the form of a factory acceptance test (FAT) at the supplier location before delivery. The FAT is generally witnessed by a representative of the purchasing company. Examples of equipment that may undergo factory acceptance testing might include such items as a wet gas compressor for a catalytic cracker, or a safety instrumented system including a verification of both the hardware and software functionality of the system.

10.4   RECEIVING

To maintain thorough QA for receiving, a facility needs to recognize all of its receipt pathways. If all parts and assets are received at a central storeroom or laydown yard, QA efforts would be focused there. However, some facilities may have multiple departments bringing in materials. In addition, the methods by which contractors receive materials vary considerably between facilities (see Section 10.12). If multiple pathways are involved, QA efforts at later stages such as during asset installation may need to be intensified, since no central source exists to verify proper receipt QA.

QA for receiving generally involves some kind of receipt inspection to verify that the parts and equipment received are the same as those that were designed, specified and purchased. In addition, receipt inspection may be used to inspect parts for damage or nonconformities. The type of inspection and the personnel involved can vary with the type of assets received, as well as with the facility culture and resources. In its simplest form, receipt inspections involve checking the packing list against the applicable purchase order. At its most complex, receipt can be a time for formal inspection, including material testing and positive material identification (PMI), as discussed in Section 10.3. Practices often vary according to the importance of the assets received. In such cases, thorough receiving procedures and/or training are important to help ensure that receiving personnel know when a formal inspection is required. Personnel performing receipt inspections, such as of spare parts, need to receive appropriate training (e.g., to conduct visual inspections).

> **Receiving incident**
>
> A vendor made a change to a gasket on a gaseous chlorine strainer. The gasket was now designed for a new strainer model. The vendor updated its specification; however, a user of the older model strainer was not notified of the change by the vendor. The difference in gaskets was not caught during receiving. Installers noticed that the new strainer gasket was different. It subsequently failed a leak test.

Many companies track nonconformities found during receipt QA and share this information with other company locations. This enables a company to more quickly identify quality issues such as poor vendor performance.

10.5   STORAGE AND RETRIEVAL

Many of the QA considerations for storage are equipment-specific, such as providing areas with humidity and/or static control for electrical components, storing pressure safety valves upright, and turning motors periodically. General considerations, such as binning, labeling, and inventory control measures (e.g., first-in, first-out; cycle count; reorder procedures) help to ensure that proper parts are available, are not confused with others, and do not spend too much of their useful life in storage, since some spare parts such as gas monitoring sensors and elastomers have a limited shelf life. In some cases, facilities supplement these measures by creating segregated storage areas, such as for exotic metals. All of these quality considerations are relatively simple to manage with procedures and/or training for storeroom and material handling personnel. Small facilities that do not have a full-time stockroom person will need to have very good procedures and practices in place for retrieval. Some additional equipment storage issues include:

- How a storeroom is set up is important, including labeling, segregation and the interface with receiving.

- Spare parts in storage can require regular inspection and testing to verify the spares are functional and can be put into service when needed. This includes maintaining the proper storage environment.

- Small parts require careful attention, since many small parts may look similar.

- Electrical and control equipment, including control valves, need to be stored and handled carefully all the way through to installation.

- Equipment not stored in the warehouse may require extra preservation measures. For example, all pressure penetrations need to be sealed up, and a nitrogen atmosphere and/or a positive pressure inside the vessel may be required until installation.

- Specific equipment such as molecular sieves will have its own particular storage requirements.

Retrieval. Developing QA steps for routine retrieval of materials from the storeroom during hours that the storeroom is attended is quite straightforward. Facilities can use part-numbering systems and/or drawings, along with work orders, to provide control and tracking of issued materials. Such retrieval systems can be set up through the computerized maintenance management system (CMMS); however, many facilities have successfully employed manual (i.e., non-computerized) systems for parts checkout.

Nonstandard storage and retrieval systems can result in a more difficult QA process. Informal storage of spare parts and materials, such as out in a process unit, can present opportunities for these materials to be misapplied. Further difficulties arise when one unit can "borrow" parts from another. Limits can be set on the amount and types of parts that can be stored locally.

Even with well-designed storage and retrieval systems, human error is always possible. Most companies have some type of requisition process that can incorporate quality control checks. Other measures to reduce the potential for human error include minimizing the number of materials of construction used and/or the types of seal or gasket material used; clearly identifying each individual item; and employing bar coding, RFID tags and/or a similar means to aid in computerized tracking of parts and materials.

Unused parts (e.g., bolts, gaskets, small valves) returned to storage can lead to misapplications when mistakes are made as materials are returned to storage. Administrative controls with associated training need to be established for parts return practices.

Few storerooms are staffed around the clock. Facilities that rely on the training of their storeroom personnel to ensure appropriate parts and materials retrieval may consider providing additional controls for the times that the storeroom is unattended. One common practice is to restrict storeroom access to selected individuals. In such cases, all selected individuals can receive appropriate training and have procedures in place to ensure that storeroom and retrieval QA is not compromised.

For some facilities, restricting parts retrieval for particular work order items is not practical. For example, an oil production field technician may have maintenance rounds that encompass many miles. Requiring trips to a central storeroom for each work order is unreasonable. In such cases, the technician's truck can be treated as another storeroom, with appropriate procedures, training and documentation of replacement parts required.

10.6 CONSTRUCTION AND INSTALLATION

Construction and installation are the last chance in the asset life cycle to detect and compensate for any QA vulnerabilities at earlier stages. Companies and facilities that do not correct vulnerabilities in the earlier stages of the life cycle need to intensify QA for construction and installation. Of course, errors made during installation can nullify a program full of good practices up to that point. Controls need to be in place to prevent and/or detect installation errors (such as mixing low-temperature valves with carbon steel valves or incorrectly aligning rotating equipment) before they lead to failures.

Construction and installation involve a significant amount of material handling, which needs to be performed with deliberate discipline to ensure all materials of construction are correct for a given service. This includes not only vessels and piping but also all other assets exposed to process materials and/or the environment including internal components or packing, bolts, flanges, gaskets, seals, instrumentation, flexible connections, sight glasses, rupture disks, welding consumables, etc. This deliberate, disciplined material handling may involve tagging and tracking of assets and construction materials from receipt on-site (Section 10.4) as well as storage and retrieval (Section 10.5) through to final installation. Techniques such as bar coding and RFID tagging may be able to aid the tracking and checking processes.

Construction and installation QA procedures need to mandate the use of qualified personnel. For many companies, this entails auditing contractor performance as well as providing training for company personnel. Installation specifications can also help provide guidance to these personnel. Most companies have contractor approval processes; however, such processes are more often developed to ensure liability coverage and/or workplace safety rather than to guarantee contractor performance. Facility management may consider adding AIM considerations to the contractor approval process. Appendix 10C contains a sample service contractor QA plan. QA activities help ensure that personnel performing construction and installation are qualified and that their finished work is acceptable. As during the asset fabrication stage, hold points are often used to have independent inspection opportunities at key times during construction.

Inspection and testing of the final installation are also very common. Particular tests may include hydrotesting of pressure vessels, instrument and interlock functional tests and water runs, as well as appropriate PMI checks similar to those done during fabrication or procurement (Section 10.3). Using procedures and/or checklists can contribute to the consistency and quality of these activities.

One of the final opportunities to identify QA issues is during a pre-startup safety review (PSSR) or operational

> **Operational readiness after maintenance incident**
>
> During a scheduled maintenance shutdown, two piping elbows that were identical except metallurgy were switched when reinstalled, exposing the carbon steel elbow to high-temperature hydrogen, resulting in catastrophic failure of the elbow. Released hydrogen immediately ignited, causing a huge fireball and US$ 30 million in property damage.

readiness review (ORR). Facilities may consider requiring a QA review as part of ORR activities. During such a review, the installed assets can be compared to the design documentation and any project-specified installation requirements can be verified. This will help avoid incidents such as the one described in the sidebar above. Typical ORR QA activities are to:

1. Document discrepancies between design and installation

2. Evaluate whether these discrepancies are tolerable (this evaluation can be similar to a change review process)

3. Make necessary corrections prior to asset startup

4. Document closure of any identified items.

Some additional considerations at this stage include the following:

- A large capital project or significant facility change will generally have a precommissioning stage. The ORR is usually conducted after pre-commissioning.

- Many QC activities such as loop checks and blowing out of lines are performed during both precommissioning and pre-startup.

- Some QM tasks such as radiography and pressure testing may affect adjacent construction activities. Simultaneous Operations (SIMOPS) reviews can help identify risk issues associated with adjacent operations.

- Late field changes during construction, such as piping system rerouting, need to be identified and receive a proper review, including potential impacts on quality management.

- Operational readiness reviews are not limited to initial startup. ORRs can be performed any time a unit is shut down, changes are made to the process, or maintenance is performed on a unit.

10.7 IN-SERVICE REPAIRS, ALTERATIONS AND RERATING

Many in-service repairs occur in response to asset deficiencies. General issues with asset deficiency resolution are discussed in Chapter 11. Occasionally, facilities need to repair, alter or rerate pressure vessels, tanks and piping. These terms are differentiated in Table 10-3. Because of the potential catastrophic consequences of, and the technical issues involved with, this type of work, special QA requirements have been defined in applicable codes and standards. Table 10-4 lists some of the codes and standards applicable to repair, alteration and rerating.

In general, these codes and standards provide requirements and guidance on the following issues:

- *Authorization* – The personnel who authorize the work before it is performed

- *Approval* – The personnel who need to approve the work once it has been performed

TABLE 10-3. Repair, Alteration and Rerating

Term	Definition
Repair	Any work necessary to restore an asset to a suitable state in accordance with the design conditions
Alteration	Any physical change in an asset that has design implications, such as those changes affecting pressure-containing capabilities
Rerating	A change in the design temperature and/or the maximum allowable working pressure of an asset

TABLE 10-4. Sample of Codes and Standards Having QA Requirements Applicable to Repair, Alteration and Rerating

Application	Applicable Code or Standard
Atmospheric Storage Tanks (API 650 tanks)	API 653, Tank Inspection, Repair, Alteration, and Reconstruction, Section 7
Electrical Systems	NFPA 70, National Electric Code
Instrumentation	IEC 61508, IEC 61511 for Safety Instrumented Systems
Low Pressure Tanks (API 620 tanks)	API 653, Tank Inspection, Repair, Alteration, and Reconstruction, Section 7
Piping	API 570, Piping Inspection Code: Inspection, Repair, Alteration, and Rerating of In-service Piping, Section 8
	ASME PCC-2, Repair of Pressure Equipment and Piping
Pressure Vessels	API 510, Section 7 or National Board (NB)-23, part RC
	ASME Boiler and Pressure Vessel Code, Section VIII-1
	ASME PCC-2, Repair of Pressure Equipment and Piping

- *Workmanship* – Details on specific aspects related to how the work is performed (e.g., welding techniques, heat treatment requirements, materials) and worker qualifications

- *Inspection and Testing* – Inspections and tests required during and after the work

- *Documentation* – All of the documents that are required.

Other codes and standards apply to specific situations. For example, ASTM A780 is a standard practice for repair of damaged hot-dip galvanized coatings. Numerous codes and standards apply to various types of welding, which repairs and alterations often require.

While having personnel who are experts on these issues is not imperative for a facility, it is important that organizations be aware of the issues and have access to a knowledgeable vessel/piping contractor who can assist with adhering to these requirements. Following these QA requirements will help to ensure the integrity of the assets and avoid any legal or regulatory problems.

Construction and service history files can be helpful for personnel who are troubleshooting performance problems. In addition, records can be critical to safety for weld repairs. For example, post-weld heat treatment (PWHT) during original construction will generally dictate PWHT requirements for weld repairs. Catastrophic incidents during startup of vessels have occurred as a result of improper heat treatment. Some plants have developed work order review protocols intended to ensure that qualified personnel develop repair plans for critical repairs such as those involving welding.

10.8 TEMPORARY INSTALLATIONS AND TEMPORARY REPAIRS

Specific problems may be presented by temporary installations and temporary repairs. An installation or repair may be designated "temporary" to avoid being subject to the requirements of a permanent repair or installation. Sometimes, this includes bypassing QA checks required of permanent assets. To help ensure that these situations do not lead to catastrophic consequences, facilities need to consider implementing a policy for regulating temporary installations and repairs. Such a policy may be integrated into a facility's management of change (MOC) procedure to help ensure that QA issues are adequately covered.

To manage temporary repairs and installations, facilities can take measures to ensure that these situations are identified, that any exceptions to specifications are noted, and the situation is reviewed to determine whether:

- Modifications to operating limits are necessary for the duration of the installation

- Procedures need to be updated

- Affected personnel need to be informed of the changes.

Depending on the type of temporary installation or repair, the asset may need to be inspected prior to startup, restart or continued

> **Temporary installation incident**
>
> At a facility in Flixborough, UK, failure of a temporary bypass line between two reactors led to a massive cyclohexane release and vapor cloud explosion resulting in 28 fatalities and loss of the entire facility. Because of the rush to resume production, the new bypass was not tested prior to startup, and engineering standards and manufacturer's recommendations were not considered.

use, with additional inspections considered and scheduled as necessary. Temporary installations and repairs need to be assigned an expiration date and procedures developed to help ensure that the installation or repair is removed, upgraded or reviewed again prior to that date and that QA checks are performed after putting the installation back to its original configuration. Two examples of QA considerations for temporary installations and repairs are given below to illustrate some of the issues involved.

Example 1 — Temporary Building. A blast-resistant portable building needs to be temporarily leased, brought on-site and located adjacent to a process area for reasons that exclude other possible alternatives. Although the temporary building is not process equipment, it is nevertheless an asset needing to meet specific design criteria for the location, even on a temporary basis. The design of the building will need to match facility siting requirements such as blast loading, ventilation and emergency egress for the specific location (References 10-5 to 10-8), as well as receive proper authorization. Procurement will need to ensure the temporary building that is selected will meet design requirements as well as be suitable for its purpose. Receiving will need to verify that the portable building provided by the supplier and shipped to the site matches design specifications. (This will likely only involve a paperwork check, and certainly not a full functional check such as verification of its blast-resistance capabilities.) Installation checks may involve an operational readiness review that provides a final verification that the asset meets all design specifications, it is placed in the proper location, and it has the proper utilities, signage, etc. It may also verify that any ITPM tasks required to be performed while the asset is in service are scheduled, that all procedural updates and training have been conducted such as emergency response plan updates, and that the temporary building has an authorized expiration date posted by which time it is to be removed from the site.

Example 2 — Pipe Clamps. Various types of clamps are employed at some facilities to stop minor leaks such as from cooling water piping systems and allow continued process operation, with the expectation that permanent repairs will be made at the next shutdown or turnaround. Issues related to the use of pipe clamps include treating them as temporary repairs that do not require (a) management of change or other evaluation of their potential impact or (b) tracking of locations or how long the clamps are in place. As a result, they often proliferate and may increase the likelihood of a more significant piping system failure. Considering pipe clamps from a QA perspective can lead to a more disciplined approach to the use of pipe clamps including having authorization, safety review, design specification, procurement, storage and retrieval, installation check and periodic inspection requirements, as well as tracking of locations and ensuring timely clamp removal and piping system repairs. This will require management emphasis and significant operating discipline, especially if a facility has in the past treated temporary repairs such as pipe clamps in a more casual manner.

10.9   DECOMMISSIONING / RE-USE

QA for decommissioning may be an AIM concern if re-use of assets is intended or possible. It may also be an AIM concern if decommissioning a large facility or one that would otherwise take an extended amount of time to decommission. (See Section 3.7 for life cycle considerations during decommissioning.)

Any asset that is not removed and disposed of upon decommissioning is subject to re-use. "Mothballed" units (taken out of service with the possibility of future use) and "boneyards" (storage areas for unused equipment) present opportunities for saving money, but they also present significant QA challenges.

Facilities that recognize these QA challenges may consider establishing decommissioning   and recommissioning procedures.   A decommissioning procedure may consider depressurization and cleaning of assets, additional measures for asset preservation and any ongoing inspections and/or preventive maintenance (PM) that needs to be performed. In addition, design and inspection documentation needs to be retained, and assets on their way to the boneyard need to be labeled or tagged. Similarly, units that are mothballed for re-use at a later date (e.g., seasonally operated assets) can have procedures to ensure that liquids are drained, systems are purged, and other measures are taken to help preserve asset life, such as maintaining a proper atmosphere to prevent corrosion.

> **Decommissioning incident**
>
> In a Texas refinery, propane leaked from cracked control station piping that had been out of service for 15 years. A huge fire resulted, injuring four people and shutting down the refinery for two months. The unused piping was only isolated by a closed valve that leaked through when a low point cracked from ice formation.

Some facilities provide QA for re-using assets through recommissioning procedures. Recommissioning procedures may include a change-of-service approval process. Variables to consider in such an approval process include (1) the length of time the asset was out of service and (2) the extent to which ongoing inspections and/or PM were performed. Generally, recommissioning involves inspections and other asset checks to verify that the used asset is suitable for the new service. This may require a fitness-for-service evaluation. If appropriate, pressure vessels may be rerated according to the practices listed in Section 10.7.

Quality Assurance While Decommissioning Larger or More Complex Facilities. Decommissioning often requires existing systems to be re-engineered or new systems installed to allow old redundant systems to be safely engineered down and decommissioned. This is particularly relevant for safety-related systems.

To illustrate this point, the decommissioning of a fully manned offshore oil platform will take an extended amount of time, with some of the relevant considerations being as follows:

- Existing asset power generation needs to be engineered down to reflect changes in fuel sources and platform-wide power requirements. New temporary generators may need to be installed to simplify requirements for future global isolations to facilitate module separation.

- Traditionally, assets may have relied on fuel gas for power supply. After production has been stopped, fuel gas will no longer be available, requiring a switch to a diesel supply, with a need to consider storage capacity during winter months, integrity of pipework and reliability of pumps.

- Alternative pumps may need to be installed or skid-mounted units brought online during decommissioning for firewater supplies.

- Decommissioning activities will increase demand for platform utilities such as service water and HVAC. Any reliability issues may impact the decommissioning schedule.

- Demand will likewise be increased on platform cranes; capacity, reliability and possible replacement may need to be considered.

During decommissioning, managing asset integrity of new systems such as these need to ensure a fit-for-purpose design is adopted and installed that reflects the remaining limited operating lifetime of each asset. The decommissioning phase of a large or complex facility will benefit from having a decommissioning strategy, setting out clear guidelines that ensure a fit-for-purpose design is adopted, including robust controls for managing change and managing deviations from existing company standards. The latter is particularly pertinent if decommissioning is not a core business, as most company design and operating standards are based on a typical asset life cycle of 25+ years, and therefore may not be cost-effective or appropriate for decommissioning without modification.

10.10 USED ASSETS

Unforeseen problems can arise when purchasing used equipment. The purchasing facility may be provided design documentation, some data about previous service and/or previous repairs. Often, however, none of this information is available. Facilities that purchase used assets may consider developing specific procedures for doing so. These procedures may include many of the same considerations listed in Section 10.9 for asset re-use, such as recommissioning procedures and fitness-for-service evaluations. In addition, procedures for obtaining used assets may identify methods for securing and maintaining equipment file information.

10.11 SPARE PARTS

Much of the information regarding QA of procurement, fabrication, receiving, storage and retrieval systems (Sections 10.2 through 10.5) is directly applicable to managing spare parts, including materials handling aspects discussed earlier. Additional considerations for spare parts management include:

- Identifying spare parts to stock for new units, installations or equipment

- Developing and implementing procedures and providing training for purchasing, receiving, labeling, storing and retrieving replacement parts

- Developing and implementing procedures and providing training for approving substitute parts in place of original equipment manufacturer (OEM) parts

- Auditing vendor-managed inventories.

Procedures for storing spare parts may need to take into account equipment-specific requirements such as maximum shelf life or packaging to minimize corrosion or other environmental effects over time. Human factors may also need to be addressed, such as providing clear markings so similar assets such as rupture disks with different burst pressures or materials of construction are not confused. Additional considerations include the possibility of having supplier-managed inventories on-site (these may need to be periodically checked or audited due to a potentially greater risk of parts or materials of construction getting changed) and managing costs by just-in-time delivery of spare parts (ensure a full QM process is followed).

10.12 CONTRACTOR-SUPPLIED ASSETS AND MATERIALS

For facilities that allow contractors to supply their own equipment and/or materials, QA of those equipment/materials may either be covered by the facility's QA procedures or delegated to a contractor QA program. In either instance, facilities need to ensure that all contractors are aware of the QA requirements. To the extent that contractors are providing QA services, facilities may consider auditing the contractors' practices. In addition, facilities need to ensure that (1) all contractor-supplied equipment and materials are identified as soon as practical in the life cycle, (2) QA activities are initiated and (3) equipment documentation is collected. If a contractor is doing certain QA tasks on behalf of a company, such as a piping installer also performing weld inspections, the operating company needs some form of QA oversight of the contractor QA activities ("QA of the QA"), which may be by a third party.

10.13 QA PROGRAM ROLES AND RESPONSIBILITIES

The roles and responsibilities for the QA program are generally assigned to Engineering, Inspection and/or Maintenance departments. Example roles and responsibilities for the quality management program are provided in Table 10-5 for an existing facility. The matrix uses the following letter designators:

R   indicates the job position(s) with primary responsibility for the activity

A   indicates approver of the work or decisions made by a responsible party

S   indicates the job position(s) that typically support the responsible party in completing the activity

I   indicates the job position(s) that are informed when the activity is completed or delayed.

CHAPTER 10 REFERENCES

10-1   U.S. Chemical Safety and Hazard Investigation Board, *Investigation Report: Chlorine Release*, Report No. 2002-04-I-MO, Washington, DC, May 2003.

10-2   Atherton, J. and F. Gil, *Incidents that Define Process Safety*, American Institute of Chemical Engineers, New York, NY, 2008.

10-3   Center for Chemical Process Safety, *Guidelines for Hazard Evaluation Procedures, Third Edition*, American Institute of Chemical Engineers, New York, NY, 2008.

10-4   API RP 578, *Material Verification Program for New and Existing Alloy Piping Systems*, American Petroleum Institute, Washington, DC.

10-5   API RP 752, *Management of Hazards Associated with Location of Process Plant Buildings*, American Petroleum Institute, Washington, DC.

10-6   API RP 753, *Management of Hazards Associated with Location of Process Plant Portable Buildings*, American Petroleum Institute, Washington, DC.

10-7   Center for Chemical Process Safety, *Guidelines for Evaluating Process Plant Buildings for External Explosions, Fires, and Toxic Releases, Second Edition*, American Institute of Chemical Engineers, New York, NY, 2012.

10-8   Center for Chemical Process Safety, *Guidelines for Facility Siting and Layout*, American Institute of Chemical Engineers, New York, NY, 2003. (*Second Edition*: 2017.)

TABLE 10-5. Example Roles and Responsibilities Matrix for the QA Program (Existing Facility)

R = Primary Responsibility
A = Approval
S = Support
I = Informed

Activity	Maintenance and Engineering Department									Other Personnel					
	Maintenance Manager	Engineering Manager	Maintenance Supervisors	Maintenance Engineers	Project Managers/Engineers	Maintenance Technicians	Maintenance Planner/Scheduler	Production/Process Engineers	Inspection Manager	Safety Manager/Process Safety Coordinator	Plant Manager	Inspectors	Storeroom Personnel	Purchasing Personnel	Contractor/Construction Supervisor
QM improvement plan development	A	A	S	R	R	I	S	S	A	I	I	I	I	I	I
QM plan development	A	A	S	R	R	I	S	S	A	I	I	I	I	I	I
Design QC/QA, including specification development	I	R	I	I	S			S	I	I	I			R	
Purchasing QC/QA	I	A		S	S			S	I			S	I	R	
Fabrication QC/QA	I	A		I	S	I		S	R		I	I	I	I	I
Receiving QC/QA	I	I	I	A	A	I	I		A		I	I	R	I	I
Storage and retrieval QC/QA	R	I	R	A	A	I	I	A	A		I	I	R	I	I
Construction and installation QC/QA	S	S	I	S	S	I		A	A		I	I			
Asset rerating	R	R		I	S			I	A		I	I			
Asset repair QC/QA	R	A	I	S	I	I		A	A	I	I	I	I	I	I
Temporary repairs and installations	A	A	I	S	S	I	S	S	I	R	I	I	I	I	I
Contractor QC/QA	A	A	I	S	S	I		I	A	I	I	I	I		R
Decommissioning/recommissioning QC/QA	R	R	I	S	S	I		I	A	I	I	I	I	I	
Used asset QC/QA	S	R	I	S	S	I		I	A	I	I	I	I	S	S

Additional Chapter 10 Resources

API 510, *Pressure Vessel Inspection Code: Maintenance Inspection, Rating, Repair and Alteration*, American Petroleum Institute, Washington, DC.

API 570, *Piping Inspection Code: Inspection, Repair, Alteration, and Rerating of In-service Piping*, American Petroleum Institute, Washington, DC.

API 610/ISO 13709, *Centrifugal Pumps for Petroleum, Petrochemical and Natural Gas Industries*, American Petroleum Institute, Washington, DC.

API 620, *Design and Construction of Large, Welded, Low-pressure Storage Tanks*, American Petroleum Institute, Washington, DC.

API 650, *Welded Steel Tanks for Oil Storage*, American Petroleum Institute, Washington, DC.

API 653, *Tank Inspection, Repair, Alteration, and Reconstruction,* American Petroleum Institute, Washington, DC.

American Society of Mechanical Engineers, *International Boiler and Pressure Vessel Code*, New York, NY.

ASME B31.3, *Process Piping*, American Society of Mechanical Engineers, New York, NY.

ASME B73.1, *Specification for Horizontal End Suction Centrifugal Pumps for Chemical Process*, American Society of Mechanical Engineers, New York, NY.

ASME B73.2, *Specifications for Vertical In-line Centrifugal Pumps for Chemical Process*, American Society of Mechanical Engineers, New York, NY.

ASME PCC-2, *Repair of Pressure Equipment and Piping*, American Society of Mechanical Engineers, New York, NY.

ASTM E1476-97, *Standard Guide for Metals Identification, Grade Verification, and Sorting*, ASTM International, West Conshohocken, PA.

IEC 61511, *Functional Safety: Safety Instrumented Systems for the Process Industry Sector - Part 1: Framework, Definitions, System, Hardware and Software Requirements*, International Electrotechnical Commission, Geneva, Switzerland.

National Board Inspection Code, National Board of Boiler and Pressure Vessel Inspectors, Columbus, OH.

NFPA 70, *National Electrical Code*, National Fire Protection Association, Quincy, MA.

Pipe Fabrication Institute, *Standard for Positive Material Identification of Piping Components Using Portable X-Ray Emission Type Equipment*, New York, NY, 2005.

UL 142, *Steel Aboveground Tanks for Flammable and Combustible Liquids*, Underwriters Laboratories Inc., Northbrook, IL.

APPENDIX 10A. SAMPLE VENDOR QA PLAN

VENDOR QUALITY ASSURANCE PLAN

For major AIM program asset items such as a new pressure vessel, our facility may require that a vendor develop and implement a QA plan to ensure the quality of the asset item before use. The following vendor QA plan can be used as an audit tool for companies approving new vendors. A vendor QA plan may have, but is not limited to, the following features.

Statement of Design and Fabrication Specifications. For assets that our personnel designed, this statement provides verification that the vendor adequately understands our specifications. For assets designed by the vendor, this statement ensures that the vendor has formally established the specifications and allows for our input into the specifications.

Definition and Scheduling of QA Activities. This section defines the specific tasks that will promote and verify the quality of the asset. These tasks may include, as appropriate:

- Material quality verifications
- Worker training and qualification verifications
- Procedure quality verifications
- Fabrication specification verifications
- Manufacturing/fabrication process quality control
- Nondestructive testing during manufacturing/fabrication/field installation (dimensional checks; visual inspections; hardness, ultrasonic, radio-graphic, performance, dye penetrant, pressure, magnetic particle tests; etc.)

Definition of Roles and Responsibilities for QA Activities. This section establishes the roles and responsibilities of each organization (i.e., the vendor, our company, subcontractors to the vendor and independent contractors) in accomplishing each of the required activities. The roles and responsibilities may include planning, performance, witnessing, evaluation, and/or documentation activities.

List of Required Documentation for QA Activities.   This section lists the documentation that our company needs to receive to have confidence that the vendor has appropriately implemented the QA plan and that the quality of the asset is acceptable. This documentation may include, as appropriate:

- Manufacturer data sheets (e.g., ASME pressure vessel forms)
- Physical data sheets
- Mill test reports
- Material verification reports
- Weld maps
- Pressure test sheets
- Reports for noted deficiencies and their resolutions
- Stress relief charts
- Hardness readings
- Charpy impact results
- Calculations
- NDT interpretation results
- As-built drawings
- Nameplate facsimiles
- Bill of materials.

The vendor is responsible for developing the QA plan (when one is required by our company) and submitting the completed plan to us for comments. Our personnel responsible for purchasing the item will review the plan and provide comments to the vendor. The responsible company personnel will also ensure that the requirements of the plan are complete before our company places the asset into service. This person also resolves any quality deficiencies with the vendor before placing the asset into service and ensures that our company receives and files all required documentation.

APPENDIX 10B. POSITIVE MATERIAL IDENTIFICATION

The consequences can be devastating when incorrect assumptions are made about materials of construction or when materials are inadvertently substituted. Serious incidents have occurred as a result of the substitution of incorrect piping materials. Many of these incidents were attributed to substituting carbon steel for chrome alloys. In one case, the catastrophic failure occurred shortly after the substitution. In another case, the catastrophic failure occurred 20 years later.

Diligent control over the materials of construction used in hazardous processes, including welding consumables, can help prevent incidents and yield economic benefits as well. Such controls are commonly referred to as positive material identification, or PMI.

A comprehensive PMI effort can be integrated into a facility's hazard management system through a variety of practices. Material control begins with the material selection phase of the design process. The material selection process involves the economic consideration of the expected performance of materials in the anticipated exposure environment. PMI covers control of the material condition (i.e., mechanical, physical, and corrosion-resistant properties for the application) and verification of the material composition (i.e., the elemental constitution).

Material Condition. Heat treatment, mechanical working, surface treatment or a combination of these actions can adversely affect the condition of a material. The condition of the material may create sensitivity to degradation with small changes in environmental conditions, thus creating equipment hazards.

Improper material condition is a common cause of a material-related asset failure because of the large number of end-state conditions resulting from combinations of heat treating, forming, and fabrication methods. Material condition may create equipment hazards by increasing the sensitivity of the material to degradation at environmental conditions different from what was anticipated in the selection process. Examples of controls for material condition include (1) controlling grain size for high-temperature applications and for fatigue applications, (2) solution annealing for cold-formed austenitic materials to improve stress corrosion cracking resistance, and (3) tempering bolts for hydrogen sulfide services.

Material condition is generally not determined through analysis for composition and physical appearance. Equipment hazards caused by material condition are managed using QA testing during design, material specification and manufacturing. Facilities sometimes have limited ability to measure material condition because the tests often require sampling for destructive testing or specialized technical analysis, such as hardness testing or metallographic examination. Generally, control of material condition is dependent on strong engineering and purchasing procedures to ensure that the correct material condition is specified, ordered and marked and that the appropriate mill test or asset fabricator documentation is provided with receipt of the materials.

Material Composition. Quantification of the material composition or identification of a specific alloy can be accomplished using several methods available to most facilities. Typical composition identification methods include:

- *Portable X-ray spectroscopy*, which determines composition of elements found in most alloys and can provide estimates of specific alloy designation.

- *Portable optical emissions spectroscopy*, which determines composition of elements found in most alloys and can provide estimates of specific alloy designation. When used in an argon atmosphere, carbon content can be determined in both ferrous and austenitic materials.

- Access to commercial or company testing laboratory to perform compositional analysis.

Sorting or classifying materials within metallurgical types can be accomplished using a variety of methods available to most facilities. Typical classification methods include:

- Classification or sorting using basic principles
  - o Color – copper alloys versus white metals
  - o Density – aluminum versus magnesium, titanium versus steel
  - o Magnetic response – strong, weak, or no magnetic response to identify ferrous metals versus austenitic or nickel-based materials

- Chemical etches can determine some metals present and some types of surface coatings; care is required when handling reagents

- Resistivity testing to determine broad types of materials, but not alloy type

- Grinding wheel spark test – an experienced operator can differentiate ferrous materials and determine if the metal can be welded based on carbon content.

These methods may be specified as part of the routine QA procedures or performed on an as-needed basis, such as for poorly labeled filler metal composition or for service materials without readily discernable markings.

Recognized and generally accepted good engineering practice (RAGAGEP) guidance for establishing formal material verification procedures as part of the facility's overall PMI practices is found in:

- API RP 578 – *Material Verification Program for New and Existing Alloy Piping*

- ASTM E1476-97 – *Standard Guide for Metals Identification, Grade Verification, and Sorting Systems.*

These documents provide guidance for establishing the basic elements of a material verification work process that includes (1) components to test and level of examination, (2) test methods, (3) acceptance criteria, (4) material marking, (5) test documentation and (6) resolution of material nonconformances.

In addition, these documents suggest good practices to justify variances, including when specific material verification of all incoming material is generally not required. This may be acceptable if mill test report documentation is provided at material receipt in conjunction with physical verification of material markings for a representative sample. The documents suggest that the need for stringent or statistically founded composition verification increases for both new and existing materials when:

- Verification of a specific alloy grade is needed, or the specificity of the material composition (such as for trace elements) is essential for performance

- The cost of the material is high

- An incident or near miss lists "incorrect material" as a causal factor

- The facility has had unfavorable experience (or lack of experience) with a supplier, or the material is used infrequently at the facility

- The consequences of failure are high

- Historical fabrication and installation practices or facility documentation suggest poor material controls and/or poor material traceability.

PMI Practices. PMI practices, in addition to composition verification, are performed to help ensure that correct materials of construction, including welding consumables, are specified and used. PMI practices are applicable during any AIM activity and are intended to complement the materials verification procedures. PMI practices include procedural controls and employee awareness activities that are usually developed and maintained by technical personnel within the facility. PMI helps to ensure the correct selection, purchase, receipt and installation of process equipment with the appropriate materials of construction. Typical PMI procedures and practices for a facility may include:

- Equipment Design Standards
  - Materials selection and fluid compatibility guidelines

- Project Engineering Procedures
  - Shop inspection
  - Component inspection and material verification

- Process Safety Management
  - Process knowledge management

- Hazard identification and risk analysis
- Management of change
- Operational readiness
- QA procedures

- Maintenance Work Controls
  - Maintenance job planning to ensure replacement in kind or MOC approval for a change

- Engineering Standards or Procedures for Pressure Equipment
  - QA manual (e.g., ASME fabrication or repair)
  - Company engineering standards

- Employee Awareness
  - Knowledge of standard material markings (color coding and ASTM designations)
  - Knowledge of fabrication codes and standards
  - Basic materials incompatibility knowledge (e.g., stainless steel with chloride solutions, brass alloys with ammonia, aluminum alloys with high pH, carbon steel in strong acid/water solutions)
  - "Something is not right" awareness – appearance, nonstandard markings, weldability or dimensional anomalies

- Weld Procedures and Weld Qualification Procedures
  - Technical approval of the weld procedure for each job
  - Control of filler metal identification and storage conditions

- Purchasing Controls
  - Order placement requires using recognized and specific material designations, such as the Unified Numbering System
  - Requirements for inclusion of appropriate documentation with receipt of materials
  - Technical approval requirements for ordering alloy or specialty materials

- Warehouse Controls
  - Receiving
    > Confirmation of material markings with purchase order
    > Confirmation of receipt of documentation, such as mill test reports
  - Storage
    > Segregation by alloy
    > Color coding or other markings
    > Control of environmental conditions that could lead to degradation (chloride or sulfur compounds in the environment, standing water, etc.)
  - Issuing
    > Issue matched to correct work order or project documentation
    > Return-to-stock procedures for unused materials.

APPENDIX 10C. SAMPLE SERVICE CONTRACTOR QA PLAN

Service Contractor QA Plans. For extensive maintenance, inspection, testing, and construction activities by service contractors such as a significant process modification during a turnaround, within areas of the plant covered by the AIM program, our company may require a contractor to develop and implement a QA plan to ensure the quality of their work. The following contractor QA plan can be used as an audit tool for companies approving new contractors. A contractor QA plan may have, but is not limited to, the following features.

Statement of the Scope of Work. For work planned by our company, this statement provides verification that the contractor adequately understands our objectives. For work planned by the contractor, this statement ensures that the contractor has formally established plans for their work and allows for our company input into these plans.

Definition and Scheduling of QA Activities. This section defines the specific tasks that will promote and verify the quality of the work. These tasks may include, as appropriate:

- Material quality verifications

- Worker training and qualification verifications

- Procedure quality verifications

- Fabrication specification verifications

- Fabrication process quality control

- Nondestructive testing during fabrication / field installation (dimensional checks; visual inspections; hardness, ultrasonic, radiographic, performance, dye penetrant, pressure and magnetic particle tests; etc.)

- Oversight of responsibilities by qualified supervisors.

Definition of Roles and Responsibilities for QA Activities. This section establishes the roles of each organization (i.e., the contractor, our company, subcontractors to the prime contractor and independent contractors) in accomplishing each of the required activities. The roles may include planning, performance, witnessing, evaluation and/or documentation activities.

List of Required Documentation for QA Activities. This section lists the documentation that our company needs to receive to have confidence that the contractor has appropriately implemented the QA plan and that the quality of the work is acceptable. This documentation may include, as appropriate:

- Documentation of QA activities for specific assets
- Work completion checklists, including completion of
  - Mechanical (e.g., all components and protective coatings/linings installed)
  - Electrical (e.g., all circuits and grounding completed)
  - Instrumentation (e.g., control loops calibrated and interlocks tested)
- Material verification reports
- System integrity and performance test reports
- Reports for noted deficiencies and their resolutions
- Stress relief charts
- Calculations
- NDT interpretation results
- As-built drawings.

The contractor, with input from all subcontractors, is responsible for developing the QA plan (when one is required by our company) and submitting the completed plan to us for comments. Our personnel responsible for hiring the contractor will review the plan and provide comments to the contractor. Our responsible personnel will also ensure that the requirements of the plan are complete before our company places newly installed/modified assets into service. This person also resolves any quality deficiencies with the contractor before placing the new or modified assets into service and ensures that our company receives/files all required documentation.

11
EQUIPMENT DEFICIENCY MANAGEMENT

Successful asset integrity management (AIM) programs include effective plans for recognizing and correcting equipment deficiencies. A deficiency is identified through the evaluation of asset condition based on AIM activity results or by the observation of substandard asset performance or condition during the course of normal operations. A *deficiency* occurs when an observed condition is outside the established limits (*acceptance criteria*) that define asset integrity.

A deficient asset condition may be discovered (1) during acceptance testing or inspection for new asset fabrication or installation; (2) in the course of performing inspection, testing and preventive maintenance (ITPM) activities; or (3) while measurements are taken when the asset is accessible during a repair. In addition, operating personnel can observe deficiencies that first appear as operating difficulties; this is often the start of the traditional work order process. Without routine assessments and subsequent evaluations of asset condition, deficiencies could remain unrecognized or ignored. Each AIM activity needs to include work processes and procedures for making observations, performing the appropriate evaluations, and documenting observations and evaluations. Likewise, operations

> **Equipment deficiency mis-management**
>
> Operators were transferring methanol from a portable chemical transporter tank on an offshore oil platform.
> When the tank was lifted by a crane to gravity feed a storage tank, methanol began spraying out from a hole in the transfer hose. The methanol ignited and the fire spread, one man receiving second-degree burns.
> After fires were extinguished, it was discovered that the transfer hose was split and had been repaired with duct tape prior to the operation.
>
> (July 2007 CCPS Beacon)

personnel need to have a procedure for documenting and reporting suspected asset deficiencies. Assessments and subsequent evaluations form the basis for determining whether the process can temporarily continue functioning while the deficient condition exists.

11.1 EQUIPMENT DEFICIENCY MANAGEMENT PROCESS

To effectively manage deficient asset conditions, facilities need to implement a process to ensure that the following actions occur:

- Acceptance criteria are established that define proper asset performance/ conditions

- Equipment condition is routinely evaluated

- Deficient conditions are identified

- Proper responses to deficient conditions are developed and implemented

- Equipment deficiencies are communicated to affected personnel

- Deficient conditions are appropriately resolved

- The AIM program is updated based on learnings from the deficiency and its correction.

The following sections discuss each of these actions, except for ensuring asset conditions are routinely evaluated. This topic has been discussed in detail in Chapters 6 and 10. Chapter 6 discussed routine evaluation of asset condition through ITPM activities. Chapter 10 discussed establishing quality assurance (QA) activities to evaluate asset conditions throughout the life of the asset. In addition to these actions, organizations can consider using failure analysis and/or root cause analysis (RCA) techniques to identify underlying causes of equipment deficiencies. Chapter 15 contains more information on failure analysis and RCA.

11.2 ACCEPTANCE CRITERIA

Acceptance criteria are established to provide (1) standards for new equipment fabrication, installation or repair activities; (2) confidence in asset integrity while the deficient condition is being resolved (i.e., during the time required to implement permanent corrective actions), accounting for any uncertainty in the assessment process; and/or (3) confirmation of function. Acceptance criteria are specific to each type of equipment and each observation method. An equipment condition that does not satisfy an acceptance criterion may not signal an immediate threat for catastrophic consequences; rather, the asset condition can indicate that attention from the integrity management system is needed.

Figure 11-1 illustrates the concept for selecting acceptance criteria for asset conditions observed of in-service equipment. Acceptance criteria values need to be conservative enough to provide time for further condition assessment and to accommodate uncertainty in the absolute condition of the asset, since the observation process may not reveal the worst condition of the asset.

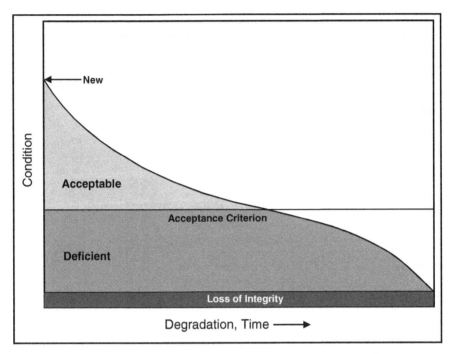

Figure 11-1. Technical evaluation condition selection

Asset acceptance criteria are based on recognized and generally accepted good engineering practice and/or on engineering calculations used as the design basis for assets. Such engineering calculations can be documented in the facility's process safety information. In addition, subject matter experts (SMEs) such as inspectors, mechanics and equipment vendors can be consulted and information from several resources can be used during the development of acceptance criteria. Table 11-1 provides examples of resources useful for determining acceptance criteria. Chapter 12 contains lists of codes and standards applicable to many common types of process equipment.

When determining acceptance criteria, personnel need to consider a variety of potential asset deficiencies that can occur and develop corresponding acceptance criteria based on measurable asset conditions or attributes. Table 11-2 provides broad categories of asset conditions/attributes that are commonly used to define acceptance criteria. The criteria can be defined quantitatively or qualitatively. The information in this table is presented for major types of assets and three AIM activity areas: (1) new asset design/fabrication/installation, (2) inspection and testing, and (3) repair.

TABLE 11-1. Acceptance Criteria Resources

Vessels and tanks	• Inspection codes and standards • As-built drawings • Vessel/tank specifications
Piping	• Inspection codes and standards • Piping specifications
Instrumentation	• Manufacturer manuals • Instrument specifications • Professional society (e.g., ISA) documents
Rotating equipment	• Manufacturer manuals • As-built drawings • Equipment specifications
General	• Manufacturer manuals • Asset specifications

These acceptance criteria are used when evaluating an observed condition to determine if the condition represents a deficiency requiring resolution. Developing all acceptance criteria prior to placing assets into service may not be necessary, but limits need to be known at or soon after the time of observation (e.g., inspection) so that a detailed evaluation can be completed. One option is to have conservative screening-level acceptance criteria in place before placing assets into service. If a screening-level criterion is exceeded, then action can be taken to either correct the condition or generate an acceptance criterion with a more formal analysis that may be less conservative but have a greater degree of precision and confidence.

Each acceptance criterion needs to also explain the actions required when the asset condition is found to exceed that criterion. For example, when the wall thickness of pressure-containing equipment is determined to be too thin, the options often include rerating, repairing, or replacing the equipment and/or performing a fitness-for-service (FFS) analysis to set new acceptance criteria. Appendix 11A contains an overview of the recommended practice API RP 579 (Reference 11-1) that addresses fitness for service. This may also apply to other equipment such as instrumentation. If a sensor is out of service, some compensating provision may need to be taken until the sensor is replaced.

TABLE 11-2. Examples of Acceptance Criteria for Common Asset Types

Asset Type	New Equipment Fabrication and Installation	Inspection and Testing	Repair
Pressure vessels, tanks and piping	• Pressure rating for design and testing • Weld quality • Dimensional/ alignment tolerances • Materials of construction • Valve leakage rates • Installation criteria for supports, pipe, and hangers	• Thickness requirements for each pressure boundary component • Requirements for assessment of support and anchoring systems • Foundation settlement limits • Tolerance to linear indications • Tolerance to distortion • Leakage	• Weld quality • Materials of construction • Dimensional/ alignment tolerances • Pressure rating for design and testing • Leakage • Installation criteria for supports, pipe and hangers
Pressure relief valves (PRVs)	• Materials of construction • Design pressure and temperature • Relief capacity • Storage conditions • Installation criteria • Leakage	• Set testing pressure limits • Pre-disassembly test limits • Set pressure and blowdown tolerances • Visual inspection (pipe and PRV) for fouling/plugging • Leakage	• Materials of construction • Dimensional tolerances • Installation criteria • Leakage
Instrumentation	• Materials of construction • Installation criteria • Component calibration • Functional performance criteria	• Component calibration • Functional performance criteria	• Materials of construction • Installation criteria • Component calibration

TABLE 11-2. *Continued*

Equipment	New Equipment Fabrication and Installation	Inspection and Testing	Repair
Rotating equipment	• Materials of construction • Performance testing criteria • Pressure testing requirements • Storage conditions • Installation criteria • Leakage	• Vibration limits • Bearing or coolant temperatures • Trip speeds • Requirements for assessment of support and anchoring systems • Leakage	• Weld quality • Materials of construction • Dimensional/alignment tolerances • Pressure testing requirements • Installation criteria • Leakage
Fire protection	• Installation criteria • Pressure testing requirements • Performance testing criteria	• Performance testing criteria • Requirements for assessment of support and anchoring systems	• Installation criteria • Weld quality • Materials of construction • Dimensional/alignment tolerances
Electrical	• Conformance to applicable National Electric Code (NEC) requirements	• Breaker testing performance	• Conformance to applicable NEC requirements
Fired heaters	• Installation criteria for supports, pipe hangers, and burners • Weld quality • Materials of construction for components: tubes, supports, refractory and instrument taps	• Tube and instrument tap dimensions and condition • Pipe hanger spring settings • Refractory condition	• Installation criteria for supports, pipe hangers, and burners • Weld quality • Materials of construction for components, tubes, supports, refractory and instrument taps

11.3   EQUIPMENT DEFICIENCY IDENTIFICATION

Personnel may identify deficiencies during the performance of AIM activities, while an asset is in service, or out of service, or prior to the asset being placed into service. Facility response to deficient conditions is time-dependent. The urgency of the response depends on (1) whether the condition is identified while the asset is in service, (2) how close the condition is to loss of integrity (see Figure 11-1), and/or (3) the increase in risk to personnel or assets from the loss of function of a safety system. For assets not in service, regulatory requirements and good practice generally dictate that a deficiency be corrected before use.

To help ensure that deficiencies are identified, an AIM program needs to include procedures for evaluating ITPM results. For each activity, the acceptance criteria and associated actions (such as further evaluation, communication and resolution) can be included in a written procedure (see Section 9.4) to support the evaluation of the observation results (e.g., QA inspection, ITPM task results) and ensure that the asset deficiency is properly managed. Some issues regarding deficiency identification that are borderline or unclear may need to get elevated to a chief inspector or corporate engineer.

Sometimes, personnel with specific technical expertise may be required to determine (1) if an asset is deficient, (2) the appropriate precautions for continued operation, and (3) the ultimate resolution of the deficiency. Typically, specific expertise and appropriate methods are needed to evaluate more complex issues (e.g., localized corrosion, higher-than-expected corrosion rates, suspect welds) related to pressure-containing equipment such as pressure vessels, tanks and piping. In addition, specific expertise and appropriate techniques are needed to perform FFS evaluations.

11.4   RESPONDING TO EQUIPMENT DEFICIENCIES

Operational and safety risks often accompany the following actions: (1) placing an asset into service having a deficient condition, (2) continuing to operate an asset with a deficient condition, or sometimes (3) immediately shutting down equipment having a deficient condition. Therefore, facility management, ideally with the support of technical assurance personnel, needs to be responsible for these decisions. In addition, facilities need to have written procedures in place and the appropriate means for ensuring that all hazards associated with the selected course of action are addressed and mitigated. The written procedures for deficiency resolution may include the following:

- Methods of hazard and risk assessment

- Specifying what deficiencies are considered near misses that warrant incident investigation and/or root cause failure analysis

- Management and engineering approval authority for temporary mitigation and corrective actions

- Tracking and closure of temporary mitigation measures

- Communication of hazards to affected parties

- Training/informing of affected parties on temporary mitigation measures

- Asset-specific documentation.

The risks associated with deficiency resolution are best assessed from both engineering and managerial perspectives. The level of detail of the risk assessment may (1) be dictated in the written procedure by the complexity and estimated risks associated with the deficiency and (2) provide the basis for establishing the timing and response for both the temporary mitigation measures and the ultimate corrective actions for the deficient condition. The response to an asset deficiency occurs in stages, beginning with the definition of the problem (i.e., the identification and evaluation of a specific deficiency), through implementation of temporary mitigation measures, culminating with the final corrective action. The facility's management of change and operational readiness review (pre-startup safety review) systems can be used to document and provide proper authorizations along the stages and activities of the response path.

Responses for In-Service Assets. Typically, resolving a deficiency for an in-service asset involves determining whether or not the asset (or its larger system such as process unit or unloading operation) needs to be shut down. The deficiency may be such that the best decision is to shut down and repair or remove the asset from service. Some deficiencies, particularly of critical assets such as firewater pumps on an offshore oil platform, may require mandatory shutdown. If not already established by regulatory requirement or company policy, such mandatory-shutdown decisions are best made ahead of time; i.e., before assets are found to be deficient.

When deciding to keep deficient assets in service until they are replaced or repaired, facility personnel need to be able to demonstrate and document that continued operation is safe; i.e., that the risk of operating with the deficiency is tolerable, including documentation of those factors upon which the decision was based.

If a decision is made to leave a deficient asset in service, a variety of non-exclusive resolution paths are possible to help ensure continued safe and reliable operation until the asset is renewed, replaced or permanently repaired:

- The decision to temporarily continue operation with a deficient asset may require ***ITPM adjustments*** such as increasing the number and/or extent of thickness measurements, using different nondestructive examination techniques, or changing the ITPM schedule (e.g., more frequent monitoring of the degradation rate and extent of damage).

- The decision to temporarily continue operation with a deficient asset may require one or more ***temporary risk-compensating provisions*** such as:
  - Replacing a lost function
  - Enhancing equipment integrity such as by reinforcement
  - Changing operating conditions so that the degradation rate is reduced or a greater safety factor is realized
  - Providing an additional layer of protection
  - Increasing surveillance or monitoring
  - Giving more advance warning.

- The decision to temporarily continue operation with a deficient asset may require *scheduling of asset repair or replacement* for permanent correction of the deficient condition. This might take into account degradation rates and P-F intervals when these are known with sufficient confidence; see Figure 11-1. Continued operation would follow technical recommendations to minimize degradation while making the asset available for a specific period of time.

- The decision to temporarily continue operation with a deficient asset may require a *fitness-for-service evaluation*. An overview of FFS, based on API RP 579 (Reference 11-1), is presented in Appendix 11A.

- The assessment performed for evaluating the risk of continued operation with a particular asset in a deficient condition may indicate that the risk is very low (e.g., by determining that the worst-case consequence of asset failure would be a non-critical product quality impact). If this is the case, then the decision might be made to schedule repair or replacement at the earliest convenient time without adjusting ITPM, adding risk-compensating provisions or performing a FFS evaluation, while including the decision, rationale and required approvals in the facility's management of change documentation.

In all cases, the burden of proof needs to be considered that it is safe to continue operation with one or more assets in a deficient condition, based on a reasonable technical evaluation of the risks and the compensating provisions.

Responses for Out-of-Service Assets. The resolution of a deficient condition for out-of-service assets involves a similar thought process. Facility management needs to decide if the asset is to be placed (back) into service before the deficiency is permanently corrected. If so, similar resolution paths are appropriate to help ensure safe, reliable operation. Following a risk assessment, the resolution path involving a deficient condition for out-of-service equipment comprises at least one of the above in-service paths and one of the following paths:

- Immediate repair or renewal for permanent correction of the deficient condition

- Delayed restart until an evaluation to confirm FFS is completed and, if necessary, the scope of repairs, de-rating of a vessel or other response paths is defined.

> **U.S. OSHA PSM requirement**
>
> The employer shall correct deficiencies in equipment that are outside acceptable limits (defined by the process safety information in paragraph (d) of this section) before further use or in a safe and timely manner when necessary means are taken to assure safe operation.
>
> 29 CFR 1910.119(j)(5)

Temporary measures for some types of deficiencies may include clamps or other forms of leak sealing, new piping bypasses, and/or alternate process lineups for relief protection. Employing temporary mitigation measures and/or other temporary changes (e.g., safety systems placed on bypass, alternate fire protection or gas detection measures, operation without spare equipment or with temporary equipment) requires appropriate technical review and documentation and includes a schedule for removal of the temporary measures or technical re-evaluation. Such technical review and/or re-evaluation can be considered part of the deficiency resolution process and needs to be tracked until the temporary measure is removed. The identification of an out-of-service deficient condition can be resolved using QA procedures and documented in the asset information file. Management of change and operational readiness procedures can help manage the deficiency response process, including updating of process safety information with any new safety limits or design parameters.

11.5   EQUIPMENT DEFICIENCY COMMUNICATION

Communicating asset deficiencies to affected personnel is critical to help ensure that incidents do not occur as a result of the deficiencies. "Affected personnel" almost always includes operators, but maintenance and contract personnel may also need to be aware of some asset deficiencies. Communication needs to occur at three points in the equipment deficiency cycle:

- *Immediate hazard and initial response.* For asset deficiencies that are discovered via asset failures such as a pump seal failure, personnel discovering the deficiency need to know to communicate the immediate hazard (e.g., fire, release of toxic materials) to potentially affected personnel, such as workers in the immediate area, and the actions needed to correct or mitigate the hazard (e.g., sound the evacuation alarm, stop the pump, close inlet and outlet block valves).

- *Status of deficient equipment.* For an asset that is shut down and for an asset operating in a deficient state (with or without temporary repairs), updated asset status needs to be promptly communicated to personnel. In addition, the precautions (if any) associated with operating the deficient asset also needs to be clearly communicated, such as running the process at lower pressures due to a vessel rerating.

- *Return of deficient equipment to normal service.* When the asset deficiency is permanently corrected, personnel need to be informed when the asset is returned to normal service.

Communication of the immediate hazard and the initial response is typically addressed by facility emergency response procedures. The status of deficient assets and return of deficient assets to normal service can be addressed via asset deficiency logs and/or the site MOC program. This includes any required operator training or re-training aspects to manage the change in equipment condition until permanent repairs are implemented.

11.6 TRACKING OF TEMPORARY REPAIRS

Sometimes, facilities have found that, without adequate controls in place, temporary repairs implemented in response to equipment deficiencies have a tendency to remain in place long enough that they effectively become "permanent." This has led to catastrophic incidents because a temporary repair, by definition, is not fit for permanent service. The best strategy to counter this tendency is to develop a system for tracking temporary repairs of deficient assets. Typically, the asset deficiency process uses the following systems to track temporary repairs:

- MOC procedures for temporary changes, including authorized time until change is reversed or a new MOC is submitted to make the change permanent

- Asset deficiency logs/forms

- Asset deficiency flags / designators in the computerized maintenance management system (CMMS).

For batch type of operations, temporary repairs often can be corrected at the end of a batch. Correcting temporary repairs on continuous operations can be more problematic. The asset deficiency resolution process for continuous operations needs to dictate that temporary repairs be scheduled for termination as soon as possible and no longer than the date of the next scheduled shutdown or turnaround. In addition, the system could allow personnel to readily identify temporary repairs so that they can be corrected sooner in the event of an unscheduled shutdown. On the other hand, extending shutdown or turnaround dates needs to include a reanalysis of risks associated with keeping temporary repairs in place for a longer period of time than originally authorized.

11.7 DEFICIENCY MANAGEMENT ROLES AND RESPONSIBILITIES

The roles and responsibilities for deficiency management can be assigned to personnel in various departments. A high-level procedure that describes the management system and approval authorities used to address asset deficiencies is recommended.

Typically, personnel in the Maintenance, Engineering and Operations departments will be involved in managing deficiencies. Outside personnel, such as contractors, may also be involved. Example roles and responsibilities for managing deficiencies are provided in Table 11-3. The matrices use the following letter designators:

R indicates the job position(s) with primary <u>responsibility</u> for the activity

A indicates <u>approver</u> of the work or decisions made by a responsible party

S indicates the job position(s) that typically <u>support</u> the responsible party in completing the activity

I indicates the job position(s) that are <u>informed</u> when the activity is completed or delayed.

CHAPTER 11 REFERENCES

11-1   API RP 579, *Recommended Practice for Fitness-For-Service,* American Petroleum Institute, Washington, DC.

Additional Chapter 11 Resource

API RP 571, *Damage Mechanisms Affecting Fixed Equipment in the Refining Industry,* American Petroleum Institute, Washington, DC.

TABLE 11-3. Example Roles and Responsibilities Matrix for Equipment Deficiency Resolution

R = Primary Responsibility
A = Approval
S = Support
I = Informed

Activity	Inspection and Maintenance Personnel						Engineering Personnel			Operations Personnel			Other Personnel				
	Maintenance Manager	Maintenance Engineers	Maintenance Supervisors	Inspectors	Maintenance Technicians	Maintenance Planner/Scheduler	Engineering Manager	Project Engineers	Process Engineers	Unit Superintendent	Production Supervisors	Operators	Plant Manager	EHS Manager	Process Safety Coordinator	Contractor	Equipment Vendor
Identifying acceptance criteria																	
• New equipment		S		S	S		A	R	S				I		I	S	S
• Operating equipment		R		S			A	S	S		I		I		I	S	
Identifying deficient equipment																	
• New equipment	S	S		S	S		S	R					I		I		
• Operating equipment	S	R		S	S	S	A		S		A	S	I		I		
Responding to equipment deficiencies																	
• New equipment	S	S		S			A	R	S	S	A	S	I		I	S	S
• Operating equipment	S	R		S			A		S	S	A	S	I	I	I		

TABLE 11-3. *Continued*

R = Primary Responsibility
A = Approval
S = Support
I = Informed

Activity	Inspection and Maintenance Personnel						Engineering Personnel			Operations Personnel			Other Personnel				
	Maintenance Manager	Maintenance Engineers	Maintenance Supervisors	Inspectors	Maintenance Technicians	Maintenance Planner/Scheduler	Engineering Manager	Project Engineers	Process Engineers	Unit Superintendent	Production Supervisors	Operators	Plant Manager	EHS Manager	Process Safety Coordinator	Contractor	Equipment Vendor
Tracking equipment deficiencies to completion																	
• New equipment				S	S			R					I		I		
• Operating equipment	R	R	S	S	S	I	S		S		R	S	I	I	I		
Equipment deficiency communication																	
• New equipment							S	R		R			I		I		
• Operating equipment	R	S	S	I	I	I	S			R	S	I	I	I	I		

Appendix 11A. Fitness for Service

This appendix provides a brief overview of API RP 579. Readers are encouraged to refer to API RP 579 when researching and performing FFS assessments.

A fitness-for-service (FFS) assessment is performed on deficient pressure-retaining equipment to justify continued service or to identify the parameters required to enter or return the equipment to service. The FFS assessment of pressure-retaining equipment typically involves the resolution of a deficiency found during an inspection. An FFS assessment is also used to define inspection acceptance criteria for a degraded condition, such as for large areas of metal depletion or flaws in the metal or welding of the pressure-retaining component. FFS assessments can also provide alternative documentation for the suitability of equipment for service if the original equipment records (such as the manufacturer's data report, Form U-1) are not available.

A deficiency occurs when the condition of an asset is outside the established limits that define asset integrity (acceptance criteria). Typically, personnel identify deficiencies in pressure equipment during inspection or repair activities. However, inspections of new asset fabrication and installation can also uncover deficiencies such as gouges or dents that may be difficult to resolve using the equipment design and fabrication codes. Such deficiencies may also be problematic because of delivery schedule and business interruption concerns. To assist facilities in evaluating these areas of concern, API RP 579 was developed to provide standardized technical evaluation methods for a wide variety of commonly identified modes of degradation in pressure-retaining equipment. API RP 579 defines FFS assessments as "engineering evaluations that are performed to demonstrate the structural integrity of an in-service component containing a flaw or damage." The FFS evaluation methodologies provide consistent engineering tools that may be applied with increasing technical rigor as necessary, based on the extent and magnitude of the asset's degraded condition.

Scope and Purpose of FFS. Facilities inspect in-service pressure-retaining equipment to characterize the condition of the equipment, which enables personnel to predict with confidence the remaining service life of the equipment. A routine inspection of pressure equipment is designed to find indications of degradation that warrant follow-up investigation, not to locate the areas of maximum degradation. Following the inspection(s), conservative evaluation criteria are initially applied to the inspection results to determine suitability for service. However, the evaluation of the areas of concern may require additional analysis beyond the conservative evaluation criteria derived from fabrication and inspection codes.

FFS evaluations using API RP 579 are applicable to pressure boundary components of metallic pressure vessels, piping, and storage tanks that have been designed and fabricated to a nationally recognized code or standard such as the ASME BPV code, ASME B31.3, API 620 or API 650. FFS evaluations are to be performed under the supervision of an engineer familiar with the requirements of the applicable code.

The FFS assessment may result in a variety of outcomes, resulting in follow-up actions ranging from continued service within existing safe operating limits to immediate removal from service for remediation or replacement. Common results from an FFS assessment include:

- Operating with reduced operating pressure or temperature rating

- Reducing maximum fill heights for tanks

- Changing service conditions or providing inherent process safeguards to prevent further damage

- Intensifying and/or increasing the frequency of on-stream condition monitoring (CM) to determine (1) if the degradation mode is active and (2) the rate of degradation

- Reducing service life or providing a time-specific remediation schedule

- Defining limited repair scopes through the inspection codes.

In addition, a FFS assessment provides the owner-user of the pressure-retaining equipment with valuable information and options for planning and scheduling inspections and maintenance work on the asset. This information includes:

- Equipment design basis documentation, including documents that are used to comply with regulatory requirements by establishing that equipment is safe for continued operation

- Acceptance criteria for future inspections

- Justification for the deferral of equipment replacement or repair

- Options for rerating, altering, repairing or replacing equipment

- Forecast of remaining useful life of equipment

- Future inspection requirements.

FFS Evaluation Methodology. The FFS methodology, as presented in API RP 579, discusses nine asset damage or flaw conditions for applicable pressure-retaining equipment, pressure vessels, piping, and storage tank shells. The damage or flaw conditions discussed in API RP 579 are:

- Brittle fracture
- General metal loss
- Local metal loss

- Pitting corrosion
- Blisters and laminations
- Weld misalignment and shell distortions
- Crack-like flaws
- Equipment operating in the creep range
- Fire damage.

For each of these damage and flaw conditions, the assessment techniques and the technical acceptance requirements for three levels of FFS assessments are described. Each level of assessment increases in rigor and complexity. The increases in rigor are associated with (1) the amount and detail of additional inspection data and (2) the need for specialized expertise. The three levels of assessment are designed to reach, if possible, acceptable outcomes with minimum analysis effort. The level of effort required to complete an FFS assessment is consistent with the severity level of the degradation, the economics of the equipment remediation and the economics of the necessary business interruption. Typically, engineers or inspectors who are familiar with the equipment fabrication codes can perform a Level 1 FFS assessment. Levels 2 and 3 require individuals with specialized knowledge in fracture mechanics, materials science, nondestructive testing (NDT), and structural engineering. A Level 2 analysis requires information similar to that required for Level 1 but uses more detailed calculations. A Level 3 analysis is generally a numerical finite element model requiring the involvement of personnel knowledgeable in Level 2 evaluations but who are also skilled in the setup, running and interpretation of the numerical model results.

For each damage or flaw condition, a damage-specific procedure for conducting an FFS assessment is discussed. API RP 579 describes an eight-step procedure for conducting an FFS assessment for each damage condition of a component:

1. Flaw and damage mechanism identification
2. Applicability and limitations of the FFS assessment procedures
3. Data requirements for conducting the FFS assessment
4. FFS assessment techniques and acceptance criteria
5. Remaining life evaluation
6. Equipment or component remediation
7. On-stream monitoring of the component or equipment
8. Documentation of the FFS assessment.

Step 1 (flaw and damage mechanism identification) is not specifically covered in API RP 579. This step involves understanding the results of the applied inspection technique in order to properly characterize the damage condition(s) to be evaluated during the FFS assessment. API RP 579 provides a list of RAGAGEPs that are useful for damage and flaw mechanism identification and characterization. API RP 571, *Damage Mechanisms Affecting Fixed Equipment in the Refining Industry*, gives additional information on fixed-equipment damage mechanisms.

API RP 579 provides specific guidance on Steps 2 through 8 for each flaw and damage mechanism identified. The data requirements and documentation requirements for these steps are similar to those required for typical in-service inspection plans.

12
EQUIPMENT-SPECIFIC INTEGRITY MANAGEMENT

The previous chapters in this book focused on the programmatic aspects, such as management activities and systems, of a facility's asset integrity management (AIM) program. This chapter provides more detailed AIM program information for common types of equipment. The information provides facility personnel with guidance and suggestions for managing the integrity of particular equipment included in the AIM program. Specifically, the information included in this chapter can assist facilities in selecting inspection, testing and preventive maintenance (ITPM) tasks (Chapter 6), identifying appropriate quality management activities (Chapter 10) and initiating appropriate repairs and corrections for deficient conditions (Chapter 11). It should be noted that some specialized applications are not included in this chapter and will warrant their own analysis, plan and procedure.

Much of the information in this chapter is documented in a series of matrices in Appendix 12A for different types of equipment that might require equipment-specific procedures, training and/or tasks. Each matrix is organized into the four phases of asset integrity introduced in Chapter 1. The objective of each phase is explained below:

1. *New equipment design, fabrication and installation.* During this phase, activities focus on ensuring that new equipment is suitable for its intended service, so many of the activities in this phase are directly related to the quality management activities for the early part of the asset life cycle as discussed in Chapter 10.

2. *Inspection and testing.* During this phase, activities focus on ensuring the ongoing integrity of equipment or functionality of safeguards for a specified inspection and testing interval. These activities are included in the ITPM plan discussed in Chapter 6.

3. *Preventive maintenance.* During this phase, activities focus on preventing premature failure of equipment and its components, and can include performing servicing tasks (e.g., lubrication) and/or inspecting

and replacing components that are subject to wear. These activities are also part of the ITPM plan discussed in Chapter 6.

4. ***Repair.*** During this phase, activities focus on responding to identified equipment deficiencies and failures, then repairing and returning equipment to service in a condition suitable for their intended use. Such repair activities are similar to the practices discussed in Chapter 11 regarding deficiency correction.

In the matrices, these four phases are on four separate pages for each equipment type, and specific information for each of the phases is provided in the rows on each page. The information has been divided into the following categories:

1. ***Example activities and typical frequencies.*** Actions listed for the equipment design, fabrication, and installation phase include QA activities for that type of equipment, as well as the timing of these activities during the design, fabrication and installation cycle. For the inspection, testing and preventive maintenance phases, the activities and frequencies provide the typical information needed for the ITPM plan for that type of equipment. The activities listed for the repair phase include common repairs as well as some repair QA activities.

2. ***Technical basis for activity and frequency.*** This row provides the jurisdictional requirement, code, standard or other rationale (e.g., common industry practice, manufacturer recommendation) typically used to justify the activity and frequency.

3. ***Sources of acceptance criteria.*** Information sources that typically contain acceptance criteria for AIM activities are provided in this row.

4. ***Typical failures of interest.*** This row lists some typical equipment failures that can affect asset integrity; however, it is not a comprehensive list of possible failures or failure modes.

5. ***Personnel qualifications.*** This row provides general and specific qualification requirements for personnel performing the activities.

6. ***Procedure requirements.*** The information in this row outlines types of procedures typically developed for the phase and/or the general content of the procedures.

7. ***Documentation requirements.*** This row lists typical documentation created from the activities, as well as documentation retention recommendations. In addition, suggestions for documenting activities by exception are provided.

The following sections of this chapter each have associated AIM matrices in Appendix 12A for these equipment classes:

Section 12.1 - Vessels, Tanks and Piping

Section 12.2 - Relief and Vent Systems

Section 12.3 - Instrumentation and Controls

Section 12.4 - Rotating Equipment

Section 12.5 - Fired Equipment

Section 12.6 - Electrical Systems.

In addition, each section in this chapter provides information about relevant codes, standards and recommended practices, as well as general information for the equipment class or detailed information beyond what is included in the matrix.

The remaining sections of this chapter present relevant information specific to various types of equipment but with no AIM activity matrices included. Matrices have not been included for fire protection systems because volumes of published information are generally already accessible to most facility personnel. In addition, the specific ITPM activities for fire protection systems are usually provided by the authority having jurisdiction (e.g., local fire marshal or company insurance carrier). Matrices for miscellaneous systems have not been included in the book because of the wide variety of designs and issues with these systems. However, information on applicable codes, standards and recommended practices and a brief discussion of AIM activities are provided.

12.1 VESSELS, TANKS AND PIPING

The design, inspection and testing requirements for most types of fixed equipment are governed by codes, standards and recommended practices. The primary organizations issuing codes and standards for vessels, tanks and piping include the American Society of Mechanical Engineers (ASME), the American Petroleum Institute (API) and the National Board of Boiler and Pressure Vessel Inspectors (NBBPVI). Tables 12-1, 12-2 and 12-3 provide brief summaries of a few of the more common codes, standards and recommended practices for pressure vessels, atmospheric and low-pressure storage tanks, and process piping, respectively.

In addition, other common practices have been developed for certain applications and specific chemicals. These include ANSI/CGA G-2.1 for processes using anhydrous ammonia (Reference 12-1), as well as guidance developed by the Chlorine Institute for chlorine and related processes (chlorineinstitute.org) and the International Institute of Ammonia Refrigeration for ammonia-based refrigeration systems (iiar.org).

TABLE 12-1. Codes/Standards/Practices for Pressure Vessels

Issuing Organization	Document Number	Document Title	Application
API	API 510	Pressure Vessel Inspection Code: Maintenance Inspection, Rating, Repair, and Alteration	Covers the maintenance, inspection, repair, alteration, and rerating procedures for pressure vessels
API	Recommended Practice (RP) 572	Inspection of Pressure Vessels	Covers the inspection of pressure vessels
API	ANSI/API 660 or ISO 16812	Shell-and-Tube Heat Exchangers for General Refinery Services	Defines the minimum requirements for the equipment design, material selection, fabrication, inspection, testing, and preparation for shipment of shell-and-tube heat exchangers
ASME	ASME Code, Section VIII	ASME Boiler and Pressure Vessel Code (BPVC), Unfired Pressure Vessels	Provides requirements applicable to the design, fabrication, inspection, testing, and certification of pressure vessels operating at either internal or external pressures exceeding 15 psig
ASME	PCC-2	Repair of Pressure Equipment and Piping	Provides methods for repair of equipment within the scope of ASME codes and standards after it has been placed in service, including relevant design, fabrication, examination, and testing practices
NBBPVI	NBIC	National Board Inspection Code	Provides rules and guidelines for in-service inspection of boilers, pressure vessels, piping, and pressure relief valves (PRVs); also provides rules for the repair, alteration, and rerating of pressure-retaining items and for the repair of PRVs
NFPA	55	Compressed Gases and Cryogenic Fluids Code	Provides general information on siting of tanks and cylinders; includes some references to maintenance practices

TABLE 12-2. Codes/Standards/Practices for Atmospheric/Low-Pressure Storage Tanks

Issuing Organization	Document Number	Document Title	Application
API	Recommended Practice RP 575	Inspection of Atmospheric and Low Pressure Storage Tanks	Covers the inspection of atmospheric and low-pressure storage tanks designed to operate at pressures from atmospheric to 15 psig
API	API 620	Design and Construction of Large, Welded, Low-pressure Storage Tanks	Covers the design and construction of large, welded, low-pressure carbon steel aboveground storage tanks, including flat-bottom tanks that have a single vertical axis of revolution; applies to tanks with pressure in their vapor spaces at not more than 15 psig
API	API 650	Welded Steel Tanks for Oil Storage	Covers the material, fabrication, erection and testing requirements for aboveground, vertical, cylindrical, closed- and open-top, welded steel storage tanks; applies to tanks with internal pressures approximating atmospheric pressure
API	API 653	Tank Inspection, Repair, Alteration, and Reconstruction	Covers the inspection, repair, alteration, and reconstruction of steel aboveground storage tanks

In general, the listed codes, standards and recommended practices contain design, inspection and testing requirements needed to ensure structural and containment integrity of the fixed-equipment items. Many times, however, these documents do not include process performance requirements that might be important for ensuring process safety. Therefore, additional design considerations, inspections and tests may be needed to ensure safety performance. Common examples include:

- Performance testing and monitoring of scrubber operations

- Performance testing and monitoring of heat-exchange equipment when failure to adequately remove heat can have process safety implications (e.g., removal of heat from an exothermic reaction)

- Lubrication of swivel joints in loading arms to ensure that the joints are sealed.

TABLE 12-3. Codes/Standards/Practices for Process Piping

Issuing Organization	Document Number	Title	Application
ASME	B31.3	Process Piping	Provides requirements for materials and components, design, fabrication, assembly, erection, examination, inspection, and testing of piping
ASME	PCC-1	Guidelines for Pressure Boundary Bolted Flange Joint Assembly	Useful for developing effective bolted flange joint assembly procedures and troubleshooting leaking joints
ASME	PCC-2	Repair of Pressure Equipment and Piping	Provides methods for repair of equipment and piping within the scope of ASME codes and standards after it has been placed in service, including relevant design, fabrication, examination, and testing practices
API	570	Piping Inspection Code: Inspection, Repair, Alteration, and Rerating of In-Service Piping Systems	Provides procedures for the inspection, repair, alteration, and rerating of metallic piping that have been in-service
API	RP 574	Inspection Practices for Piping System Components	Covers inspection practices for piping, tubing, valves (not including control valves), and fittings; this document is a supplement to API 570
NBBPVI	NBIC	National Board Inspection Code	Provides rules and guidelines for in-service inspection of boilers, pressure vessels, piping, and PRVs; also provides rules for the repair, alteration, and rerating of pressure-retaining items and for the repair of PRVs

Fixed Equipment of FRP Construction. Design, fabrication, and testing requirements for fiberglass-reinforced plastic (FRP) vessels, tanks and equipment are contained in several documents published by API, ASME and ASTM International. However, these documents generally do not provide specific guidance on the inspection and testing of in-service equipment. Common practices for FRP-constructed equipment include:

- ASME Section X — Fiberglass-Reinforced Plastic Pressure Vessels

- API Spec 12P.— Specification for Fiberglass-Reinforced Plastic Tanks

- ASTM D3299 — Filament-Wound Fiberglass-Reinforced Plastic Chemical Resistant Tanks

- ASTM D2563 — Recommended Practice for Classifying Visual Defects in Laminates

- ASTM D2583 — Indentation Hardness of Rigid Plastics Using a Barcol Impressor.

Transportation Containers. Transportation equipment that remains stationary at a facility (e.g., tank trucks, railcars) needs to be included in the AIM program as a special category of "fixed" equipment. In the United States, the design, inspection, and testing requirements for this equipment are defined by Department of Transportation (DOT) regulations (49 CFR Chapter 1, Subchapter C). For most organizations, the company shipping the materials and/or the owner of the tank truck/railcar is responsible for ensuring the integrity of this equipment. Therefore, most facilities will need to ensure only that the equipment and its installation is in compliance with applicable regulations. This may warrant some means of verification that the equipment is being inspected, tested and maintained as prescribed by the design code to which the equipment has been manufactured.

Intermodal tanks are particularly likely to be installed or connected for a lengthy period of time. For this reason, intermodal tanks carry transport markings, labels, and placards as well as a globally harmonized system (GHS) label with signal words, hazard warnings, precautionary phrases and pictograms.

AIM Activity Matrices for Vessels, Tanks and Piping. Tables 12A-1, 12A-2 and 12A-3 in Appendix 12A depict AIM activity matrices for pressure vessels (including columns, filters, and heat exchangers), storage tanks, and piping systems, respectively. As described in Section 6.1.1 of this document, selecting ITPM tasks for fixed equipment will be based on specific identified damage mechanisms and will likely cover a broad range of activities, from visual inspections to detailed condition monitoring such as by acoustic emission and eddy current techniques. It should be noted that since some piping system components such as flexible hoses can be pressure tested and routinely inspected and then are periodically replaced, nevertheless even these components need to be appropriately evaluated to determine their as-found condition prior to replacement.

12.2   RELIEF AND VENT SYSTEMS

The design, inspection and testing requirements for most types of pressure relief equipment are governed by codes and standards, as well as guidance such as provided by the AIChE Design Institute for Emergency Relief Systems (DIERS). These requirements are often included in the code or standard for the fixed equipment on which the relief device is installed (e.g., pressure vessel, atmospheric storage tank).   ASME, API, NBBPVI, and the National Fire Protection Association (NFPA) are the primary organizations issuing codes and standards for these pressure relief devices. Brief summaries of some common codes, standards and practices for pressure relieving devices (i.e., pressure relief valves, rupture disks and pressure-vacuum relief devices) are presented in Table 12-4. In addition, the Center for Chemical Process Safety document *Emergency Relief Systems Design Using DIERS Technology* (Reference 12-2) provides guidance on sizing relief systems involving reactive materials and two-phase flow situations.

Other codes, standards and practices have been developed for certain applications and chemicals. These include ANSI/CGA G-2.1 for processes using anhydrous ammonia (Reference 12-1) and standards developed by the Compressed Gas Association and Underwriters Laboratories Inc. (UL) for selected applications. Several NPFA standards provide guidance on pressure relief device design and maintenance; however, these standards often reference requirements in other documents.

As with other types of equipment, success/failure criteria will need to be established for pressure relief devices as they are tested and inspected. For example, the ASME Boiler and Pressure Vessel Code, in Section VIII-1 paragraph UG-126(d), gives the following functional test criteria for pressure relief valves:

The set pressure tolerances, plus or minus, of pressure relief valves shall not exceed 2 psi (15 kPa) for pressures up to and including 70 psi (500 kPa) and 3% for pressures above 70 psi (500 kPa).

For pressure relief valves that comply with paragraph UG-125(c)(3) of the same reference, the deviation is limited to not less than 0% or greater than +10% of the set pressure, per API RP 576, *Inspection of Pressure Relieving Devices.*

Codes and standards for other relief system equipment such as flame arresters, detonation arresters, emergency vents, vent headers and thermal oxidizers are not as widely published. Many times, the design criteria are based on the process and on manufacturer-supplied design criteria. In addition, ITPM activities for this equipment are typically based on a combination of manufacturer recommendations and operating experience and knowledge. Although thermal oxidizers may be considered part of a vent system for destruction of off-gases, their ITPM requirements are covered by the information in Section 12.5 on fired equipment.

TABLE 12-4. Codes/Standards/Practices for Pressure-Relieving Devices

Issuing Organi- zation	Document Number	Document Title	Application
ASME	B31.3	Process Piping	Provides requirements for materials and components, design, fabrication, assembly, erection, examination, inspection, and testing of piping
API	510	Pressure Vessel Inspection Code: Maintenance, Inspection, Rating, Repair and Alteration	Covers the maintenance, inspection, repair, alteration, and rerating procedures for pressure vessels used by the petroleum and chemical process industries; also covers inspection and testing of pressure relief devices
API	RP 520	Sizing, Selection, and Installation of Pressure-Relieving Devices in Refineries, Part I and Part II	Covers the sizing, selection, and installation of pressure relief devices for equipment that has a maximum allowable working pressure of 15 psig or greater
API	RP 576	Inspection of Pressure Relieving Devices	Describes the inspection and repair practices for automatic pressure-relieving devices, including pressure safety valves (PSVs), pilot-operated PRVs, rupture disks, and weight-loaded pressure vacuum vents
ASME	ASME B&PV Code, Section I	ASME BPVC – Power Boilers	Provides requirements for all methods of construction and relief protection of power, electric and miniature boilers, as well as high-temperature water boilers used in stationary service
ASME	ASME B&PV Code, Section IV	ASME BPVC – Heating Boilers	Provides requirements for design, fabrication, installation and inspection of steam-generating and hot water boilers intended for low-pressure service that are directly fired by oil, gas, elec-tricity or coal; also covers methods of checking safety valve and safety relief valve capacity
ASME	ASME B&PV Code, Section VIII	ASME BPVC – Pressure Vessels	Provides requirements applicable to the design, fabrication, inspection, testing and certification of pressure vessels operating at either internal or external pressures above 15 psig; also provides requirements for pressure vessel relief devices
NBBPVI	NBIC	National Board Inspection Code	Provides rules and guidelines for in-service inspection of boilers, pressure vessels, piping and PSVs; also gives rules for the repair, alteration and rerating of pressure-retaining items and for the repair of PSVs
NFPA	68	Standard on Explosion Prevention by Deflagration	Covers design, location, installation, maint-enance and use of devices and systems that vent combustion gases and pressures resulting from an enclosed deflagration
CGA	S-1.3	Pressure Relief Device Standards - Part 3 - Stationary Storage Contain-ers for Compressed Gases	Establishes design basis for pressure relief devices and defines their installation practices. References ASME codes

Factory Mutual Insurance Company (FM), UL, and ASTM International have established standards/guidelines for testing and approving flame arresters and detonation arresters. In addition, the United States Coast Guard (USCG) has developed regulations for marine vapor control that (1) provide requirements for the installation of a flame/detonation arrester for marine applications and (2) contain guidelines for testing and approving arresters. The CCPS book *Deflagration and Detonation Flame Arresters* (Reference 12-3) also provides information on designing, installing, inspecting and maintaining arresters.

Piping associated with relief and vent systems (e.g., vent piping, relief inlet and discharge piping, thermal oxidizer piping) is typically designed, fabricated and installed in accordance with the process piping standard ASME B31.3 (Reference 12-4). Special design issues, such as allowable pressure drop and support requirements, as well as relief piping concerns such as plugging, are identified in applicable relief device codes and standards such as API RP 572, Part II (Reference 12-5). The piping inspection code API 570 (Reference 12-6) has requirements for the inspection of relief and vent system piping.

It should be noted that since some relief system components, notably rupture disks and explosion vents, cannot be functionally tested and still remain in service, ITPM of these components focuses on periodic inspections and component replacement. Nevertheless, even these components need to be appropriately tested and/or inspected to determine their as-found condition prior to replacement.

Tables 12A-4 through 12A-12 in Appendix 12A present AIM activity matrices for the following types of relief and vent systems:

- Table 12A-4: Pressure Relief Valves

- Table 12A-5: Rupture Disks

- Table 12A-6: Conservation Vents and Other Low Pressure/Vacuum Relief Devices

- Table 12A-7: Flame/Detonation Arresters

- Table 12A-8: Emergency Vents

- Table 12A-9: Vent Headers

- Table 12A-10: Thermal Oxidizers

- Table 12A-11: Flare Systems

- Table 12A-12: Explosion Vents.

12.3 INSTRUMENTATION AND CONTROLS

Instrumentation and controls provide three broad functions in process operations:

- Actively controlling and/or sequencing the process operation to run within intended process limits based on sensor and operator inputs.

- Measuring and monitoring process parameters for production and quality control purposes.

- Providing on-demand safety functions.

ITPM activities for instrumentation and controls will vary depending on the function provided, as described in the following paragraphs. Note that these are not hard and fast categories; for example, safety functions are often implemented in the basic process control system (BPCS). Note also that some instrumentation and controls will need to have ITPM activities that pertain to both their functionality and the ability to contain hazardous materials and energies.

Controlling. The primary AIM objective for BPCS components is to keep the controls operating normally and reliably over time. In terms of hazard identification and risk analysis, this objective can be expressed as reducing the frequency of process upset initiating events that are associated with control system failures. All control system components within the scope of the AIM program need to be appropriately maintained. For some components, this may mean making periodic adjustments to compensate for wear. For others, it can involve valve stroking or checking the functionality of on/off valves. Sensors may need regular calibration and logic solvers may need temperature and humidity controls to be maintained.

Measuring and Monitoring. Measuring and monitoring instrumentation might include elements such as mass or volumetric flowmeters, sampling systems and on-line analyzers. Such instrumentation may or may not be included as part of the AIM program, depending on the program philosophy and scope. The primary ITPM activity associated with measuring and monitoring instrumentation is calibration, which is generally performed according to manufacturer recommendations. Some components may need to be replaced on a regular basis or when proper calibration cannot be achieved.

Providing Safety Functions. Safety functions need to reliably function on demand. Most equipment providing safety functions will be within the scope of the AIM program and thus need to be appropriately maintained. Since most safety systems operate in a standby, on-demand mode, ITPM activities include functionally testing the safety system on a periodic basis. For some safety functions, this may be as straightforward as verifying an alarm system from the sensor through to the annunciator. Safety systems that are considered safety instrumented systems (SISs) work automatically without the need for operator action to avoid a loss event. SISs complying with IEC 61508 or IEC 61511 as described in Section 7.6 will have the frequency of functional tests specified as part of the requirements for meeting specific reliability goals. Other instrumentation and controls may provide safety functions by way of interlocks or permissives to avoid hazardous equipment or process configurations.

Codes, Standards and Practices. Codes and standards published by API and the International Society for Automation (ISA) govern the design, inspection and testing requirements for some instrumentation and controls. In addition, ASME and NFPA have published codes that are applicable to burner management systems. Table 12-5 provides a brief summary of common practices for instrumentation and controls. Other documents also contain information on instrumentation and controls. For example, the Chlorine Institute has issued a publication outlining requirements for chlorine detection systems (Reference 12-7). ANSI/CGA G-2.1 (Reference 12-1) and IIAR standards provide information on instrumentation and controls for anhydrous ammonia installations and ammonia refrigeration systems, respectively. References 12-8 to 12-11 are codes/standards/practices pertaining to emergency shutdown valves for offshore installations, in addition to regulatory requirements for such valves.

In addition to the practices listed in Table 12-5, ISA issues other publications that provide information on the design and maintenance of instrumentation and control systems. The CCPS document *Guidelines for Safe Automation of Chemical Processes* (Reference 12-12) also contains information on designing and maintaining safety instrumented systems (SISs) as well as maintaining BPCS components.

Table 12A-13 in Appendix 12A is an AIM activity matrix for SISs and emergency shutdowns (ESDs). The supplemental materials accompanying this document include additional matrices for the following types of instrumentation and controls:

- Critical process controls

- Critical alarms and interlocks

- Toxic chemical monitors and detection systems

- Flammable area monitors and detection systems

- Conductivity, pH, and other process analyzers

- Burner management systems.

TABLE 12-5. Codes/Standards/Practices for Instrumentation and Controls

Issuing Organization	Document		Application
	Number	Title	
API	RP 551	Process Measurement Instrumentation	Provides procedures for installation of the more generally used measuring and control instruments and related accessories
API	RP 554	Process Instrumentation and Control	Covers performance requirements and considerations for the selection, specification, installation, and testing of process instrumentation and control systems
API	555	Process Analyzers	Addresses the associated systems, installation, and maintenance of analyzers
ASME	CSD-1	Controls and Safety Devices for Automatically Fired Boilers	Covers requirements for the assembly, maintenance, and operation of controls and safety devices installed on automatically operated boilers that are directly fired with gas, oil, gas-oil, or electricity, subject to certain service limitations and exclusions
International Electro-technical Commission (IEC)	61508	Functional Safety of Electrical/Electronic/Programmable Electronic Safety-Related Systems	Sets out a generic approach for all safety life-cycle activities for systems comprising electrical, electronic, and/or programmable electronic components that are used to perform safety functions
IEC	61511	Safety Instrumented Systems for the Process Industry Sector	Provides a safety life cycle and requirements for design, installation, and maintenance of SISs
ISA	ISA-TR-84.00.02 Parts 1 through 5	Safety Instrumented Functions (SIFs) – Safety Integrity Level (SIL) Evaluation Techniques	This series covers the different evaluation techniques that can be used to determine if a specific SIS design satisfies the SIL requirements defined in the SIF
ISA	ISA-91.00.01	Identification of Emergency Shutdown (ESD) Systems and Controls that Are Critical to Maintaining Safety in the Process Industries	Provides general requirements for determining safety-critical ESDs and controls, and for maintaining the identified instrumentation
NFPA	85	Boiler and Combustion Systems Hazards Code	Discusses the fundamentals, maintenance, inspection, training, and safety for the reduction of combustion system hazards

12.4   ROTATING EQUIPMENT

Most of the codes and standards available for rotating equipment cover the design, fabrication and installation of the equipment. Very few codes and standards are available for ITPM activities for rotating equipment. As a result, many of the inspection and testing requirements are derived from the manufacturer recommendations, common industry practices (e.g., vibration analysis), operating experience and history. In addition, some organizations use analysis techniques such as reliability-centered maintenance (RCM) to determine ITPM activities for rotating equipment. Note that some rotating equipment will need to have ITPM activities that pertain to both the functionality of the equipment and the ability to contain hazardous materials and energies.

ANSI, API and the Hydraulic Institute (HI) publish codes and standards for pumps, compressors, fans, turbines, and gearboxes. Brief summaries of some common codes, standards and practices from ANSI and API for pumps, compressors, turbines, and fans and gearboxes are presented in Tables 12-6 through 12-9, respectively.

Other available guidance includes standards and bulletins published by IIAR (iiar.org) that contain specific information on the design, installation, inspection and maintenance of ammonia refrigeration equipment. Also, design, inspection and test requirements for electric motors are provided in NPFA 70, *National Electric Code*, and NFPA 70B, *Recommended Practice for Electrical Equipment Maintenance* (References 12-13 and 12-14).

Table 12A-14 in Appendix 12A provides an AIM activity matrix for pumps. The supplemental materials accompanying this document include additional matrices for the following types of rotating equipment:

- Reciprocating compressors
- Centrifugal compressors, including specific protection systems (e.g., pressure cutouts)
- Process fans and blowers
- Agitators and mixers
- Electric motors
- Gas turbines
- Steam turbines
- Gearboxes.

TABLE 12-6. Codes/Standards/Practices for Pumps

Issuing Organization	Document Number	Document Title	Application
ANSI	ANSI/ASME B73.1	Specification for Horizontal End Suction Centrifugal Pumps for Chemical Process	Covers centrifugal pumps of horizontal, end suction single stage, centerline discharge design; includes dimensional interchangeability requirements and certain design features to facilitate installation and maintenance
ANSI	ANSI/ASME B73.2	Specification for Vertical In-Line Centrifugal Pumps for Chemical Process	Covers motor-driven centrifugal pumps of vertical shaft, single stage design with suction and discharge nozzles in-line; includes dimensional interchange-ability requirements and certain design features to facilitate installation and maintenance
ANSI	ANSI/ASME B73.3	Specification for Sealless Horizontal End Suction Centrifugal Pumps for Chemical Process	Covers sealless centrifugal pumps of horizontal, end suction single stage, centerline discharge design; includes dimensional interchangeability and features to facilitate installation and maintenance
API	610	Centrifugal Pumps for Petroleum, Petrochemical, and Natural Gas Industries	Specifies requirements for centrifugal pumps, including running in reverse as hydraulic power turbines
API	674	Positive Displacement Pumps - Reciprocating	Covers the minimum requirements for reciprocating positive displacement pumps
API	675	Positive Displacement Pumps - Controlled Volume	Covers the minimum requirements for controlled volume positive displacement pumps
API	676	Positive Displacement Pumps - Rotary	Covers the minimum requirements for rotary positive displacement pumps
API	681	Liquid Ring Vacuum Pumps and Compressors	Defines the minimum requirements for the basic design, inspection, testing, and preparation for shipment of liquid ring vacuum pumps and compressors
ANSI	ANSI/ASME B73.1	Specification for Horizontal End Suction Centrifugal Pumps for Chemical Process	Covers centrifugal pumps of horizontal, end suction single stage, centerline discharge design; includes dimensional interchangeability requirements and certain design features to facilitate installation and maintenance
API	682	Pumps - Shaft Sealing Systems for Centrifugal and Rotary Pumps	Specifies requirements and provides suggestions for sealing systems for centrifugal and rotary pumps

TABLE 12-7. Codes/Standards/Practices for Compressors

| Issuing Organization | Document | | Application |
	Number	Title	
API	617	Axial and Centrifugal Compressors and Expander-compressors for Petroleum, Chemical, and Gas Industry Services	Covers the minimum requirements for centrifugal compressors that handle air or gas
API	618	Reciprocating Compressors for Petroleum, Chemical, and Gas Industry Services	Covers the minimum requirements for reciprocating compressors and their drivers handling process air or gas with either lubricated or nonlubricated cylinders
API	681	Liquid Ring Vacuum Pumps and Compressors	Defines the minimum requirements for the basic design, inspection, testing, and preparation for shipment of liquid ring vacuum pumps and compressors

TABLE 12-8. Codes/Standards/Practices for Turbines

| Issuing Organization | Document | | Application |
	Number	Title	
API	616	Gas Turbines for the Petroleum, Chemical, and Gas Industry Services	Covers the minimum requirements for open, simple, and regenerative-cycle combustion gas turbine units for services of mechanical drive, generator drive or process gas generation
API	611	General Purpose Steam Turbines for Petroleum, Chemical, and Gas Industry Service	Covers the minimum requirements for the basic design, materials, related lubrication systems, controls, auxiliary equipment, and accessories for general-purpose stream turbines
API	612	Petroleum, Petrochemical and Natural Gas Industries - Steam Turbines - Special-purpose Applications	Specifies requirements and gives recommendations for the design, materials, fabrication, inspection, testing, and preparation for shipment of special-service steam turbines

TABLE 12-9. Codes/Standards/Practices for Fans and Gearboxes

Issuing Organization	Document Number	Document Title	Application
API	673	Special Purpose Fans	Covers the minimum requirements for centrifugal fans intended for continuous duty
API	613	Special Purpose Gear Units for Petroleum, Chemical, and Gas Industry Services	Covers the minimum requirements for special-purpose, enclosed, precision, single- and double-helical, one- and two-stage speed increasers and reducers of parallel-shaft design
API	677	General-purpose Gear Units for Petroleum, Chemical, and Gas Industry Services	Covers the minimum requirements for general-purpose, enclosed, single- and multi-stage gear units incorporating parallel-shaft helical and right-angle bevel gears

12.5 FIRED EQUIPMENT

The applicable codes, standards and practices for fired equipment are grouped based on application: (1) boilers and (2) fired heaters and furnaces including thermal oxidizers. Boiler design requirements are governed by Sections I and IV of the ASME Boiler and Pressure Vessel Code (BPVC). Inspection and testing requirements are governed by the National Board Inspection Code published by NBBPVI (Reference 12-15). API publishes most of the codes and standards for fired heaters and furnaces in process industry service. Table 12-10 lists commonly employed codes, standards and recommended practices for fired heaters and furnaces.

NFPA and some industrial insurers have also published documents that provide information on the design, installation and maintenance of fired heaters and furnaces (References 12-16 through 12-22). Specifically, NFPA has written several codes that address furnaces for special applications. The burner control system is a key element of boilers, fired heaters and furnaces; information on these systems is provided in Section 12.3 of this chapter.

Table 12A-15 in Appendix 12A presents an AIM activity matrix for fired heaters, furnaces and boilers.

TABLE 12-10. Codes/Standards/Practices for Fired Heaters and Furnaces

Issuing Organization	Document		Application
	Number	Title	
API	673	Special Purpose Fans	Covers the minimum requirements for centrifugal fans intended for continuous duty
API	613	Special Purpose Gear Units for Petroleum, Chemical, and Gas Industry Services	Covers the minimum requirements for special-purpose, enclosed, precision, single- and double-helical, one- and two-stage speed increasers and reducers of parallel-shaft design
API	677	General-purpose Gear Units for Petroleum, Chemical, and Gas Industry Services	Covers the minimum requirements for general-purpose, enclosed, single- and multi-stage gear units incorporating parallel-shaft helical and right-angle bevel gears
API	535	Burners for Fired Heaters in General Refinery Service	Provides guidelines for selection and/or evaluation of burners installed in fired heaters
API	560	Fired Heaters for General Refinery Service	Covers minimum requirements for the design, materials, fabrication, inspection, testing, and preparation for shipment for fired heaters
API	RP 573	Inspection of Fired Boilers and Heaters	Covers inspection practices for fired boilers and process heaters (furnaces).
NFPA	85	Boiler and Combustion Systems Hazards Code	Discusses the fundamentals, maintenance, inspection, training, and safety for the reduction of combustion system hazards

12.6 ELECTRICAL SYSTEMS

NFPA is a primary source of codes and standards governing the design, inspection and testing of electrical equipment. NFPA 70, which is the *National Electric Code* (Reference 12-13), provides design information. NFPA 70B, *Recommended Practice for Electrical Equipment Maintenance* (Reference 12-14), provides information on inspection and testing activities for most types of electrical equipment. In addition, NFPA 111, *Standard on Stored Electric Energy,*

Emergency, and Standby Power Systems (Reference 12-23), provides design, inspection, and testing information for emergency generators and uninterruptible power supply (UPS) systems. IEEE 446, *Recommended Practice for Emergency and Standby Power Systems for Industrial and Commercial Applications* (Reference 12-24) also provides information on the uses, power sources, design and maintenance of emergency and standby power systems. API RP 651 and API RP 1652 (References 12-25 and 12-26) provide guidance related to cathodic protection of aboveground petroleum storage tanks and underground petroleum storage tanks and piping systems, respectively. European Union Directive 94/9/EC covers equipment and protective systems intended for use in potentially explosive atmospheres (ATEX).

Table 12A-16 in Appendix 12A presents an AIM activity matrix for switchgear. The supplemental materials accompanying this document include additional matrices for the following types of electrical equipment:

- Transformers

- Motor controls

- Uninterruptible power supplies

- Emergency generators

- Lightning protection

- Grounding systems.

12.7   FIRE PROTECTION AND SUPPRESSION SYSTEMS

NFPA codes provide extensive information regarding equipment design and installation as well as ITPM information for fire protection systems. The information presented in Table 12-11 documents some of the NFPA codes that contain design and installation requirements as well as ITPM requirements. Note that fireproofing and fire walls are discussed in Section 12.10 along with other passive mitigation systems.

The NFPA publication *Fire Protection Systems: Inspection, Test and Maintenance Manual* (Reference 12-27) provides a thorough examination and presentation of inspection, testing and maintenance requirements. AIM activities related to fire protection and suppression systems, such as inspection and testing of deluge systems, are generally performed by fire prevention and firefighting specialists. The authority having jurisdiction (e.g., local fire marshal) can be consulted when evaluating inspection, testing and maintenance requirements, and may have additional requirements that need to be met for local or country-specific jurisdictions.

TABLE 12-11. Summary of Commonly Used NFPA Codes for Fire Protection Systems

Fire Protection Systems	NFPA Code	
	Design and Installation Requirements	Inspection, Test and Maintenance Requirements
Fire detection and alarm systems	NFPA 72	NFPA 72
Automatic sprinkler systems	NFPA 13	NFPA 25
Water spray systems	NFPA 15	NFPA 25
Foam-water sprinkler systems	NFPA 16	—
Foam systems	NFPA 11	NFPA 25
Standpipe and hose systems	NFPA 14	NFPA 25 and 1962
Fire pumps	NFPA 20	NFPA 25
Water supply systems	NFPA 22 and 24	NFPA 25
Fire hydrants	NFPA 24	NFPA 25
Portable fire extinguishers	NFPA 10	NFPA 10
Fire doors and dampers	NFPA 80 and 90A	—.
Halon systems	NFPA 12A	NFPA 12A
Carbon dioxide systems	NFPA 12	NFPA 12
Clean agent systems	NFPA 2001	NFPA 2001
Dry chemical extinguishing systems	NFPA 17	NFPA 17
Buildings and structures	NFPA 101	NFPA 101

12.8 VENTILATION AND PURGE SYSTEMS

Ventilation and purge systems are typically needed for (1) industrial hygiene purposes, (2) places of safe refuge from toxic equipment releases, and (3) electrical classification. The design, inspection and testing requirements for ventilation and purge systems in industrial hygiene applications are dependent on many variables, such as the chemical(s) of concern, the activity being performed (e.g., sampling, connecting of railcars) and the equipment configuration.

The American Conference of Governmental Industrial Hygienists (ACGIH) and other organizations have published various standards and guidelines documents, including *Industrial Ventilation: A Manual of Recommended Practice for Design* (Reference 12-28), that provide design requirements for various situations. These publications also contain general information on inspection and testing activities. In general, inspection and testing activities involve functional testing and/or performance testing of the system as a whole and/or of specific system components such as air handling equipment.

Ventilation and purge systems for electrical classification typically involve pressurizing a room or building (e.g., motor control center, analyzer room) or an electrical equipment enclosure (e.g., electrical panel) to prevent the ingress of flammable vapors and gases and/or combustible dusts. NFPA 496, *Standard for Purged and Pressurized Enclosures for Electrical Equipment* (Reference 12-29), provides design requirements for different (1) electrical classifications, (2) types of electrical enclosures, (3) control rooms, and (4) analyzer rooms. In addition, NFPA 70 provides additional information on the types of electrical equipment permitted and the installation of equipment in different electrical classification areas.

Ventilation for places of safe refuge typically includes (1) a pressurized room or building and (2) a means to prevent the ingress of contaminated outside air. After initial construction, the building or enclosure can be tested for tightness and for positive-pressure ventilation. UFC 4-024-01 (Reference 12-30) addresses procedures for designing airborne chemical protection for buildings.

Design codes usually do not specify ITPM requirements for building ventilation, enclosure ventilation and purge systems. Typical ITPM activities include (1) visual inspection of the installation; (2) testing and calibration of instrumentation used to detect loss of enclosure pressure and/or presence of a flammable gas/vapor; (3) functional testing of alarms, automatic power disconnects and/or air intake dampers; and (4) preventive maintenance of ventilation equipment such as fans used to pressurize rooms.

12.9 PROTECTIVE SYSTEMS

This section discusses common protective systems, such as systems for maintaining vapor space concentrations, gas fuel purge systems, exothermic reaction stopper systems, and chemical water curtains, which are used in the process industries.

Purging and Inerting Systems. Systems for maintaining vapor space concentrations are typically installed on storage tanks and/or reactors to maintain the vapor space of the vessel or enclosure below the lower flammable limit or above the upper flammable limit. These systems generally involve introducing an inert gas, typically nitrogen or carbon dioxide, or a flammable gas such as natural gas, into the vapor space. The specifics of these systems such as the type of blanketing or purging (sweep-through, pressure, vacuum, siphon or a combination) employed determine the design and operation of the system. NFPA 69, *Standard on Explosion Prevention Systems* (Reference 12-31), describes and provides design requirements for different purging systems. However, this standard does not contain significant information on the inspection and testing requirements of these systems. Typically, the ITPM activities for these systems include visual inspection, functional testing (e.g., of alarms) and/or PM of system components such as calibration of gas concentration monitors and rebuilding of supply gas regulators. Boilers and other fired equipment need to be purged of fuel gas prior to initial ignition of the burner; this activity requires a specific type of protective purge system. NFPA 54, *National Fuel Gas Code* (Reference 12-32), provides design requirements and inspection and testing guidelines for such systems.

Chemical Reaction Last-Resort Safety Systems. Facilities that involve exothermic reactions often have systems installed to stop potential runaway reactions. These protective systems may (1) use a chemical injection to stop a reaction (e.g., a polymerization chain inhibitor), (2) open valve(s) for emergency de-inventory and/or depressuring of the reactor and/or (3) "quench" the reaction with a compatible liquid (often water), providing a heat sink to quickly lower the temperature of the reactor contents. These automated systems typically involve the movement of material into and/or out of the reaction vessel. The design requirements for these systems are process-dependent, and no consensus codes and standards exist for these systems. ITPM activities typically include a functional test of system operation and maintenance of the critical system/components as described in the applicable sections above.

While no consensus codes and standards exist for exothermic reaction safety systems, several publications provide information on the design, installation, inspection, testing and maintenance of these systems. The information in the following CCPS publications can be considered when developing AIM activities for these types of systems (References 12-33 to 12-36):

- *Essential Practices for Managing Chemical Reactivity Hazards*

- *Guidelines for Chemical Reactivity Evaluation and Application to Process Design*

- *Guidelines for Safe Storage and Handling of Reactive Materials*

- *Guidelines for Process Safety in Batch Reaction Systems*

The U.K. Health and Safety Executive document *Designing and Operating Safe Chemical Reaction Processes* (Reference 12-37) can also be consulted.

Water Curtains. Water curtains are installed in some facilities to knock down released chemical vapors and/or protect escaping personnel from thermal radiation. The design requirements for these systems are usually governed by the codes and standards for the chemical involved. ANSI/CGA G-2.1 (Reference 12-1) provides design requirements for water curtains for ammonia systems. API RP 751, *Safe Operation of Hydrofluoric Acid Alkylation Units* (Reference 12-38), contains design information and guidance on inspection and testing of water curtains used in HF acid alkylation units. The CCPS publication *Guidelines for Post-Release Mitigation in the Chemical Process Industry* (Reference 12-39) also provides information on water curtains.

The ITPM activities for water curtains typically include a functional test of the water flow and observation of the flow pattern, as well as calibration and testing of any detection and activation systems. In addition, the ITPM plan may include periodic inspections and any necessary tests to verify the availability and integrity of the containment system for the spray water and absorbed chemical.

12.10 PASSIVE MITIGATION SYSTEMS

Passive mitigation systems are facility design and construction features that are intended to reduce the severity of consequences of a major incident without needing to detect or actively respond to the loss event (fire, explosion, toxic release, etc.). Examples of passive mitigation systems are blast-resistant building construction, blast walls, fire walls, fireproofing, dikes and other secondary containment features, drainage, and remote impoundment. If properly designed and constructed, these passive mitigation systems can provide significant risk reduction against loss events that are within their design limits. For example, if an explosion occurs, the blast wave generated by the explosion will create a blast overpressure and impulse that will decay with distance from the origin of the blast. If a blast-resistant control building is designed to withstand a certain blast overpressure and impulse level, and the blast overpressure and impulse at the control building location that is generated by an explosion are less than the design limit, then people located inside the control building at the time of the blast are protected from direct blast or building damage effects.

However, in addition to being properly designed and constructed, passive mitigation systems also need to be maintained over time for them to protect against loss events if they occur. For example, if fireproofing insulation is physically damaged or allowed to deteriorate and is not repaired, then it cannot be expected to be effective if a major fire occurs. Passive mitigation systems are easy to overlook since they are not part of the primary containment system and have no moving parts or electrical systems requiring more typical maintenance.

The type of ITPM activities to be performed on passive mitigation systems within the AIM program will vary but will mostly involve periodic inspection by personnel who are trained to recognize deficiencies and incipient failures, including structural weaknesses and deterioration. Some of these inspections can be included as part of operator rounds. Many passive mitigation systems have no established codes and standards associated with them, so company-specific maintenance procedures may need to be developed for them. Secondary containment features such as double-walled piping will have special requirements and procedures and will overlap with fixed-equipment inspections.

12.11 SOLIDS-HANDLING SYSTEMS

A primary objective of the AIM program for solids-handling equipment is to prevent conditions that can result in fires and explosions caused by combustible dusts. This typically involves (1) reducing the amount of dust generated, (2) controlling the dust and (3) preventing ignition sources. NFPA 654, *Standard for Prevention of Fire and Dust Explosion for the Manufacturing, Processing, and Handling of Combustible Particulate Solids* (Reference 12-40), provides design, inspection and testing requirements for solids-handling systems. The design information includes criteria for (1) explosion and fire protection equipment, (2) control of explosion hazards and (3) appropriate process equipment, such as material transfer systems (e.g., mechanical conveyors, pneumatic conveying systems), duct systems, pressure protection systems, air-moving devices, air-material separators, gates and dampers, size reduction devices, particle size separation equipment, mixers and blenders, and dryers.

NFPA 654 also includes inspection, testing and maintenance guidelines for solids-handling systems. Specifically, the code instructs facilities to establish a program to inspect, test and maintain the following:

- Fire and explosion protection and prevention equipment in accordance with the applicable NFPA standards

- Dust control equipment

- Housekeeping procedures

- Potential ignition sources

- Electrical, process and mechanical equipment including process interlocks

In addition, NFPA 654 requires (1) lubricating material-feeding device bearings (e.g., conveyor drives), air-moving device bearings (e.g., fans, blowers), air-separation devices as applicable, and gates/dampers; (2) periodically checking the bearings on material-feeding devices and air-moving devices for excessive wear; (3) periodically cleaning the material-feeding devices (if the conveyed material has a tendency to adhere to the device); (4) periodically checking air-moving devices for heat and vibration; (5) performing PM on fans and blowers and (6) periodically inspecting the filter media on air-separation devices.

Additional guidance on preventing fires and explosions can be found in NFPA 69, *Standard on Explosion Prevention Systems* (Reference 12-41). References 12-42 to 12-44 also provide information on safe design and operation of solids-handling equipment.

12.12 REFRIGERATION SYSTEMS

AIM program objectives for refrigeration equipment includes both containment of refrigeration fluids under pressure (many of which are toxic and/or flammable) and reliable operation of the refrigeration systems on a continuous basis. Some codes, standards and recommended practices for refrigeration systems are listed in Table 12-12.

12.13 UTILITIES

In many processes, loss of a utility such as cooling water or electrical power can potentially result in a process safety incident. For facilities with such processes, the design, inspection and testing activities for utilities need to be included in the AIM program. Selection of utilities to be included in the AIM program follows the same approach described in Chapter 5. Depending on the strategy used for determining asset criticality, some utility systems and components may also be designated as critical assets, as discussed in Section 5.4. For example, processing of thermally sensitive organic peroxides might designate the cooling water system, electrical power, plant air, nitrogen, and refrigeration systems as critical utilities.

The QA program for an AIM program utility system needs to provide assurance of utility system and component reliability. Ensuring reliability often involves providing redundant equipment items such as redundant cooling water supply pumps and/or installing backup systems such as emergency electrical generators. The ITPM activities for AIM utility systems typically include (1) functional testing of the system redundancy and/or backup supply systems, (2) calibration and testing of instrumented systems used, and (3) performing tasks necessary to maintain the reliability of utility components, such as vibration analysis of cooling water pumps and infrared analysis of electrical switchgear.

12.14 SAFETY EQUIPMENT

AIM programs may also include some or all of the following safety equipment:

- Safety showers and eyewash stations
- Employee alarm systems such as evacuation alarms
- Community alarm systems
- Emergency lighting and escape path lighting
- Emergency response equipment such as self-contained breathing apparatus, firefighting equipment and spill containment supplies.

TABLE 12-12. Codes/Standards/Practices for Refrigeration Equipment

Issuing Organi-zation	Document		Application
	Number	Title	
ASHRAE	15	Safety Standard for Refrigeration Systems	Specifies safe design, construction, test, installation and operation of mechanical and absorption refrigeration systems, including heat-pump systems used in stationary applications
ASHRAE	34	Designation and Safety Classification of Refrigerants	Establishes a uniform system for assigning reference numbers, safety classifications and concentration limits to refrigerants
IIAR	2	Equipment, Design, and Installation of Closed-Circuit Ammonia Mechanical Refrigerating Systems	Specifies equipment, design and installation of closed circuit ammonia refrigerating systems
IIAR	3	Ammonia Refrigeration Valves	Specifies criteria for materials, design parameters, marking and testing for valves and strainers used in closed circuit ammonia refrigerating systems
IIAR	4	Installation of Closed-Circuit Ammonia Refrigeration Systems	Specifies minimum requirements for the safe installation of closed-circuit ammonia mechanical refrigeration systems and overpressure device piping when used in conjunction with a closed-circuit ammonia refrigeration system
IIAR	5	Start-Up and Commissioning of Closed Circuit Ammonia Refrigeration Systems	Specifies minimum criteria for the start-up and commissioning of ammonia mechanical refrigerating systems
IIAR	7	Developing Operating Procedures for Closed-Circuit Ammonia Mechanical Refrigerating Systems	Defines the minimum requirements for developing operating procedures for closed-circuit ammonia mechanical refrigerating systems
IIAR	8	Decommissioning of Closed-Circuit Ammonia Refrigeration Systems	Specifies minimum criteria and procedures for decommissioning of closed-circuit ammonia refrigeration systems
ICC	IFC	International Fire Code	Model code that establishes minimum regulations for fire prevention and fire protection systems using prescriptive and performance-related provisions
ICC	IMC	International Mechanical Code	Model code that establishes minimum regulations for mechanical systems using prescriptive and performance-related provisions
IAPMO	UMC	Uniform Mechanical Code	Model code that governs the installation, inspection and maintenance of HVAC and refrigeration systems

The design, inspection and testing requirements for much of this equipment are governed by ANSI standards (typically for the design), OSHA or other safety and health regulations and NFPA codes. Table 12-13 provides a list of some standards and regulations applicable to certain safety equipment. Typical ITPM activities for safety equipment include inventory maintenance, visual inspection, functional testing and periodic refurbishing or replacement.

TABLE 12-13. Standards and Regulations for Selected Safety Equipment

Safety Equipment	Applicable Standard or Regulation		
	ANSI	U.S. OSHA	NFPA
Eyewashes	ANSI Z358.1	29 CFR 1910.151(c)	–
Safety showers	ANSI Z358.1	29 CFR 1910.151(c)	–
Employee alarm systems		29 CFR 1910.165	NFPA 72
• Firefighting equipment • Protective equipment • SCBAs • Fire apparatus	ANSI Z88.2	29 CFR 1910.156	NFPA 600, 1851, 1911, 1915, 1971, 1981, and 1991
Respiratory protection equipment	ANSI Z88.2	29 CFR 1910.134	NFPA 1981 and 1991

CHAPTER 12 REFERENCES

(Consult the latest version of each document)

12-1   ANSI/CGA G-2.1, *Requirements for the Storage and Handling of Anhydrous Ammonia*, American National Standards Institute, Washington, DC.

12-2   *Emergency Relief Systems Design Using DIERS Technology: The Design Institute for Emergency Relief Systems (DIERS) Project Manual*, American Institute of Chemical Engineers, New York, NY, 1993.

12-3   Grossel, S.S., *Deflagration and Detonation Flame Arresters*, American Institute of Chemical Engineers, New York, NY, 2002.

12-4   ASME B31.3, *Process Piping*, American Society of Mechanical Engineers, New York, NY.

12-5   API RP 572, *Inspection of Pressure Vessels*, American Petroleum Institute, Washington, DC.

12-6   API 570, *Piping Inspection Code: Inspection, Repair, Alteration, and Rerating of In-service Piping*, American Petroleum Institute, Washington, DC.

12-7   Pamphlet 73, *Atmospheric Monitoring Equipment for Chlorine*, Chlorine Institute, Arlington, VA.

12-8   U.K. Health and Safety Executive, "Riser emergency shut down valve (ESDV) leakage assessment," Technical offshore operational guidance SPC/TECH/OSD/49, www.hse.gov.uk/foi/internalops/hid_circs/technical_osd/spc_tech_osd_49.htm.

12-9   U.K. HSE, "Safety instrumented systems for the overpressure protection of pipeline risers," Technical offshore operational guidance SPC/TECH/OSD/31, www.hse.gov.uk/foi/internalops/hid_circs/technical_osd/spc_tech_osd_31.htm.

12-10   BS EN 10418, *Petroleum and natural gas industries – Offshore production installations – Basic surface safety systems*, UK Standard.

12-11   API RP 14 C, *Recommended Practice for Analysis, Design, Installation, and Testing of Basic Surface Safety Systems for Offshore Production Platforms*, American Petroleum Institute, Washington, DC.

12-12   Center for Chemical Process Safety, *Guidelines for Safe Automation of Chemical Processes, Second Edition*, American Institute of Chemical Engineers, New York, NY, 2016.

12-13   NFPA 70, *National Electrical Code*, National Fire Protection Association, Quincy, MA.

12-14   NFPA 70B, *Recommended Practice for Electrical Equipment Maintenance*, National Fire Protection Association, Quincy, MA.

12-15   *National Board Inspection Code*, National Board of Boiler and Pressure Vessel Inspectors, Columbus, OH.

12-16   NFPA 31, *Standard for the Installation of Oil-burning Equipment*, National Fire Protection Association, Quincy, MA.

12-17   NFPA 54, *National Fuel Gas Code*, National Fire Protection Association, Quincy, MA.

12-18   NFPA 85, *Boiler and Combustion Systems Hazard Code*, National Fire Protection Association, Quincy, MA.

12-19   NFPA 86, *Ovens and Furnaces*, National Fire Protection Association, Quincy, MA.

12-20   NFPA 86C, *Standard for Industrial Furnaces Using a Special Processing Atmosphere*, National Fire Protection Association, Quincy, MA.

12-21 NFPA 86D, *Standard for Industrial Furnaces Using Vacuum as an Atmosphere*, National Fire Protection Association, Quincy, MA.

12-22 NFPA 8505, *Standard for Stoker Operation*, National Fire Protection Association, Quincy, MA.

12-23 NFPA 111, *Standard on Stored Electric Energy, Emergency, and Standby Power Systems*, National Fire Protection Association, Quincy, MA.

12-24 IEEE 446, *Recommended Practice for Emergency and Standby Power Systems for Industrial and Commercial Applications*, Institute of Electrical and Electronics Engineers, Piscataway, NJ.

12-25 API RP 651, *Cathodic Protection of Aboveground Petroleum Storage Tanks*, American Petroleum Institute, Washington, DC.

12-26 API RP 1652, *Cathodic Protection of Underground Petroleum Storage Tanks and Piping Systems*, American Petroleum Institute, Washington, DC.

12-27 NFPA, *Fire Protection Systems: Inspection, Test and Maintenance Manual*, National Fire Protection Association, Quincy, MA.

12-28 ACGIH, *Industrial Ventilation: A Manual of Recommended Practice for Design, 28th Edition*, American Conference of Governmental Industrial Hygienists, Cincinnati, Ohio, 2013.

12-29 NFPA 496, *Standard for Purged and Pressurized Enclosures for Electrical Equipment*, National Fire Protection Association, Quincy, MA.

12-30 UFC 4-024-01, *Security Engineering: Procedures for Designing Airborne Chemical, Biological, and Radiological Protection for Buildings*, United Facilities Criteria, U.S. Department of Defense, 10 June 2008.

12-31 NFPA 69, *Standard on Explosion Prevention Systems*, National Fire Protection Association, Quincy, MA.

12-32 NFPA 54, *National Fuel Gas Code*, National Fire Protection Association, Quincy, MA.

12-33 Johnson, R.W., S.W. Rudy and S.D. Unwin, *Essential Practices for Managing Chemical Reactivity Hazards,* American Institute of Chemical Engineers, New York, NY, 2003.

12-34 Center for Chemical Process Safety, *Guidelines for Chemical Reactivity Evaluation and Application to Process Design,* American Institute of Chemical Engineers, New York, NY, 1995.

12-35 Center for Chemical Process Safety, *Guidelines for Safe Storage and Handling of Reactive Materials,* American Institute of Chemical Engineers, New York, NY, 1995.

12-36 Center for Chemical Process Safety, *Guidelines for Process Safety in Batch Reaction Systems,* American Institute of Chemical Engineers, New York, NY, 1999.

12-37 U.K. Health and Safety Executive, *Designing and Operating Safe Chemical Reaction Processes*, ISBN 0-7176-1051-9, HSE Books, 2000.

12-38 API RP 751, *Safe Operation of Hydrofluoric Acid Alkylation Units*, American Petroleum Institute, Washington, DC.

12-39 Center for Chemical Process Safety, *Guidelines for Postrelease Mitigation Technology in the Chemical Process Industry*, American Institute of Chemical Engineers, New York, NY, 1996.

12-40 NFPA 654, *Standard for Prevention of Fire and Dust Explosion for the Manufacturing, Processing, and Handling of Combustible Particulate Solids*, National Fire Protection Association, Quincy, MA.

12-41 NFPA 69, *Standard on Explosion Prevention Systems*, National Fire Protection Association, Quincy, MA.

12-42 Center for Chemical Process Safety, *Guidelines for Safe Handling of Powders and Bulk Solids*, American Institute of Chemical Engineers, New York, NY, 2004.

12-43 Barton, J., *Dust Explosions, Prevention and Protection; A Practical Guide*, Institution of Chemical Engineers, Rugby, Warwickshire, UK, 2002.

12-44 Eckoff, R.K., *Dust Explosions in the Process Industries, Third Edition,* Gulf Professional Publishing, Amsterdam, The Netherlands, 2003.

APPENDIX 12A. ASSET INTEGRITY ACTIVITIES BY EQUIPMENT TYPE

This Appendix contains asset integrity activity matrices for the following types of equipment:

TABLE	Equipment Type
12A-1	Pressure vessels
12A-2	Storage tanks
12A-3	Piping systems
12A-4	Pressure relief valves
12A-5	Rupture disks
12A-6	Conservation vents and other low pressure / vacuum relief devices
12A-7	Flame/detonation arresters
12A-8	Emergency vents
12A-9	Vent headers
12A-10	Thermal oxidizers
12A-11	Flare systems
12A-12	Explosion vents
12A-13	Safety instrumented systems and emergency shutdowns
12A-14	Pumps
12A-15	Fired heaters, furnaces and boilers
12A-16	Switchgear

Each activity matrix is four pages long, with one of the AIM activity phases described in the beginning of this chapter covered on each page; namely:

- *New Equipment Design, Fabrication and Installation*
- *Inspection and Testing*
- *Preventive Maintenance*
- *Repair.*

Additional activity matrices are included in the supplemental materials accompanying this document, as described earlier in Chapter 12.

TABLE 12A-1. Asset Integrity Activities for Pressure Vessels

New Equipment Design, Fabrication and Installation		
Example activities and typical frequencies	Activity	Frequency
	• Equipment specification, vessel data sheet • Process design requirements • Materials selection • Vendor/Shop qualification • Equipment design by manufacturer • Design approval by owner • Welding/quality control (QC) plan approval • Equipment fabrication • Inspection • Documentation preparation • Installation/commissioning • Acceptance and turnover	As required for fabrication and installation
Technical basis for activity and frequency	QA practices for pressure vessels	
Sources of acceptance criteria	ASME PV codes for design and fabrication, in conjunction with more stringent requirements in company engineering standards and in facility-specific or jurisdictional requirements for the pressure boundary	
Typical failures of interest	Incorrect material or weld metal, incorrect heat treatment, incorrect dimensions, misalignment or out-of-square flanges, leak during testing, weld defects, high hardness readings, use of unqualified welder or welding procedures; if applicable, inadequate bolting (improper materials, less than full thread engagement, missing bolts)	
Personnel qualifications	Company requirements and documented skills, NDE qualifications, inspection certifications or technical training for inspection and acceptance activities	
Procedure requirements	Written procedures describing: • Engineering standards for specification of equipment • Project management (including hazard and design review schedules) • Vendor qualification • Documentation requirements • Project acceptance and turnover requirements	
Documentation requirements	Company documentation requirements typically include U1 form, welding qualifications, design calculations, material certifications, QC results, heat treating records, as-built fabrication drawings and nameplate rubbing	

TABLE 12A-1. Asset Integrity Activities for Pressure Vessels *(Continued)*

Inspection and Testing		
Example activities and typical frequencies	Activity	Frequency
	External visual inspection	5-year maximum
	Condition (e.g., thickness) monitoring	½ corrosion life or 10-year maximum
	Internal inspection or alternatively on-stream inspection (as applicable)	½ corrosion life or 10-year maximum, thickness measurement suffices if corrosion rate is less than 5 mils per year
	Additional inspections for specific degradation modes (e.g., corrosion under insulation)	As required by service conditions, condition of equipment and rate of degradation
Technical basis for activity and frequency	Scheduled with intervals set by the results of previous activity or at fixed intervals based on inspection code (API 510 or National Board Inspection Code [NBIC]) or jurisdictional requirements	
Sources of acceptance criteria	• Acceptance criteria from inspection codes API 510, NBIC and/or jurisdictional requirements • Acceptance criteria for damage from specific degradation modes per API RP 579	
Typical failures of interest	• Distortion of pressure boundary, leakage from cracks (e.g., fatigue, environmentally induced, stress corrosion cracking or caustic cracking), or holes in pressure boundary • Corrosion of pressure boundary, including corrosion under insulation or fireproofing • Lack of grounding and excessive corrosion of structural support and anchoring systems	
Personnel qualifications	Documented qualifications, industry inspection certifications (API 510 or NBIC), or specific technical training to analyze results	
Procedure requirements	Written procedures describing the inspection or test activity, including: • The manner, the extent, the location and date the inspection or test is performed and by whom • The documentation and analysis of results • The resolution of functions or condition not meeting acceptance criteria	
Documentation requirements	• Results and analysis of each inspection are documented for the life of the equipment • Inspection dates are tracked and technical deferral is required for late tests with alternate means of protection to be considered; deficient conditions are identified and resolved by the date recommended	

TABLE 12A-1. Asset Integrity Activities for Pressure Vessels *(Continued)*

Preventive Maintenance		
Example activities and typical frequencies	Activity	Frequency
	Activities identified from RCM or similar work planning initiatives, such as: • Routine visual surveillance • Process conditions monitoring/ tracking • Process performance monitoring	As required to meet preventive maintenance schedule or process monitoring needs
Technical basis for activity and frequency	Company or jurisdictional requirements	
Sources of acceptance criteria	Company requirements and good engineering practices, coupled with upper and lower safe limits for process conditions as defined in the process safety information (such as pressure, temperature, fluid composition and velocity limits)	
Typical failures of interest	• Distortion of pressure boundary, leakage from cracks (fatigue or environmentally induced), or holes in pressure boundary • Corrosion of pressure boundary, including corrosion under insulation or fireproofing • Lack of grounding and excessive corrosion of structural support and anchoring systems	
Personnel qualifications	Tasks usually require craft-specific skills or operator-specific skills that are addressed within their respective training programs	
Procedure requirements	These activities generally do not require task-specific procedures	
Documentation requirements	Results are usually recorded by exception in equipment history files	

TABLE 12A-1. Asset Integrity Activities for Pressure Vessels *(Continued)*

Repair		
Example activities and typical frequencies	Activity	Frequency
	• Equipment replacement-in-kind • Unique vessel repair activities such as weld overlay, alterations, hot taps, or welding attachments to the pressure boundary • Painting • Insulation/Fireproofing repair • Chemical cleaning • Structural support and anchoring systems repair or renewal	As required by the condition of the equipment based on recommendations from ITPM activities or observations from normal operations
Technical basis for activity and frequency	Performed when indicated by failure during normal operations or by the results of ITPM activities	
Sources of acceptance criteria	Design and fabrication codes: ASME PV codes, in conjunction with more stringent requirements in company engineering standards, or facility or jurisdictional requirements. In general, repairs and alterations are performed in accordance with ASME "R" stamp requirements	
Typical failures of interest	Incorrect material or heat treatment, incorrect dimensions, misalignment or out-of-square flanges, leak during testing, weld defects, high hardness readings, use of unqualified welder or welding procedures; if applicable, inadequate bolting (improper materials, less than full thread engagement, missing bolting)	
Personnel qualifications	Welders qualified per Section IX of the ASME Code. NDE technicians qualified in appropriate techniques. Industry inspection certifications (API 510 or NBIC) or specific technical training for pressure vessel engineering	
Procedure requirements	• Craft skill procedures for typical tasks encountered in repairs (e.g., welding, gasket installation, bolt tightening, pressure testing) • Job-specific procedures developed for repairs or alterations to the pressure boundary • Job-specific procedures for unique or complex repairs or jobs with specialized technical content (e.g., retraying, modifications to internals, catalyst handling) • Job-specific procedures with process engineering input for chemical cleaning	
Documentation requirements	Repair history is typically maintained with equipment inspection history	

TABLE 12A-2. Asset Integrity Activities for Storage Tanks

New Equipment Design, Fabrication and Installation		
Example activities and typical frequencies	Activity	Frequency
	Verify equipment specifications are met: • Process design requirements • Materials selection • Equipment design or reconstruction plan by constructor/ manufacturer • Design approval by owner • Welding/QC plan approval • Tank fabrication/ construction • Acceptance inspection and testing • Documentation preparation • Installation/commissioning • Acceptance and turnover	As-required for fabrication and installation
Technical basis for activity and frequency	QA practices for storage tanks fabrication and construction	
Sources of acceptance criteria	API 650 codes for design, fabrication and reconstruction in conjunction with more stringent requirements in Company Engineering Standards, facility-specific standards	
Typical failures of interest	Incorrect material or weld metal, incorrect dimensions, misaligned piping or out-of-square foundation, leak during testing, weld defects, high hardness readings, use of unqualified welder or welding procedures, incorrect internal coating or application of coating; if applicable, inadequate bolting (improper materials, less than full thread engagement, missing bolts)	
Personnel qualifications	Company requirements, and documented skills, NDE qualifications, inspection certifications, or technical training for inspection and acceptance activities	
Procedure requirements	Written procedures describing: • Engineering standards for specification of equipment • Project management (including hazard and design review schedules) • Vendor qualification • Documentation requirements • Project acceptance and turnover requirements	
Documentation requirements	Company documentation requirements typically include: construction certificate, welding qualifications, design calculations, material certifications, QC results and as-built drawings	

TABLE 12A-2. Asset Integrity Activities for Storage Tanks *(Continued)*

Inspection and Testing		
Example activities and typical frequencies	**Activity**	**Frequency**
	Routine in-service inspection	No greater than one month, or environmental monitoring requirements if less
	External visual inspection	Five-year maximum or ¼ life based on measured shell thickness and calculated corrosion rate
	TM inspection	Five-year maximum if corrosion rates are not known
	Internal inspection	If corrosion rates are known, maximum interval is the minimum of: ½ life based on measured shell thickness and calculated corrosion rate or 15 years
	Alternative internal inspection	10-year maximum if bottom plate corrosion rates cannot be estimated
Technical basis for activity and frequency	Scheduled with intervals set by the results of previous activity or at fixed intervals based on inspection code (API-653) or jurisdictional requirements if hazardous waste service.	
Sources of acceptance criteria	Acceptance criteria from inspection codes API-653 or jurisdictional requirements. Acceptance criteria for damage from specific degradation modes per API-579	
Typical failures of interest	Distortion of shell, roof, or nozzle, hole in roof, leakage from bottom, holes or cracks in shell or roof joints, lack of grounding, corrosion of structural roof supports and anchoring systems, excessive foundation settlement, internal coating failure, external corrosion including corrosion under insulation or fireproofing	
Personnel qualifications	Documented qualifications, industry inspection certifications (API-653), or specific technical training for storage tank engineering for analysis of results	
Procedure requirements	Written procedures describing the test or inspection activity, including: • The manner, the extent, the location and the timing for the inspection or test and by whom • The documentation and analysis of results • The resolution of functions or conditions not meeting acceptance criteria	
Documentation requirements	• Results and analysis of each inspection are documented for the life of the equipment • Inspection dates are tracked and technical deferral required for late tests with alternate means of protection to be considered; deficient conditions are identified and resolved by the date recommended	

TABLE 12A-2. Asset Integrity Activities for Storage Tanks *(Continued)*

Preventive Maintenance		
Example activities and typical frequencies	Activity	Frequency
	Activities identified from FMEA or other analysis techniques for RCM or similar working planning initiatives, such as: • Process conditions monitoring/tracking • Floating roof seal gap monitoring • Conservation vent monitoring • Water phase draining	As-required to meet PM schedule or environmental monitoring requirements
Technical basis for activity and frequency	Company or jurisdictional requirements	
Sources of acceptance criteria	Company requirements and good engineering practice coupled with upper and lower safe limits for process conditions defined in the process safety information, such as safe fill height, temperature, fluid density and inflow and outflow rates	
Typical failures of interest	Leakage, shell or roof distortion and excessive gaps in floating roof seals, stuck conservation vent, tank in standing water, standing fluid on floating roof tank	
Personnel qualifications	Tasks usually require craft-specific skills or operator-specific skills that are addressed within their respective training programs	
Procedure requirements	Task-specific procedures with appropriate personnel safety procedures for working on in-service tanks	
Documentation requirements	Results are usually recorded by exception in the equipment history file	

TABLE 12A-2. Asset Integrity Activities for Storage Tanks *(Continued)*

Repair		
Example activities and typical frequencies	Activity	Frequency
	• Equipment/component replacement-in-kind • Unique tank liquid boundary repairs such as weld overlay, hot taps, floor plate patching or replacement • Unique repairs for fixed roof and structural supports • Unique repairs to floating roof components • External painting • Internal coating repair • Insulation repair • Cleaning • Foundation and anchoring systems repair or renewal	As required by the condition of the equipment based on recommendations from the inspection and testing or PM activities
Technical basis for activity and frequency	Performed when indicated by failure, by the results of PM activities or by the results of inspection and testing activities	
Sources of acceptance criteria	Design and fabrication codes: API 653 codes in conjunction with more stringent requirements in company engineering standards, facility or jurisdictional requirements	
Typical failures of interest	Incorrect material or heat treatment, incorrect dimensions, misalignment or out-of-square flanges, leak during testing, weld defects, high hardness readings, use of unqualified welder or welding procedures; if applicable, inadequate bolting (improper materials, less than full thread engagement, missing bolts)	
Personnel qualifications	Welders qualified per ASME Code, NDE technicians qualified in appropriate techniques, industry inspection certifications (API-653), or specific technical training for storage tank engineering	
Procedure requirements	• Craft skill procedures for typical tasks encountered in repairs – such as welding, gasket installation, bolt tightening, etc. • Job-specific procedures developed for repairs or alterations to the pressure boundary • Job specific procedures for unique or complex repairs, or jobs with specialized technical content – such as for floating roof seal repairs while in-service • Job specific procedures with process engineering input for chemical cleaning	
Documentation requirements	Repair history is maintained with equipment inspection history	

TABLE 12A-3. Asset Integrity Activities for Piping Systems

New Equipment Design, Fabrication and Installation		
Example activities and typical frequencies	Activity	Frequency
	• Design/fluid service requirements • Pressure rating • Materials selection • Fabrication contractor qualification • Design approval by owner • Welding/QC plan approval • Fabrication/storage/ shipping • Installation • Acceptance inspection and testing • Documentation preparation • Acceptance and turnover • Commissioning	As required for fabrication and installation
Technical basis for activity and frequency	Quality assurance practices for piping fabrication and installation	
Sources of acceptance criteria	ANSI/ASME B31 codes for piping design and fabrication and in conjunction with more stringent requirements in Company Engineering Standards or facility-specific standards; ASME PCC-1 for bolted flange joint assemblies	
Typical failures of interest	Dimensional errors, incorrect material or weld metal, incorrect dimensions, misaligned or out-of-square flanges, incorrect pressure rating for a component, leak during testing, weld defects outside of acceptance criteria, high hardness readings, use of unqualified welder or welding procedures; inadequate bolting, supports (improper materials, less than full thread engagement, missing bolts)	
Personnel qualifications	Company requirements, and documented craft skills for installation, NDE qualifications and ASME Section IX welding requirements for welders	
Procedure requirements	Written procedures describing: • Engineering standards for specification of equipment • Project management (including hazard and design review schedules) • Vendor qualification • Documentation requirements • Project acceptance and turnover requirements	
Documentation requirements	Company documentation requirements typically include welding qualifications, weld map, design calculations, material certifications, QC results, as-built drawings and pressure test reports	

TABLE 12A-3. Asset Integrity Activities for Piping Systems *(Continued)*

Inspection and Testing		
Example activities and typical frequencies	Activity	Frequency
	External visual inspection	Default interval values in API 570
	Condition (e.g., thickness) monitoring	Lesser of default interval values in API 570 or half-life based on measured wall thickness and calculated corrosion rates
	RBI assessment	Adjustment of intervals and extent with RBI assessment, plan to be reviewed at default inspection intervals
	• Special emphasis inspection • Injection point and soil-to-air interface	• Injection point inspection: Lesser of 3 years max. or half-life based on measured wall thickness and calculated corrosion rates • Soil-to-air interface inspection: default interval values in API 570
	Additional inspections for specific degradation modes	As required by service conditions, condition of equipment and rate of degradation
Technical basis for activity and frequency	Scheduled with intervals set by the results of previous inspection or default maximum intervals listed in the inspection code (API 570)	
Sources of acceptance criteria	• Acceptance criteria from inspection code API 570 or jurisdictional requirements • Acceptance criteria for damage from specific degradation modes per API RP 579	
Typical failures of interest	Leakage from cracks (e.g., fatigue, environmentally induced, stress corrosion cracking, caustic cracking), internal or external corrosion, corrosion under insulation, excessive vibration, unsupported or bound piping, permanent distortion, piping component not meeting pressure rating, inadequate bolting, failed piping supports	
Personnel qualifications	Documented NDE qualifications, industry inspection certifications (API 570), or specific technical training for piping engineering for analysis of results	
Procedure requirements	Written procedures describing the inspection or test activity that includes: • The extent and location of the activity, how and when the inspection or test is performed and by whom • How the results are documented and when the results are analyzed • How a function or condition not meeting the acceptance criteria is resolved	
Documentation requirements	• Results and analysis of each inspection are documented for the life of the equipment • Inspection dates are tracked and technical deferral required for late tests with alternate means of protection to be considered; deficient conditions are identified and resolved by the date recommended	

TABLE 12A-3. Asset Integrity Activities for Piping Systems *(Continued)*

Preventive Maintenance		
Example activities and typical frequencies	Activity	Frequency
	• Activities identified from failure modes and effects analysis (FMEA) or other analysis techniques for RCM, risk-based inspection (RBI), or similar work planning initiatives • Process conditions monitoring/ tracking	As required to meet preventive maintenance schedule
Technical basis for activity and frequency	Company or jurisdictional requirements	
Sources of acceptance criteria	Upper and lower safe limits for process conditions, such as pressure, temperature, fluid composition and velocity, as defined in the process safety information	
Typical failures of interest	Process conditions exceed safe upper or lower limit	
Personnel qualifications	Operator training	
Procedure requirements	Operating procedures	
Documentation requirements	Results are usually recorded by exception in the equipment history file	

TABLE 12A-3. Asset Integrity Activities for Piping Systems *(Continued)*

Repair		
Example activities and typical frequencies	Activity	Frequency
	• Piping/component replacement-in-kind • Commissioning activities • Temporary clamps • Hot taps/stopples, etc. • Painting • Insulation repair • Cleaning • Support, hanger and anchoring systems repair or renewal	As required by the condition of the equipment based on recommendations from the inspection and testing or preventive maintenance activities
Technical basis for activity and frequency	Performed when indicated by failure, by the results of PM activities or by the results of inspection and testing activities	
Sources of acceptance criteria	ANSI/ASME B31 Design and Fabrication Code; in conjunction with more stringent requirements in Company Engineering Standards, facility, or jurisdictional requirements; ASME PCC-2 for repair of pressure piping	
Typical failures of interest	Dimensional errors, incorrect material or weld metal, misaligned or out-of-square flanges, incorrect pressure rating for a component, leak during testing, weld defects outside of acceptance criteria, high hardness readings, use of unqualified welder or welding procedure, leakage during hot tap/stopple operations, inability to remove hot tap/stopple machines, inadequate bolting (improper materials, full thread engagement, missing bolts), inadequate or missing piping supports	
Personnel qualifications	Welders qualified per ASME Section IX Code, NDE technicians qualified to appropriate techniques, industry inspection certifications (API 570), or specific technical training for storage tank engineering.	
Procedure requirements	• Craft skill procedures for typical tasks encountered in repairs, such as welding, gasket installation, bolt tightening, etc. • Job-specific procedures developed for repairs or alterations to the pressure boundary • Job-specific procedures for unique or complex repairs, or jobs with specialized technical content, such as line lifting, hot taps, stopples and clamp installations • Job-specific procedures with process engineering input for chemical cleaning	
Documentation requirements	Repair history is maintained with equipment inspection history	

TABLE 12A-4. Asset Integrity Activities for Pressure Relief Valves

New Equipment Design, Fabrication and Installation		
Example activities and typical frequencies	Activity	Frequency

	• Design requirements and process specifications
	  - Service requirements
	  - Component materials
	  - Sizing design basis and sizing calculations
	• Vendor/shop qualification
	• Equipment design by manufacturer
	• Equipment fabrication
	• Inspection and testing
	• Documentation preparation
	• Installation and commissioning
	• Acceptance and turnover
		As required for fabrication and installation

Technical basis for activity and frequency	QA practices for pressure relief valve fabrication, testing and installation	
Sources of acceptance criteria	• Codes and standards for design and fabrication of pressure relief devices (ASME BPVC- Section VIII, NBIC, NFPA 30, API RP 520, or other standards applicable for the specific application [e.g., ammonia, LPG]), in conjunction with requirements for pressure vessels • Company engineering standards, facility-specific and/or jurisdictional requirements	
Typical failures of interest	Incorrect materials or internal components, incorrect pressure rating for a component, weld defects outside of acceptance criteria, dimensional errors, misalignment or out-of-square flanges, leak during testing, inadequate bolting (improper materials, full thread engagement, missing bolts), inadequate discharge piping structural supports, improper use of intervening valves	
Personnel qualifications	• Manufacturer requirements, documented skills, inspection certifications, or technical training for inspection and acceptance activities during manufacture and installation • Training on the sizing, selection and specification of relief devices in accordance with applicable codes and standards	
Procedure requirements	Written procedures describing: • Engineering standards for specification of equipment • Project management (including hazard and design review schedules) • Vendor qualification • Documentation requirements • Project acceptance and turnover requirements • Proper installation requirements	
Documentation requirements	Company documentation requirements typically include manufacturer's data forms, design and sizing calculations, material certifications, initial pop test results, QC results and device drawings	

TABLE 12A-4. Asset Integrity Activities for Pressure Relief Valves *(Continued)*

Inspection and Testing		
Example activities and typical frequencies	Activity	Frequency
	External visual inspection	Annual
	Process conditions, including positions of upstream and downstream valves and pressure indications between rupture disks and pressure relief valves	As part of every startup, then at least weekly thereafter (if not continuously monitored)
	Pop testing of pressure relief valve	As required by the service conditions
	Inspection of inlet and outlet piping for fouling and plugging	Whenever the device is replaced or removed for testing
	Additional inspections for specific degradation modes	As required by service conditions, condition of equipment and rate of degradation
Technical basis for activity and frequency	Scheduled with intervals determined by the results of previous activities or at fixed intervals based on inspection codes (ASME, NBIC, or API-RP-576) or jurisdictional requirements	
Sources of acceptance criteria	• Acceptance criteria from inspection codes (ASME, NBIC, or API) or jurisdictional requirements • Company standards for evaluating the condition of the device and process conditions	
Typical failures of interest	• Leakage from the valve resulting from internal component fatigue, corrosion, leaking gaskets • Failing to open at set pressure • Failing to fully reseat after opening • Fouling or plugging of pressure relief inlet • Plugging in the discharge piping by animals or water/ice resulting from loss of covers/flappers • Process valves closed that prevent the device from functioning • Fouling or plugging of the vent header • Failure of the inert gas purge system	
Personnel qualifications	Specific technical training on pressure relief valve inspection, testing, handling and installation procedures	
Procedure requirements	Written procedures describing the inspection or test activity including: • The manner, the extent, the location and the date the inspection or test is performed and by whom • The documentation and analysis of results • The resolution of functions or conditions not meeting acceptance criteria	
Documentation requirements	• Results and analysis of each inspection are documented for the life of the equipment • Inspection dates are tracked and technical deferral is required for late tests with alternate means of protection to be considered; deficient conditions are identified and resolved by the date recommended	

TABLE 12A-4. Asset Integrity Activities for Pressure Relief Valves *(Continued)*

Preventive Maintenance		
Example activities and typical frequencies	Activity	Frequency
	Activities identified from RCM or similar work planning initiatives, such as • Routine visual surveillance • Process conditions monitoring/ tracking	As required to meet preventive maintenance schedule or process monitoring needs
Technical basis for activity and frequency	Company or jurisdictional requirements	
Sources of acceptance criteria	Upper safe limits for process conditions (pressure), company requirements and good engineering practice for process conditions defined in the process safety information	
Typical failures of interest	• Leakage from the valve caused by failure to reset after functioning • Process conditions in excess of the design criteria for the device • Fouling or plugging of the vent header • Failure of the inert gas purge system	
Personnel qualifications	Specific technical training on pressure relief valve inspection, testing, handling and installation procedures	
Procedure requirements	These activities generally do not require task-specific procedures	
Documentation requirements	Results are usually recorded by exception in the equipment history file	

TABLE 12A-4. Asset Integrity Activities for Pressure Relief Valves *(Continued)*

Repair		
Example activities and typical frequencies	**Activity**	**Frequency**
	• Equipment replacement in kind • Mounting locations - repair or renewal • Piping conditions - internal restrictions • Visual inspection after the PSV operates	• 5 years, or to date stamped on the nameplate • As required by the condition of the equipment based on recommendations from ITPM activities or observations from normal operations • API RP 576
Technical basis for activity and frequency	Performed when required by the qualification period for the device or as indicated by failure during normal operations or the results or ITPM activities	
Sources of acceptance criteria	• Design codes (ASME, NBIC, or API) • Company engineering standards, facility or jurisdictional requirements • Some jurisdictions require an ASME "VR" stamp for repairs	
Typical failures of interest	• Incorrect materials or internal components, incorrect pressure rating for a component, weld defects outside of acceptance criteria, dimensional errors, misalignment or out-of-square flanges • Leak during testing • Incorrect bolting (improper materials, full thread engagement, missing bolts), inadequate discharge piping structural supports • Leakage from the valve caused by internal component fatigue or leaking gaskets	
Personnel qualifications	Specific technical training on pressure relief valve inspection, testing, handling and installation procedures	
Procedure requirements	• Craft skill procedures for typical tasks encountered in repairs and replacements (e.g., gasket installation, bolt tightening, pressure testing) • Job-specific procedures developed for repairs or replacements	
Documentation requirements	Repair history is maintained with equipment inspection history	

TABLE 12A-5. Asset Integrity Activities for Rupture Disks

New Equipment Design, Fabrication and Installation		
Example activities and typical frequencies	**Activity**	**Frequency**

	• Design requirements and process specifications
	  - Material(s) in the process
	  - Component materials selection
	  - Vessel limitations
	  - Sizing design basis and sizing calculations
	• Vendor/shop qualification
	• Equipment design by manufacturer
	• Equipment fabrication
	• Inspection and testing
	• Documentation preparation
	• Installation and commissioning
	• Acceptance and turnover
		As-required for fabrication and installation

Technical basis for activity and frequency	• Quality assurance practices for rupture disks fabrication, testing and installation • Pressure equipment design codes	
Sources of acceptance criteria	• Codes and standards for design and use of rupture disks (e.g., ASME Boiler and Pressure Vessel Code - Section VIII, NB-23, or API RP-520) in conjunction with requirements for pressure vessels • Company engineering standards, or facility-specific or jurisdictional requirements	
Typical failures of interest	Incorrect materials, incorrect pressure rating for a component, dimensional errors, misalignment or out-of-square flanges, failure during testing, incorrect installation (upside down), inadequate bolting (improper materials, full thread engagement, missing bolts), inadequate discharge piping structural supports, improper use of intervening valves	
Personnel qualifications	• Craft skills and knowledge required by the individual procedures • Training on the specific procedures for the inspection and testing activities	
Procedure requirements	Written procedures describing: • Engineering standards for specification of equipment • Vendor qualification • Documentation requirements • Project acceptance and turnover requirements • Proper installation requirements • Materials of construction records	
Documentation requirements	Company documentation requirements typically include: manufacturers data forms, design and sizing calculations, material certifications, QC results and device drawings	

TABLE 12A-5. Asset Integrity Activities for Rupture Disks *(Continued)*

Inspection and Testing		
Example activities and typical frequencies	Activity	Frequency
	• External visual inspection, including examination of tell-tale device • Process condition monitoring, including positions of upstream and downstream valves	A s part of every startup, then at least weekly thereafter
	Inspection of inlet and outlet piping for fouling and plugging	Whenever the device is replaced or removed for testing
	Additional inspections for specific degradation modes (e.g., coating of disc surface)	As required by service conditions, condition of equipment and rate of degradation
Technical basis for activity and frequency	Scheduled with intervals set by the results of previous activity or at fixed intervals based on inspection codes (ASME, NBIC, or API) or jurisdictional requirements	
Sources of acceptance criteria	• Acceptance criteria from inspection codes (ASME, NB, or API) or jurisdictional requirements • Company standards for evaluating the condition of the device and process conditions	
Typical failures of interest	• Leaking gaskets • Failing to open at set pressure • Fouling or plugging of pressure relief inlet • Plugging in the discharge piping by animals or water/ice resulting from loss of covers/flappers • Fouling or plugging of the vent header • Failure of the inert gas purge system • Failure of the rupture disk due to high pressure or corrosion • Abnormal vacuum/pressure cycling • Process valves closed that prevent the device from functioning	
Personnel qualifications	• Craft skills and knowledge required by the individual procedures • Training on the specific procedures for the inspection and testing activities	
Procedure requirements	Written procedures describing the inspection or test activity including: • The manner, the extent, the location and the timing for the inspection or test and by whom • The documentation and analysis of results • The resolution of functions or conditions not meeting acceptance criteria	
Documentation requirements	• Results and analysis of each inspection are documented and retained long enough to establish trends • Inspection dates are tracked and technical deferral required for late tests with alternate means of protection to be considered; deficient conditions are identified and resolved by the date recommended	

TABLE 12A-5. **Asset Integrity Activities for Rupture Disks** *(Continued)*

Preventive Maintenance	
Example activities and typical frequencies	Generally, the only PM for rupture disks are those activities listed here under inspection and testing
Technical basis for activity and frequency	Not applicable
Sources of acceptance criteria	Not applicable
Typical failures of interest	Not applicable
Personnel qualifications	Not applicable
Procedure requirements	Not applicable
Documentation requirements	Not applicable

TABLE 12A-5. Asset Integrity Activities for Rupture Disks *(Continued)*

Repair		
Example activities and typical frequencies	Activity	Frequency
	Replacement of rupture disk	As required
Technical basis for activity and frequency	Common repair activity	
Sources of acceptance criteria	• Design codes (ASME, NBIC, or API) • Company engineering standards, facility or jurisdictional requirements.	
Typical failures of interest	• Incorrect materials, incorrect pressure rating for a component, dimensional errors, misalignment or out-of-square flanges, failure during testing, incorrect installation (upside down) • Leaking gaskets • Incorrect bolting (improper materials, full thread engagement, missing bolts), inadequate discharge piping structural supports • Failure of the rupture disk due to high pressure or corrosion	
Personnel qualifications	• Craft skills and knowledge required by the individual procedures • Training on the specific procedures for the repair activities • Training on the specific procedures for the inspection and testing activities	
Procedure requirements	• Craft skill procedures for typical tasks encountered in repairs and replacements, such as proper installation, gasket installation and bolt tightening • Job-specific procedures developed for rupture disk replacements	
Documentation requirements	Replacement history is maintained with equipment inspection history	

TABLE 12A-6. Asset Integrity Activities for Conservation Vents and Other Low Pressure/Vacuum Relief Devices

New Equipment Design, Fabrication and Installation		
Example activities and typical frequencies	Activity	Frequency

	• Design requirements and process specifications
	  - Service requirements
	  - Vessel limitations
	  - Material(s) selection
	  - Sizing design basis
	• Vendor/shop qualification
	• Equipment design by manufacturer
	• Equipment fabrication
	• Inspection and testing
	• Documentation preparation
	• Installation and commissioning
	• Acceptance and turnover
		As required for fabrication and installation

Technical basis for activity and frequency	Quality assurance practices for pressure/vacuum relief device fabrication, testing and installation	
Sources of acceptance criteria	• Codes and standards for tank design and fabrication (API) in conjunction with requirements for the relief device application • Company engineering standards, or facility-specific or jurisdictional requirements	
Typical failures of interest	Incorrect materials of construction, incorrect pressure rating, improper sizing, weld defects outside of acceptance criteria, dimensional errors, misalignment or out-of-square flanges, leak during testing, incorrect installation, inlet/outlet blockage	
Personnel qualifications	• Manufacturer or company requirements, documented skills, inspection certifications or technical training for inspection and acceptance activities during manufacture, installation and testing • Trained in the sizing, selection and specification of pressure/vacuum relief systems and components in accordance with the company or manufacturers' standards and/or applicable codes or standards	
Procedure requirements	Written procedures describing: • Engineering standards for specification of equipment • Project management (including hazard and design review schedules) • Vendor qualification • Documentation requirements • Project acceptance and turnover requirements • Proper installation requirements	
Documentation requirements	Company documentation requirements typically include: manufacturers data forms, design and sizing calculations, material certifications, welding certifications and documentation (if applicable), QC results and fabrication drawings	

TABLE 12A-6. Asset Integrity Activities for Conservation Vents and Other Low Pressure/Vacuum Relief Devices *(Continued)*

Inspection and Testing		
Example activities and typical frequencies	Activity	Frequency
	External visual inspection	Annual
	Internal inspection	As appropriate for the service
	Inspection of inlet and outlet piping for fouling and plugging	Whenever the device is replaced or removed for testing
	Additional inspections for specific degradation modes	As required by service conditions, condition of equipment and rate of degradation
Technical basis for activity and frequency	Scheduled with intervals set by the results of previous activity or at fixed intervals based on inspection codes (e.g., API 653) and/or jurisdictional requirements (e.g., environmental permit)	
Sources of acceptance criteria	• Acceptance criteria from inspection codes (e.g., API 653) and/or jurisdictional requirements (e.g., environmental permit) • Company standards for evaluating the condition of the device and process conditions	
Typical failures of interest	• Leakage from the relief device resulting from internal component fatigue, corrosion, leaking gaskets, foreign material • Failing to open at set pressure • Failing to reseat after opening • Fouling or plugging of inlet • Plugging in the discharge piping by insects, animals or water/ice • Fouling or plugging of the vent header, if applicable • Failure of the inert gas purge system, if applicable • Cover left in place after vessel maintenance	
Personnel qualifications	Industry inspection certifications; specific technical training on low pressure/vacuum relief device inspection procedures	
Procedure requirements	Written procedures describing the inspection or test activity including: • The manner, the extent, the location and the timing for the inspection or test and by whom • The documentation and analysis of results • The resolution of functions or conditions not meeting acceptance criteria	
Documentation requirements	• Results and analysis of each inspection are documented for the life of the equipment • Inspection dates are tracked and technical deferral required for late tests with alternate means of protection to be considered; deficient conditions are identified and resolved by the date recommended	

TABLE 12A-6. Asset Integrity Activities for Conservation Vents and Other Low Pressure/Vacuum Relief Devices *(Continued)*

Preventive Maintenance		
Example activities and typical frequencies	Activity	Frequency
	Activities identified from FMEA or other analysis techniques for RCM or similar working planning initiatives, such as: • Routine visual surveillance • Process performance monitoring	As required to meet PM schedule or process monitoring needs
Technical basis for activity and frequency	Company or jurisdictional requirements	
Sources of acceptance criteria	Company requirements and good engineering practices, coupled with upper and lower safe limits for process conditions as defined in the process safety information	
Typical failures of interest	• Leakage through the relief device caused by failure to reset after functioning • Process conditions in excess of the design criteria for the device • Fouling or plugging of the atmospheric vent • Failure of the inert gas purge system, if applicable	
Personnel qualifications	• Craft-specific skills or operator-specific skills that are addressed within their respective training programs • Specific technical training on pressure/vacuum relief device inspection, testing, handling and installation procedures	
Procedure requirements	Task-specific procedures with appropriate personnel safety procedures for working on in-service tanks	
Documentation requirements	Results are usually recorded by exception in the equipment history file	

TABLE 12A-6. Asset Integrity Activities for Conservation Vents and Other Low Pressure/Vacuum Relief Devices *(Continued)*

Repair		
Example activities and typical frequencies	Activity	Frequency
	• Equipment replacement-in-kind • Mounting locations - repair or renewal • Piping conditions - internal restrictions	As required by the condition of the equipment based on recommendations from ITPM activities or observations from normal operations
Technical basis for activity and frequency	Performed when indicated by failure, by the results of PM activities or by the results of inspection and testing activities	
Sources of acceptance criteria	• Codes for design and fabrication in conjunction with requirements for the vessel being protected • Company engineering standards, or facility-specific or jurisdictional requirements	
Typical failures of interest	• Incorrect materials, incorrect pressure rating for a component, dimensional errors, misalignment or out-of-square flanges, failure during testing, incorrect installation, inlet/outlet blockage • Leaking gaskets • Failure of the relief device due to high pressure, high vacuum or corrosion	
Personnel qualifications	• Craft skills and knowledge required by the individual procedures • Training on the specific procedures for the repair activities • Training on the specific procedures for the inspection and testing activities	
Procedure requirements	• Craft skill procedures for typical tasks encountered in repairs and replacements, such as proper installation, gasket installation and bolt tightening • Job-specific procedures developed for relief device replacements	
Documentation requirements	Replacement history is maintained with equipment inspection history	

TABLE 12A-7. Asset Integrity Activities for Flame/Detonation Arresters

New Equipment Design, Fabrication and Installation		
Example activities and typical frequencies	**Activity**	**Frequency**
	Verify equipment specifications are met: • Service requirements • Materials selection • Sizing design basis and sizing calculations	As required for fabrication and installation
	Fabrication shop/contractor qualifications • Design approval by plant • Welding/QC plan	
	Fabrication and installation	
	Inspection	
	Documentation / Commissioning and testing	
	Acceptance and turnover	
Technical basis for activity and frequency	QA practices for flame/detonation arrestor fabrication, testing and installation	
Sources of acceptance criteria	• Codes for design and fabrication in conjunction with requirements for the vessel being protected • Company engineering standards, or facility-specific or jurisdictional requirements	
Typical failures of interest	Incorrect materials or internal components, incorrect dimensions, inadequate flow rate provided, misalignment or out-of-square flanges, or improper installation	
Personnel qualifications	• Manufacturer requirements, documented skills, inspection certifications or technical training for inspection and acceptance activities during manufacture and installation • Trained in the sizing, selection and specification of the device in accordance with the manufacturers' recommendations or other applicable code or standard	
Procedure requirements	Written procedures describing: • Engineering standards for specification of equipment • Project management (including hazard and design review schedules) • Vendor qualification • Documentation requirements • Project acceptance and turnover requirements • Proper installation requirements	
Documentation requirements	Company documentation requirements typically include: design calculations, material certifications, factory test results/certification, QC results, fabrication drawings and installation documentation	

TABLE 12A-7. Asset Integrity Activities for Flame/Detonation Arresters *(Cont'd)*

Inspection and Testing		
Example activities and typical frequencies	Activity	Frequency
	External visual inspection	Annual
	Internal inspection	As appropriate for the service
	Inspection of inlet and outlet piping for fouling and plugging	Whenever the device is replaced or removed for testing; take into consideration the type of service
	Additional inspections for specific degradation modes	As required by service conditions, condition of equipment and rate of degradation
Technical basis for activity and frequency	Scheduled with intervals set by the results of previous activity or at intervals based on inspection code or jurisdictional requirements	
Sources of acceptance criteria	• Acceptance criteria from inspection codes or jurisdictional requirements • Company standards for evaluating the condition of the device and process conditions	
Typical failures of interest	Plugging of the device by process materials, insects or dirt or by corrosion	
Personnel qualifications	Specific technical training on flame/detonation arrester relief valve inspection, testing, handling and installation procedures	
Procedure requirements	Written procedures describing the inspection or test activity including: • The manner, the extent, the location and the timing for the inspection or test and by whom • The documentation and analysis of results • The resolution of functions or conditions not meeting acceptance criteria	
Documentation requirements	• Results and analysis of each inspection are documented for long enough to establish trends • Inspection dates are tracked and technical deferral required for late tests with alternate means of protection to be considered. Deficient conditions are identified and resolved by the date recommended	

TABLE 12A-7. Asset Integrity Activities for Flame/Detonation Arresters *(Cont'd)*

Preventive Maintenance		
Example activities and typical frequencies	**Activity**	**Frequency**
	Activities identified from FMEA or other analysis techniques for RCM or similar working planning initiatives, such as: • Routine visual surveillance • Process performance monitoring	As required to meet PM schedule or process monitoring needs
Technical basis for activity and frequency	Company or jurisdictional requirements	
Sources of acceptance criteria	Company requirements and good engineering practice for process conditions defined in the process safety information	
Typical failures of interest	Plugging of the device	
Personnel qualifications	Specific technical training on flame/detonation arrester inspection, testing, handling and installation procedures	
Procedure requirements	These activities generally do not require task-specific procedures	
Documentation requirements	Results are usually recorded by exception in the equipment history file	

TABLE 12A-7. Asset Integrity Activities for Flame/Detonation Arresters *(Cont'd)*

Repair		
Example activities and typical frequencies	Activity	Frequency
	• Equipment replacement-in-kind • Mounting locations - repair or renewal • Cleaning of device removal of plugging materials	As required by the condition of the equipment based on recommendations from the ITPM activities
Technical basis for activity and frequency	Performed when indicated by failure, by the results of PM activities or by the results of inspection and testing activities	
Sources of acceptance criteria	• Codes for design and fabrication in conjunction with requirements for the vessel being protected • Company engineering standards, or facility-specific or jurisdictional requirements	
Typical failures of interest	• Incorrect materials or internal components • Incorrect dimensions • Inadequate flow rate provided • Improper installation; misalignment or out-of-square flanges • Incorrect bolting (improper materials, full thread engagement, missing bolts)	
Personnel qualifications	Specific technical training on flame/detonation arrester relief valve inspection, testing, handling and installation procedures	
Procedure requirements	• Craft skill procedures for typical tasks encountered in replacements or repairs (e.g., gasket installation, bolt tightening) • Job-specific procedures developed for replacements, repairs, or alterations	
Documentation requirements	Repair history is maintained with equipment inspection history	

TABLE 12A-8. Asset Integrity Activities for Emergency Vents

New Equipment Design, Fabrication and Installation		
Example activities and typical frequencies	Activity	Frequency
	Verify equipment specifications are met: • Service requirements • Materials selection • Sizing design basis and sizing calculations	As required for fabrication and installation
	Fabrication shop/contractor qualifications • Design approval by plant • Welding/QC plan and sizing calculations	
	Fabrication and installation	
	Inspection	
	Documentation / Commissioning and testing	
	Acceptance and turnover	
Technical basis for activity and frequency	QA practices for the equipment associated with the emergency vent system fabrication, testing and installation	
Sources of acceptance criteria	• Codes for tank design and fabrication (UL and ASME) in conjunction with requirements for the emergency venting application • Company engineering standards, or facility-specific or jurisdictional requirements	
Typical failures of interest	Incorrect materials or components used, dimensional errors, incorrect emergency venting requirements for the vessel used, welding defects outside of acceptance criteria, use of unqualified welders or welding procedures, incorrect bolting, inadequate discharge piping structural supports, inlet/outlet blockage	
Personnel qualifications	• Contractor or company requirements and documented skills for installation, welding, NDE examinations and technical training for inspection and acceptance activities • Trained in the sizing, selection and specification of emergency vent systems and components in accordance with the company or manufacturers' standards and/or applicable codes or standards	
Procedure requirements	Written procedures describing: • Engineering standards for specification of equipment • Project management (including hazard and design review schedules) • Vendor qualification • Documentation requirements • Project acceptance and turnover requirements • Proper installation requirements	
Documentation requirements	Company documentation requirements typically include: design calculations, material certifications, welding certifications and documentation, QC results and fabrication drawings	

TABLE 12A-8. Asset Integrity Activities for Emergency Vents *(Continued)*

Inspection and Testing		
Example activities and typical frequencies	**Activity**	**Frequency**
	External visual inspection	Annual
	Inspection of piping for fouling and plugging	Whenever the piping is opened to replace or remove a device (e.g., control valve)
	Additional inspections for specific degradation modes	As required by service conditions, condition of equipment and rate of degradation
Technical basis for activity and frequency	Scheduled with intervals set by the results of inspections, intervals allowed by process operation, or jurisdictional requirements	
Sources of acceptance criteria	• Acceptance criteria from vessel codes or jurisdictional requirements • Company standards for evaluating the condition of the device and process conditions	
Typical failures of interest	• Leakage from cracks or gaskets • Corrosion (internal or external) • Failed supports or restraining devices • Obstructions to movement of the relief device (including paint) • Fouling or plugging of inlet • Plugging in the discharge piping or vent header	
Personnel qualifications	Documented NDE qualifications, industry inspection certifications, or specific technical training for analysis of inspection and testing results	
Procedure requirements	Written procedures describing the inspection or test activity, including: • The manner, the extent, the location and the timing for the inspection or test and by whom • The documentation and analysis of results • The resolution of functions or conditions not meeting acceptance criteria	
Documentation requirements	• Results and analysis of each inspection are documented for long enough to establish trends • Inspection dates are tracked and technical deferral required for late tests with alternate means of protection to be considered. Deficient conditions are identified and resolved by the date recommended	

TABLE 12A-8. Asset Integrity Activities for Emergency Vents *(Continued)*

Preventive Maintenance		
Example activities and typical frequencies	Activity	Frequency
	Activities identified from FMEA or other analysis techniques for RCM or similar working planning initiatives, such as: • Routine visual surveillance • Routine performance monitoring and tracking	As required to meet PM schedule
Technical basis for activity and frequency	Company or jurisdictional requirements	
Sources of acceptance criteria	• Safe upper and lower limits for process conditions • Company requirements and good engineering practice for process conditions defined in the process safety information	
Typical failures of interest	• Leakage through the device • Process conditions in excess of the design criteria for the device • Fouling or plugging of the inlet or outlet piping or vent header	
Personnel qualifications	Documented NDE qualifications, industry inspection certifications, or specific technical training for analysis of inspection and testing results	
Procedure requirements	These activities generally do not require task-specific procedures	
Documentation requirements	Results are usually recorded by exception in the equipment history file	

TABLE 12A-8. Asset Integrity Activities for Emergency Vents *(Continued)*

Repair		
Example activities and typical frequencies	Activity	Frequency
	Replacement-in-kind	As required by the condition of the equipment based on recommendations from the ITPM activities
	Mounting locations and vent path– repairs and upkeep	
	Cleaning of device surfaces and gaskets	
Technical basis for activity and frequency	Performed when indicated by failure, by the results of PM activities or, by the results of inspection and testing activities	
Sources of acceptance criteria	• Codes for tank design and fabrication (UL and ASME) in conjunction with requirements for the emergency venting application • Company engineering standards, or facility-specific or jurisdictional requirements	
Typical failures of interest	• Incorrect materials or components used • Dimensional errors • Incorrect emergency venting requirements for the vessel used • Welding defects outside of acceptance criteria or use of unqualified welders or welding procedures • Incorrect pressure rating for a component • Misalignment or out-of-square flanges • Incorrect bolting (improper materials, full thread engagement, missing bolts) • Inadequate discharge piping structural supports	
Personnel qualifications	Contractor or company requirements and documented skills for installation, welding, NDE examinations and technical training for inspection and acceptance activities	
Procedure requirements	• Craft skill procedures for typical tasks encountered in repairs and replacement (e.g., welding, gasket installation, bolt tightening) • Job-specific procedures developed for repairs or alterations	
Documentation requirements	Repair history is maintained with equipment inspection history	

TABLE 12A-9. Asset Integrity Activities for Vent Headers

New Equipment Design, Fabrication and Installation		
Example activities and typical frequencies	**Activity**	**Frequency**

	• Verify equipment specifications are met:
	  - Service requirements
	  - Materials selection
	• Fabrication shop / contractor qualification, particularly:
	  - Design approval by plant
	  - Welding/QA plan
	• Fabrication and installation
	• Inspection
	• Documentation
	• Commissioning and testing
	• Acceptance and turnover
		As required for fabrication and installation

	External visual inspection	After installation
	NDT (e.g., weld testing, hydrostatic testing)	After fabrication
	Verification of materials of construction	When materials are received
Technical basis for activity and frequency	QA practices for piping and vent header fabrication, testing and installation	
Sources of acceptance criteria	• ANSI B31.3 codes for piping design and fabrication in conjunction with requirements for the venting application • Company engineering standards, or facility-specific or jurisdictional requirements	
Typical failures of interest	Incorrect materials used, dimensional errors, misalignment or out-of-square flanges, incorrect pressure requirements for the vessel used, leaks during testing, welding defects outside of acceptance criteria, use of unqualified welders or welding procedures, inadequate bolting (improper materials, less than full thread engagement, missing bolts), inadequate supports	
Personnel qualifications	Contractor or company requirements and documented skills for installation, welding (certifications per ASME Section IX), NDE examinations and technical training for inspection and acceptance activities	
Procedure requirements	Written procedures describing: • Engineering standards for specification of equipment • Project management (including hazard and design review schedules) • Vendor qualification • Documentation requirements • Project acceptance and turnover requirements • Proper installation requirements	
Documentation requirements	Company documentation requirements typically include: design calculations, material certifications, welding certifications and documentation (maps), QC results and fabrication drawings	

TABLE 12A-9. Asset Integrity Activities for Vent Headers *(Continued)*

Inspection and Testing		
Example activities and typical frequencies	Activity	Frequency
	External visual inspection	Intervals as specified by API standards
	Thickness measurement and/or internal inspections	Intervals as specified by API standards (e.g., ½ of calculated remaining life)
	Additional inspections for specific degradation modes	As required by condition of equipment and rate of degradation
	Verification of block valve positions	Monthly or quarterly
Technical basis for activity and frequency	Scheduled with intervals set by the results of inspections, or default intervals listed in the inspection (API-570) or jurisdictional requirements	
Sources of acceptance criteria	• Acceptance criteria from inspection codes API-570 or jurisdictional requirements • Acceptance criteria for damage from specific degradation modes per API-579	
Typical failures of interest	Leakage from welds, cracks, or gaskets, corrosion (internal or external), inadequate bolting, failed piping supports, distortion of piping, or piping component not meeting pressure rating	
Personnel qualifications	Documented NDE qualifications, industry inspection certifications (API-570), or specific technical training for analysis of inspection and testing results	
Procedure requirements	Written procedures describing the inspection or test activity including: • The manner, the extent, the location and the timing for the inspection or test and by whom • The documentation and analysis of results • The resolution of functions or conditions not meeting acceptance criteria	
Documentation requirements	• Results and analysis of each inspection are documented for the life of the equipment • Inspection dates are tracked and technical deferral required for late tests with alternate means of protection to be considered. Deficient conditions are identified and resolved by the date recommended	

TABLE 12A-9. Asset Integrity Activities for Vent Headers *(Continued)*

Preventive Maintenance		
Example activities and typical frequencies	Activity	Frequency
	• Activities identified from FMEA or other analysis techniques for RCM or similar working planning initiatives, such as routine performance monitoring and tracking • Tightening of valve packing	As required to meet PM schedule
Technical basis for activity and frequency	Company or jurisdictional requirements	
Sources of acceptance criteria	• Safe upper and lower limits for process conditions • Company requirements and good engineering practice for process conditions defined in the process safety information	
Typical failures of interest	Process conditions in excess of safe upper or lower design limits	
Personnel qualifications	Documented NDE qualifications, industry inspection certifications (API-570), or specific technical training for analysis of inspection and testing results	
Procedure requirements	These activities generally do not require task-specific procedures	
Documentation requirements	Results are usually recorded by exception in the equipment history file	

TABLE 12A-9. Asset Integrity Activities for Vent Headers *(Continued)*

Repair		
	Activity	Frequency
Example activities and typical frequencies	• Piping replacement-in-kind • Painting • Insulation repairs • Supports and hangers, repair and replacement	As required by the condition of the equipment based on recommendations from the inspection and testing or PM activities
Technical basis for activity and frequency	Performed when indicated by failure, by the results of PM activities or by the results of inspection and testing activities	
Sources of acceptance criteria	• ANSI B31.3 codes for piping design and fabrication in conjunction with requirements for the venting application • Company engineering standards, or facility-specific or jurisdictional requirements	
Typical failures of interest	Dimensional errors, incorrect material or weld metal, misaligned or out-of-square flanges, incorrect pressure rating for a component, leak during testing, weld defects outside of acceptance criteria, use of unqualified welder or welding procedure, inadequate bolting (improper materials, full thread engagement, missing bolts), inadequate or missing piping supports	
Personnel qualifications	Contractor or company requirements and documented skills for installation, welding (certifications per ASME Section IX), NDE examinations and technical training for inspection and acceptance activities	
Procedure requirements	• Craft skill procedures for typical tasks encountered in repairs and replacement (e.g., welding, gasket installation, bolt tightening) • Job-specific procedures developed for repairs or alterations	
Documentation requirements	Repair history is maintained with equipment inspection history	

TABLE 12A-10. Asset Integrity Activities for Thermal Oxidizers

New Equipment Design, Fabrication and Installation		
Example activities and typical frequencies	Activity	Frequency
	• Verify equipment specifications are met: - Process design requirements - Materials selection • Vendor/Shop qualification • Equipment design by manufacturer • Equipment fabrication • Inspection and testing • Documentation preparation • Installation and commissioning • Acceptance and turnover	As required for fabrication and installation
Technical basis for activity and frequency	QA practices for thermal oxidizer system fabrication, testing and installation	
Sources of acceptance criteria	• Codes and standards for design and fabrication of the thermal oxidizer and the burner controls and injection lines (ASME or API) • Company engineering standards, or facility-specific or jurisdictional requirements	
Typical failures of interest	• Incorrect materials or components, incorrect design or capacity for materials to be handled, weld defects outside of acceptance criteria, dimensional errors, leaks during testing, refractory material or installation problems, toxic or hazardous material releases • Ignition system and burner failures and material flow control problems • Blower/fan, motor, bearings or support installation errors	
Personnel qualifications	Manufacturer or company requirements, documented skills, inspection certifications or technical training for inspection and acceptance activities during manufacture, installation and testing	
Procedure requirements	Written procedures describing: • Engineering standards for specification of equipment • Project management (including hazard and design review schedules) • Vendor qualification • Documentation requirements • Project acceptance and turnover requirements	
Documentation requirements	Company documentation requirements typically include: manufacturers' data forms, design and sizing calculations, material certifications, QC results, fabrication drawings and testing documentation	

TABLE 12A-10. Asset Integrity Activities for Thermal Oxidizers *(Continued)*

Inspection and Testing		
Example activities and typical frequencies	Activity	Frequency
	External visual inspection of the various system components	Annual
	Additional inspections for specific degradation modes	As required by service conditions, condition of equipment and rate of degradation
	Inspection and testing of the burner system and controls	As required based on plant experience and in accordance with applicable codes and standards
Technical basis for activity and frequency	Scheduled with intervals set by the results of previous activity or at fixed intervals based on inspection codes (ASME or API) or jurisdictional requirements	
Sources of acceptance criteria	• Acceptance criteria from inspection codes (ASME or API) or jurisdictional requirements • Company standards for evaluation the condition of the various thermal oxidizer system components	
Typical failures of interest	• Obstructions/plugging in the transfer lines or in the injection nozzles • Pilot and burner flame detection or ignition systems failure • Corrosion • Refractory damage	
Personnel qualifications	Specific technical training on thermal oxidizers and on burner control and ignition systems	
Procedure requirements	Written procedures describing the inspection or test activity including: • The manner, the extent, the location and the timing for the inspection or test and by whom • The documentation and analysis of results • The resolution of functions or conditions not meeting acceptance criteria	
Documentation requirements	• Results and analysis of each inspection are documented for long enough to establish trends • Inspection dates are tracked and technical deferral required for late tests with alternate means of protection to be considered. Deficient conditions are identified and resolved by the date recommended	

TABLE 12A-10. Asset Integrity Activities for Thermal Oxidizers *(Continued)*

Preventive Maintenance		
Example activities and typical frequencies	Activity	Frequency
	Activities identified from FMEA or other analysis techniques for RCM or similar working planning initiatives, such as: • Routine visual surveillance • Process performance monitoring	As required to meet PM schedule or process monitoring needs
Technical basis for activity and frequency	Company or jurisdictional requirements	
Sources of acceptance criteria	Upper and lower safe limits for process conditions, company requirements and good engineering practice for process conditions defined in the process safety information	
Typical failures of interest	• Obstructions/plugging in the transfer lines or in the injection nozzles • Pilot and burner flame detection or ignition systems failure • Corrosion • Refractory damage • Blower/fan, motor or bearings failure	
Personnel qualifications	Specific technical training on thermal oxidizers and on burner control and ignition systems	
Procedure requirements	These activities generally do not require task-specific procedures	
Documentation requirements	Results are usually recorded by exception in the equipment history file	

TABLE 12A-10. **Asset Integrity Activities for Thermal Oxidizers** *(Continued)*

Repair		
Example activities and typical frequencies	Activity	Frequency
	• Equipment repairs and replacement-in-kind • Structural support and anchoring systems – repair and replacement	As required by the condition of the equipment based on recommendations from the ITPM activities
Technical basis for activity and frequency	Performed when indicated by failure, by the results of PM activities or by the results of inspection and testing activities	
Sources of acceptance criteria	• Design codes (ASME and API) • Company engineering standards, facility or jurisdictional requirements	
Typical failures of interest	• Incorrect materials or components, incorrect design or capacity for materials to be handled, weld defects outside of acceptance criteria, dimensional errors, leaks during testing, refractory material problems, toxic or hazardous material releases • Ignition system and burner failures and material flow control problems • Blower/fan, motor or bearings failure	
Personnel qualifications	Specific technical training on thermal oxidizers and on burner control and ignition systems	
Procedure requirements	• Craft skill procedures for typical tasks encountered in replacements and repairs (e.g., system isolation, purging, component removal/installation) • Job-specific procedures developed for repairs or alterations	
Documentation requirements	Repair history is maintained with equipment inspection history	

TABLE 12A-11. Asset Integrity Activities for Flare Systems

New Equipment Design, Fabrication and Installation		
Example activities and typical frequencies	Activity	Frequency
	Verify equipment specifications are met: • Service requirements • Materials selection • Sizing design basis and sizing calculations	As required for fabrication and installation
	Fabrication shop/contractor qualifications • Design approval by plant • Welding/QC plan	
	Fabrication and installation	
	Inspection	
	Documentation / Commissioning and testing	
	Acceptance and turnover	
Technical basis for activity and frequency	• API 537 • QA practices for flare system fabrication, testing and installation	
Sources of acceptance criteria	• Codes and standards for design and fabrication of vent lines and flare systems (ASME or API) • Company Engineering Standards, or facility-specific or jurisdictional requirements	
Typical failures of interest	• Incorrect materials or components, incorrect design or capacity for all vent sources, weld defects outside of acceptance criteria, dimensional errors, leaks during testing, incorrect seal design or seal failure, incorrect heat radiation calculations, toxic or hazardous material releases • Ignition system and burner failures, flow control problems, inadequate sweep gas system, air ingress, flashback protection failures • Improper additions or connections to the system • Structural support or foundation inadequacies	
Personnel qualifications	• Manufacturer or company requirements, documented skills, inspection certifications or technical training for inspection and acceptance activities during manufacture, installation and testing • Trained in the sizing, selection and specification of flare systems and components in accordance with the company or manufacturers' standards and/or applicable codes or standards	
Procedure requirements	Written procedures describing: • Engineering standards for specification of equipment • Project management (including hazard and design review schedules) • Vendor qualification • Documentation requirements • Project acceptance and turnover requirements • Proper installation requirements	
Documentation requirements	Company documentation requirements typically include: manufacturers data forms, design and sizing calculations, material certifications, QC results, fabrication drawings and testing documentation	

TABLE 12A-11. Asset Integrity Activities for Flare Systems *(Continued)*

Inspection and Testing		
Example activities and typical frequencies	Activity	Frequency
	External visual inspection of the various system components	• Annual for those that are accessible during operation • During each shutdown of the flare system
	Inspection of stack and associated support systems	As required based on plant experience
	Inspection and testing of the ignition system and controls	As required based on plant experience
	Inspection and verification of sweep gas system, liquid seal level and positions of valves in the system	Weekly
	Inspection of piping for fouling and plugging	Whenever the piping is opened to replace or remove a device (e.g., control valve)
	Additional inspections for specific degradation modes	As required by service conditions, condition of equipment and rate of degradation
Technical basis for activity and frequency	Scheduled with intervals set by the results of previous activity or at fixed intervals based on inspection codes (ASME or API) or jurisdictional requirements	
Sources of acceptance criteria	• Acceptance criteria from inspection codes (ASME or API) or jurisdictional requirements • Company standards for evaluation the condition of the various flare system components	
Typical failures of interest	• Obstructions/plugging in the transfer lines or in the flare stack • Pilot flame detection or ignition system failure • Corrosion • Process valves closed that prevent the flare system from functioning	
Personnel qualifications	Specific technical training on flares, vent systems and ignition systems	
Procedure requirements	Written procedures describing the inspection or test activity including: • The manner, the extent, the location and the timing for the inspection or test and by whom • The documentation and analysis of results • The resolution of functions or conditions not meeting acceptance criteria	
Documentation requirements	• Results and analysis of each inspection are documented for long enough to establish trends • Inspection dates are tracked and technical deferral required for late tests with alternate means of protection to be considered. Deficient conditions are identified and resolved by the date recommended	

TABLE 12A-11. Asset Integrity Activities for Flare Systems *(Continued)*

Preventive Maintenance		
Example activities and typical frequencies	Activity	Frequency
	Activities identified from FMEA or other analysis techniques for RCM or similar working planning initiatives, such as: • Routine visual surveillance • Process performance monitoring	As required to meet PM schedule or process monitoring needs
Technical basis for activity and frequency	Company or jurisdictional requirements	
Sources of acceptance criteria	Upper and lower safe limits for process conditions, company requirements and good engineering practice for process conditions defined in the process safety information	
Typical failures of interest	• Obstructions/plugging in the transfer lines or in the flare stack • Pilot flame detection or ignition system failure • Corrosion • Process valves closed that prevent the flare system from functioning • Structural support or foundation inadequacies	
Personnel qualifications	Specific technical training on flares, vent systems and ignition systems	
Procedure requirements	These activities generally do not require task-specific procedures	
Documentation requirements	Results are usually recorded by exception in the equipment history file	

TABLE 12A-11. Asset Integrity Activities for Flare Systems *(Continued)*

Repair		
Example activities and typical frequencies	Activity	Frequency
	• Equipment repairs and replacement-in-kind • Structural support and anchoring systems – repair and replacement	As required by the condition of the equipment based on recommendations from the ITPM activities
Technical basis for activity and frequency	Performed when indicated by failure, by the results of PM activities or by the results of inspection and testing activities	
Sources of acceptance criteria	• Design codes (ASME and API) • Company engineering standards, facility or jurisdictional requirements	
Typical failures of interest	• Incorrect materials or components, incorrect design or capacity for all vent sources, weld defects outside of acceptance criteria, dimensional errors, leaks during testing, incorrect seal design or seal failure, incorrect heat radiation calculations, toxic or hazardous material releases • Ignition system and burner failures, flow control problems, inadequate sweep gas system, air ingress, flashback protection failures • Improper additions or connections to the system • Structural support or foundation inadequacies	
Personnel qualifications	Specific technical training on flares, vent systems and ignition systems	
Procedure requirements	• Craft skill procedures for typical tasks encountered in replacements and repairs (e.g., system isolation, purging, lockout/tagout, component removal/installation) • Job-specific procedures developed for repairs or alterations	
Documentation requirements	Repair history is maintained with equipment inspection history	

TABLE 12A-12. Asset Integrity Activities for Explosion Vents

New Equipment Design, Fabrication and Installation		
	Activity	Frequency

Example activities and typical frequencies	• Design requirements and process specifications
	  - Service requirements
	  - Vessel limitations
	  - Material(s) selection
	  - Sizing design basis and sizing calculations
	• Vendor/shop qualification
	• Equipment design by manufacturer
	• Equipment fabrication
	• Inspection and testing
	• Documentation preparation
	• Installation and commissioning
	• Labeling ("WARNING: Explosion relief device")
	• Acceptance and turnover
		As required for fabrication and installation

Technical basis for activity and frequency	• Quality assurance practices for explosion vent fabrication, testing and installation • NFPA 68	
Sources of acceptance criteria	• Codes and standards for design and use of explosion vents (NFPA 68) in conjunction with requirements for protected vessels and equipment • Company engineering standards, or facility-specific or jurisdictional requirements	
Typical failures of interest	Incorrect materials of construction, incorrect pressure rating, improper sizing, weld defects outside of acceptance criteria, dimensional errors, misalignment or out-of-square flanges, leak during testing, improper installation	
Personnel qualifications	• Manufacturer or company requirements, documented skills, inspection certifications or technical training for inspection and acceptance activities during manufacture, installation and testing • Trained in the sizing, selection and specification of explosion venting systems and components in accordance with the company or manufacturers' standards and/or applicable codes or standards	
Procedure requirements	Written procedures describing: • Engineering standards for specification of equipment • Project management (including hazard and design review schedules) • Vendor qualification • Documentation requirements • Project acceptance and turnover requirements • Proper installation requirements	
Documentation requirements	Company documentation requirements typically include: manufacturers data forms, design and sizing calculations, material certifications, factory test results/certification, QC results, fabrication drawings and installation documentation	

TABLE 12A-12. Asset Integrity Activities for Explosion Vents *(Continued)*

Inspection and Testing		
Example activities and typical frequencies	**Activity**	**Frequency**
	Visual inspection	At least annually; frequency can be increased or decreased based on documented operating experience
	Visual inspection	After each process maintenance turnaround
	Visual inspection	After each actuation, before being placed back into service
	Visual inspection for physical damage or obstruction	After every act of nature or process upset condition
	Additional inspections for specific degradation modes (e.g., buildup of process material on vent enclosure)	As required by service conditions, condition of equipment and rate of degradation
Technical basis for activity and frequency	Scheduled with intervals set by the results of previous activity or at fixed intervals based on standards (NFPA 68), inspection codes and/or jurisdictional requirements	
Sources of acceptance criteria	• Acceptance criteria from standards (NFPA 68), inspection codes and/or jurisdictional requirements • Company standards for evaluation the condition of the device and process conditions	
Typical failures of interest	• Leakage from the explosion vent resulting from internal component fatigue, corrosion, leaking gaskets • Failing to open at set pressure • Damage from relief event or from external force • Failing to fully reclose after opening • Discharge blockage	
Personnel qualifications	Industry inspection certifications; specific technical training on explosion vent inspection procedures	
Procedure requirements	Written procedures describing the inspection or test activity including: • The manner, the extent, the location and the timing for the inspection or test and by whom • The documentation and analysis of results • The resolution of functions or conditions not meeting acceptance criteria	
Documentation requirements	• Results and analysis of each inspection are documented for the life of the equipment • Inspection dates are tracked and technical deferral required for late tests with alternate means of protection to be considered; deficient conditions are identified and resolved by the date recommended	

TABLE 12A-12. Asset Integrity Activities for Explosion Vents *(Continued)*

Preventive Maintenance		
Example activities and typical frequencies	Activity	Frequency
	Lubrication of closure hinges (if provided)	At least annually; frequency can be increased or decreased based on documented operating experience
	Activities identified from FMEA or other analysis techniques for RCM or similar working planning initiatives, such as: • Routine visual surveillance • Process performance monitoring	As required to meet PM schedule or process monitoring needs
Technical basis for activity and frequency	Company or jurisdictional requirements	
Sources of acceptance criteria	Company requirements and good engineering practices, coupled with upper and lower safe limits for process conditions as defined in the process safety information	
Typical failures of interest	Failure of the device to open on demand	
Personnel qualifications	Specific technical training on explosion vent inspection, testing, handling and installation procedures	
Procedure requirements	These activities generally do not require task-specific procedures	
Documentation requirements	Results are usually recorded by exception in the equipment history file	

TABLE 12A-12. Asset Integrity Activities for Explosion Vents *(Continued)*

Repair		
Example activities and typical frequencies	Activity	Frequency
	Visual inspection for physical damage or obstruction	After every act of nature or process upset condition
	Non-reclosing explosion vent replacement	As required
Technical basis for activity and frequency	Performed when indicated by failure, by the results of PM activities or by the results of inspection and testing activities	
Sources of acceptance criteria	• Codes for design and fabrication in conjunction with requirements for the vessel being protected • Company engineering standards, or facility-specific or jurisdictional requirements	
Typical failures of interest	• Incorrect materials or components, incorrect pressure rating for a replaced vent, misalignment or out-of-square flanges, failure during testing, incorrect installation • Leaking gaskets • Failure of the explosion vent due to high pressure or corrosion	
Personnel qualifications	• Craft skills and knowledge required by the individual procedures • Training on the specific procedures for the repair activities • Training on the specific procedures for the inspection and testing activities	
Procedure requirements	• Craft skill procedures for typical tasks encountered in repairs and replacements, such as proper installation, gasket installation and bolt tightening • Job-specific procedures developed for explosion vent replacements	
Documentation requirements	Replacement history is maintained with equipment inspection history	

TABLE 12A-13. Asset Integrity Activities for Safety Instrumented Systems and Emergency Shutdowns

New Equipment Design, Fabrication and Installation		
	Activity	Frequency
Example activities and typical frequencies	Identification of materials of construction	When received
	Calibration of field devices (e.g., transmitter, switches)	Before or at initial installation
	Visual inspection of field devices and installation	Initial installation
	SIS loop check	Initial installation
	SIS functional test	Initial installation
	Manufacturer tests of logic solver, operator interface and engineering interface	During fabrication
	Factory acceptance tests of logic solver, operator interface and engineering interface	Before delivery and confirmation tests after delivery
Technical basis for activity and frequency	• API RP-554, API RP-551 • Manufacturers' recommendations • Industrial insurers' recommendations • Common industry practices	
Sources of acceptance criteria	• Device specifications • SIS specifications • Manufacturers' recommendations • Industrial insurers' recommendations • Company engineering and/or maintenance standards	
Typical failures of interest	• Leakage resulting from improper installation or materials of construction (e.g., incorrect gasket, incorrect metallurgy of wetted parts) • Failure to operate on demand caused by improper installation (e.g., incorrectly wired), incorrect configuration of control system and/or incorrect calibration of the device • Potential ignition source or electrical shock due to improper installation • Improper rating for the electrical classification of the installed location • Improper management of intervening valves between the instrument and the equipment being protected	
Personnel qualifications	• Craft skills and knowledge required by the individual procedures • Training on the use and operation of special tools (e.g., signal simulators) required by procedures	
Procedure requirements	• Procurement and receiving procedures to ensure proper materials of construction • Device-specific testing, calibration and installation procedures • Manufacturers' manuals • Special tool (e.g., signal simulator) use and operation procedures	
Documentation requirements	• Vendor material of construction reports • Calibration record • Loop check sheet • Functional test record • Manufacturers' test report • Factory acceptance test report • Facility acceptance test report	Special notes: • Records retained for the life of the equipment • Installation documentation to support PSSR requirements

TABLE 12A-13. Asset Integrity Activities for Safety Instrumented Systems and Emergency Shutdowns *(Continued)*

Inspection and Testing		
	Activity	Frequency
Example activities and typical frequencies	• Calibration of field devices (e.g., transmitters, switches) • Loop check • Functional test • Checking/running of logic solver diagnostics	As specified to meet the safety performance requirements
Technical basis for activity and frequency	• API RP-554 • API RP-551 • IEC 61508 • Manufacturers' recommendations • Industrial insurers' recommendations • Common industry practices	
Sources of acceptance criteria	• Device specifications • SIS specifications • Manufacturers' recommendations • Industrial insurers' recommendations • Company engineering and/or maintenance standards	
Typical failures of interest	• Failure to operate on demand or spurious trip of system as a result of wiring failure (e.g., loose connection, short), failure of input device (e.g., sensor electronic failure), failure of controller/local solver (e.g., I/O card failure), or unauthorized change to controller/logic solver configuration and/or bypassing/forcing of the interlock/alarm • Failure to operate on demand caused by isolation/plugging of input device connection or other process conditions (e.g., buildup on a temperature probe) that render input device inoperable or incapable of accurately measuring process conditions (e.g., pressure, temperature) • Potential ignition source if the device or connection shorts • Leakage at process connection as a result of overpressurization of the joint • Improper management of intervening valves between the instrument and the equipment being protected	
Personnel qualifications	• Craft skills and knowledge required by the individual procedures • Training on the specific procedures for the inspection and testing activities • Training on the use and operation of special tools (e.g., signal simulators) required by procedures	
Procedure requirements	Written procedures describing the inspection or test activity including: • The manner, the extent, the location and the timing for the inspection or test and by whom • The documentation and analysis of results • The resolution of functions or conditions not meeting acceptance criteria	
Documentation requirements	• Calibration record • Loop check sheet • Functional test record • Diagnostic check sheet	Special notes: • Records retained long enough to establish trends • Documentation by exception may be acceptable (for certain tasks and selected electric motors)

TABLE 12A-13. Asset Integrity Activities for Safety Instrumented Systems and Emergency Shutdowns *(Continued)*

Preventive Maintenance		
Example activities and typical frequencies	Activity	Frequency
	• Replacement of logic solver battery • Replacement/cleaning of logic solver, operator interface and engineering interface cabinet air filters • Replacement of other components before end of life / wear-out interval	Manufacturers' recommendation
Technical basis for activity and frequency	Manufacturers' recommendations	
Sources of acceptance criteria	Manufacturers' recommendations	
Typical failures of interest	• Failure to operate on demand or spurious trip of system as a result of wiring failure (e.g., loose connection, short), failure of input device (e.g., sensor electronic failure), or failure of controller/local solver (e.g., I/O card failure) • Failure to operate on demand or spurious trip as a result of an unauthorized change to controller/logic solver configuration and/or bypassing/forcing of the interlock/alarm • Failure to operate on demand caused by isolation/plugging of input device connection or other process conditions (e.g., buildup on a temperature probe) that render input device inoperable or incapable of accurately measuring process conditions (e.g., pressure, temperature) • Potential ignition source if the device or connection shorts • Leakage at process connection as a result of overpressurization of the joint • Improper management of intervening valves between the instrument and the equipment being protected		
Personnel qualifications	Craft skills and knowledge required by the individual procedures	
Procedure requirements	Written procedures describing the PM activity including: • The manner, the extent, the location and the timing of the activity and by whom • The documentation and analysis of results • The resolution of functions or conditions not meeting acceptance criteria	
Documentation requirements	• Completed/closed work order • Equipment PM record	Special notes: • Records retained long enough to establish trends • Documentation by exception may be acceptable (for certain tasks)

TABLE 12A-13. Asset Integrity Activities for Safety Instrumented Systems and Emergency Shutdowns *(Continued)*

Repair		
Example activities and typical frequencies	Activity	Frequency
	• Troubleshooting • Replacement of field devices (e.g., transmitters, switches) • Replacement of logic solver, operator interface and engineering interface cabinet components (e.g., integrated circuit boards	As required
	• Identification of materials of construction	When new devices are installed as part of a repair
Technical basis for activity and frequency	Common repair activity	
Sources of acceptance criteria	• Device specifications • SIS specifications • Manufacturers' recommendations	
Typical failures of interest	• Leakage at process connection resulting from improper installation or materials of construction (e.g., incorrect gasket, incorrect metallurgy of wetted parts) • Failure to operate on demand as a result of improper installation (e.g., incorrectly wired), incorrect configuration of control system and/or incorrect calibration of the device • Potential ignition source or electrical shock to personnel as a result of improper installation • Improper electrical classification rating for the electrical classification of the area where the device is installed	
Personnel qualifications	• Craft skills and knowledge required by the individual procedures • Training on the specific procedures for the repair activities • Training on the use and operation of special tools (e.g., signal simulators) required by procedures	
Procedure requirements	• Generic written repair procedures for troubleshooting of SISs and replacement of integrated circuit boards that include references to the Manufacturers' manuals • Device-specific installation procedures • Manufacturers' manuals • Special tool (e.g., signal simulator) use and operation procedures • Procurement and receiving procedures to ensure proper materials of construction	
Documentation requirements	• Completed/closed work order • Work order or storeroom records of parts/materials used • Return to service check sheet • Vendor material of construction reports	Special notes: • Repair data (e.g., condition found, parts used, repairs made, condition left) are usually recorded in equipment history files • Repair history is retained long enough to establish trends

TABLE 12A-14. Asset Integrity Activities for Pumps

New Equipment Design, Fabrication and Installation		
Example activities and typical frequencies	Activity	Frequency
	• Identification of materials of construction • Performance test • Pressure test	Initial fabrication
	Alignment	Initial installation and any time components are loosened, removed, or replaced
	Rotational check	Initial installation and any time the driver is connected
	Vibration analysis (baseline)	Initial startup
Technical basis for activity and frequency	• Various codes, standards, or recommended practices from various organizations (e.g., API, HI, ANSI, ISO). (Application depends on industry and type of pump) • Manufacturers' recommendations • Common industry practices • Industrial recommendations	
Sources of acceptance criteria	• Applicable code, standard, or recommended practice • Pump specifications • Manufacturers' recommendations • Company engineering and/or maintenance standards • Industrial insurers' recommendations	
Typical failures of interest	• Leakage from the seal/packing assembly as a result of improper installation or incorrect materials of construction • Inadequate flow/pressure as a result of improper assembly or improper installation of internal components (e.g., inadequate impeller clearance) • Leakage from pump casing as a result of improper assembly, installation, or materials of construction • Damage to seal/packing/associated equipment due to excessive vibration • Bolting or foundation inadequacies	
Personnel qualifications	• Craft skills and knowledge required by the individual procedures • Training on the specific procedures for the inspection and testing activities • Training on the use and operation of special tools required by procedures (e.g., laser alignment equipment)	
Procedure requirements	• Procurement and receiving procedures to ensure proper materials of construction • Manufacturer testing procedures • Pre-commissioning and/or commissioning testing procedures • Manufacturers' manuals • Pump alignment procedure • Alignment tool (e.g., laser) procedure and/or manufacturers' manuals • Pump installation procedure • Vibration analysis procedure • Vibration analyzer operation procedure and/or Manufacturers' manuals	
Documentation requirements	• Vendor material of construction reports • Non-destructive testing records • Performance and/or pressure test reports • Alignment report • Pump installation record • Vibration analysis data and report	Special notes: • Records retained for the life of the equipment • Installation documentation to support PSSR requirements

TABLE 12A-14. Asset Integrity Activities for Pumps *(Continued)*

Inspection and Testing		
Example activities and typical frequencies	**Activity**	**Frequency**
	Visual inspection of sealing system	Each shift to weekly (depending on criticality)
	Vibration analysis	• Continuous for pumps with large motors (e.g., 10,000 hp) • Weekly to quarterly (depending on criticality and horsepower)
	Performance testing	Depends on service conditions and criticality
	Alternating redundant pumps	Weekly to monthly
	Operational/ functional check of standby pump	Weekly to monthly
Technical basis for activity and frequency	• Manufacturers' recommendations • Common industry practices • Industrial recommendations	
Sources of acceptance criteria	• Pump specifications • Manufacturers' recommendations • Company engineering and/or maintenance standards	
Typical failures of interest	• Leakage from the seal/packing assembly as a result of overpressurization of the assembly, inadequate lubrication, bearing failure, or wear • No or inadequate flow/pressure resulting from failure of a drive component • No or inadequate flow/pressure resulting from failure, corrosion, erosion, or wear of an internal component (e.g., impeller) • Damage to pump casing from loose or broken internal component (e.g., impeller) • Damage to seal/packing and/or associated equipment caused by excessive vibration • Bolting or foundation inadequacies	
Personnel qualifications	• Craft skills and knowledge required by the individual procedures • Training on the specific procedures for the inspection and testing activities • Training on the use and operation of special tools required by procedures (e.g., vibration analysis equipment)	
Procedure requirements	Written procedures describing the inspection or test activity including: • The manner, the extent, the location and the date the inspection or test is performed and by whom • The documentation and analysis of results • The resolution of functions or condition not meeting acceptance criteria	
Documentation requirements	• Inspection check sheet • Vibration analysis data and report • Performance test records	Special notes: • Records retained long enough to establish trends • Documentation by exception may be acceptable (for certain tasks and select pumps)

TABLE 12A-14. Asset Integrity Activities for Pumps *(Continued)*

Preventive Maintenance		
Example activities and typical frequencies	**Activity**	**Frequency**
	Bearing housing and/or gearbox oil/lubricant level check	Each shift to weekly
	Changing of bearing housing and/or gearbox oil/lubricant	Manufacturers' recommendations
	Analysis of bearing and/or gearbox oil/lubricant	Monthly to semi-annually (depending on criticality and history)
	Lubrication of metal couplings (e.g., Falk Steelflex, Fast gear or similar types)	Manufacturers' recommendations
	Internal inspection and rebuilding	As required depending on history, service conditions and criticality
Technical basis for activity and frequency	• Manufacturers' recommendations • Common industry practices • Industrial recommendations	
Sources of acceptance criteria	• Manufacturers' and/or lubricant/oil vendor's recommendations • Company engineering and/or maintenance standards	
Typical failures of interest	• Leakage from the seal/packing assembly as a result of overpressurization of the assembly, inadequate lubrication, bearing failure, or wear • No or inadequate flow/pressure resulting from failure of a drive component • No or inadequate flow/pressure resulting from failure, corrosion, erosion, or wear of an internal component (e.g., impeller) • Damage to pump casing from loose or broken internal component (e.g., impeller) • Damage to seal/packing and/or associated equipment caused by excessive vibration • Bolting or foundation inadequacies	
Personnel qualifications	• Craft skills and knowledge required by the individual procedures • Training on the specific procedures for the preventive maintenance activities	
Procedure requirements	Written procedures describing the PM activity including: • The manner, the extent, the location and the date the inspection or test is performed and by whom • The documentation and analysis of results • The resolution of functions or condition not meeting acceptance criteria	
Documentation requirements	• Lube route check sheet • Completed/closed work order • Equipment PM record • Oil/lubricant analysis report	Special notes: • Records retained long enough to establish trends • Documentation by exception may be acceptable (for certain tasks and select pumps)

TABLE 12A-14. Asset Integrity Activities for Pumps *(Continued)*

Repair		
Example activities and typical frequencies	Activity	Frequency
	• Mechanical seal replacement • Pump disassembly and assembly	As required
	Identification of materials of construction	When parts are received and/or at time of installation
Technical basis for activity and frequency	Common repair activity	
Sources of acceptance criteria	• Pump specifications • Manufacturers' recommendation • Company engineering and/or maintenance standards	
Typical failures of interest	• Leakage from the seal/packing assembly as a result of improper installation or incorrect materials of construction • Inadequate flow/pressure resulting from improper assembly or improper installation of internal components • Leakage from pump casing as a result of improper assembly, installation, or materials of construction • Damage to seal/packing and/or associated equipment caused by excessive vibration • Bolting or foundation inadequacies	
Personnel qualifications	• Craft skills and knowledge required by the individual procedures • Training on the specific procedures for the repair activities • Training on the specific procedures for the inspection and testing activities • Training on the use and operation of special tools required by procedures (e.g., laser alignment equipment)	
Procedure requirements	• Generic written repair procedures that include references to the manufacturers' manuals • Pump alignment procedure • Alignment tool (e.g., laser alignment) procedure and/or manufacturers' manuals • Pump installation procedure • Procurement and receiving procedures to ensure proper materials of construction for replacement parts • Testing procedures covering operation of testing equipment and/or performance of non-destructive testing	
Documentation requirements	• Completed/closed work order • Work order or storeroom records of parts/materials used • Return to service check sheet • Alignment report • Pump installation record • Vendor material of construction reports • Non-destructive testing records	Special notes: • Repair data (e.g., condition found, parts used, repairs made, condition left) are usually recorded in the equipment history files • Repair history retained long enough to establish trends

TABLE 12A-15. Asset Integrity Activities for **Fired Heaters, Furnaces and Boilers**

New Equipment Design, Fabrication and Installation		
Example activities and typical frequencies	Activity	Frequency

	• Heat or steam rate duty/fluid service requirements
	  - Pressure rating
	  - Materials selection
	• Fabrication contractor qualification
	  - Design approval by owner
	  - Welding/QC plan approval
	• Fabrication/storage/ shipping
	• Installation
	• Acceptance inspection and testing
	• Documentation preparation
	• Acceptance and turnover
	• Commissioning
		As required for fabrication and installation

Technical basis for activity and frequency	QA practices for equipment fabrication and installation	
Sources of acceptance criteria	• ASME Boiler and Pressure Vessel Code and NBIC • Company engineering standards, or facility-specific standards for heaters and furnaces	
Typical failures of interest	Incorrect material or weld metal, incorrect pressure rating for a component, leak during testing, weld defects outside of acceptance criteria, high hardness readings, use of unqualified welder or welding procedures, dimensional errors, incorrect refractory material, incorrect refractory installation or cure, incorrect spring hanger setting, damage to finned tube, improper combustion safeguards	
Personnel qualifications	• Boilers: ASME and National Board fabrication requirements • Heaters and furnaces: Company requirements, and documented craft skills for installation, NDE qualifications and ASME Section IX welding requirements for welders	
Procedure requirements	Written procedures describing: • Engineering standards for specification of equipment • Project management (including hazard and design review schedules) • Vendor qualification • Documentation requirements • Project acceptance and turnover requirements	
Documentation requirements	Company documentation requirements typically include: Manufacturers' data forms (boilers), welding qualifications, weld map, design calculations, material certifications, QC results, as-built drawings and pressure test reports	

TABLE 12A-15. Asset Integrity Activities for Fired Heaters, Furnaces and Boilers *(Continued)*

Inspection and Testing		
Example activities and typical frequencies	**Activity**	**Frequency**
	Inspection of heater or furnace firebox and tubes	Scheduled by user, usually in conjunction with planned unit shutdown or when an imminent failure condition is visible
	• Inspection of boilers, steam drums, tubes and firebox • Waste heat boiler inspection • Hazardous waste incinerator inspection	Inspection schedule is not code-derived, but is generally mandated by jurisdictional requirements; frequency is also impacted by inspection results / trending of operating conditions
	Functional testing of combustion safeguard systems	As required by codes/standards and/or as required to meet designated Safety Integrity Level
Technical basis for activity and frequency	Jurisdictional requirements for boilers and reliability issues for heaters and furnaces	
Sources of acceptance criteria	• Acceptance criteria from NBIC or jurisdictional requirements. • Company standards for evaluating tube life and refractory damage	
Typical failures of interest	Tube ruptures, tube bulging and distortions, creep damage, steam leaks, process fluid leaks, refractory failure, structural support failure	
Personnel qualifications	• National Board inspection certifications for boilers per jurisdictional requirements • Optional industry certifications (API-570, 510) or company-specific training for heater and furnace inspection	
Procedure requirements	Written procedures describing the inspection or test activity including: • The manner, the extent, the location and the date the inspection or test is performed and by whom • The documentation and analysis of results • The resolution of functions or condition not meeting acceptance criteria	
Documentation requirements	Results and analysis of each inspection are documented for the life of the equipment	

TABLE 12A-15. Asset Integrity Activities for **Fired Heaters, Furnaces and Boilers**
(Continued)

	Preventive Maintenance	
Example activities and typical frequencies	**Activity**	**Frequency**
	Activities identified from RCM or similar work planning initiatives, such as: • Process conditions monitoring/ tracking • Metal temperature monitoring • Water quality testing • Efficiency performance monitoring	As required to meet preventive maintenance schedule and boiler operating permit requirements
	Steam purity monitoring (where required)	As required to meet preventive maintenance schedule and boiler operating permit requirements
	Preventive maintenance of control system and combustion safeguard system components	Manufacturers' recommendations
Technical basis for activity and frequency	Company or jurisdictional requirements	
Sources of acceptance criteria	Upper and lower safe limits for process conditions defined in the process safety information, such as pressure, temperature, fluid composition and velocity	
Typical failures of interest	Process conditions (such as outlet temperatures, stack temperatures, combustion efficiency, heat input, flow balances, steam drum level, and tubular meal temperatures) exceed safe operating limit	
Personnel qualifications	Company requirements for tasks	
Procedure requirements	Written procedures describing the PM activity including: • The manner, the extent, the location and the date the inspection or test is performed and by whom • The documentation and analysis of results • The resolution of functions or condition not meeting acceptance criteria	
Documentation requirements	Results are usually recorded by exception in the equipment history file	

TABLE 12A-15. Asset Integrity Activities for **Fired Heaters, Furnaces and Boilers** *(Continued)*

Repair		
Example activities and typical frequencies	Activity	Frequency
	• Tube replacement • Burner replacement or repair • Insulation/refractory repair • Tube or piping support, hanger and anchoring systems repair or renewal • Commissioning activities following repair	As required by the condition of the equipment, based on recommendations from inspection and testing or preventive maintenance activities
Technical basis for activity and frequency	Performed when indicated by failure, or based on recommendations from PM or inspection and testing activities	
Sources of acceptance criteria	• ASME Boiler and Pressure Vessel Code and NBIC • Company engineering standards, or facility-specific standards for heaters and furnaces	
Typical failures of interest	Incorrect material or weld metal, incorrect pressure rating for a component, leak during testing, weld defects outside of acceptance criteria, high hardness readings, use of unqualified welder or welding procedures, dimensional errors, incorrect refractory material, incorrect refractory installation or cure, incorrect spring hanger setting, damage to finned tube, low water resulting in overheating, tube rupture and steam explosion	
Personnel qualifications	• Boilers: ASME and National Board fabrication requirements • Heaters and furnaces: Company requirements, and documented craft skills for repairs, NDE qualifications and ASME Section IX welding requirements for welders, optional industry certifications (API-570, 510), or company-specific training for inspection of firebox piping	
Procedure requirements	• Craft skill procedures for typical tasks encountered in repairs (e.g., welding, gasket installation bolt tightening) • Job-specific procedures for unique or complex repairs, or for jobs with specialized technical content (e.g., coil or tube replacement, new spring hanger settings, refractory repairs or burner adjustments) • Job-specific procedures with process engineering input for chemical cleaning	
Documentation requirements	Repair history is maintained with the equipment inspection history files	

TABLE 12A-16. Asset Integrity Activities for Switchgear

New Equipment Design, Fabrication and Installation		
Example activities and typical frequencies	Activity	Frequency
	• Visual inspection	Initial installation
	• Verification / testing of overcurrent protection	
Technical basis for activity and frequency	• NFPA 70	
	• Manufacturers' recommendations	
	• Industrial insurers' recommendations	
	• Common industry practices	
Sources of acceptance criteria	• NFPA 70	
	• Electric distribution system specifications	
	• Manufacturers' recommendations	
	• Industrial insurers' recommendations	
	• Company engineering and/or maintenance standards	
Typical failures of interest	• Failure to provide electrical power to safety-critical devices (e.g., controls) as a result of incorrect installation (e.g., incorrectly connected or improper sizing of the unit)	
	• Potential ignition source or electrical shock to personnel as a result of improper installation or incorrect overcurrent protection device (e.g., incorrect fuse)	
Personnel qualifications	• Craft skills and knowledge required by the individual procedures	
	• Training on the use and operation of special tools required by procedures (e.g., electrical test equipment)	
Procedure requirements	• Manufacturers' manuals	
	• Electric distribution system installation procedures	
Documentation requirements	• Record of overcurrent protection devices (e.g., fuse ratings) and settings	Special notes:
		• Records retained for the life of the equipment
	• As-built drawings	• Installation documentation to support PSSR requirements

TABLE 12A-16. Asset Integrity Activities for Switchgear *(Continued)*

Inspection and Testing		
Example activities and typical frequencies	Activity	Frequency
	Visual inspection	• Monthly to quarterly (outdoor installations) • Quarterly to semiannually (indoor installation)
	Infrared analysis	Annually to 3 years, depending on criticality, history and starter size
	Calibration and testing of protective relays; tripping of breakers; and testing insulation resistance of controls, meters and protective devices	Voltage, condition and criticality dependent; check current requirements
	• Inspection and maintenance of circuit breaker • Electrical testing of circuit breaker	3 years maximum (air-break and oil-immersed circuit breakers)
	System testing of installed circuit breaker	After completion of circuit breaker electrical testing
	Manufacturers' recommended inspection, maintenance and testing of vacuum and gas-filled circuit breakers	Manufacturers' recommendations
Technical basis for activity and frequency	• NFPA 70B • Manufacturers' recommendations • Common industry practices • Industrial insurers' recommendations	
Sources of acceptance criteria	• Electric distribution system specifications • Manufacturers' recommendations • Industrial insurers' recommendations • Company engineering and/or maintenance standards	
Typical failures of interest	• Failure to provide electrical power to safety-critical devices (e.g., controls) as a result of battery, connection, transfer switch and/or charging system failure • Potential ignition source or electrical shock to personnel as a result of a short in device or connection, incorrect overcurrent protection device (e.g., incorrect fuse), or operating equipment with too high a load (e.g., operates at too high a temperature)	
Personnel qualifications	• Craft skills and knowledge required by the individual procedures • Training on the specific procedures for the inspection and testing activities • Training on the use and operation of special tools required by procedures (e.g., electrical test equipment)	
Procedure requirements	Written procedures describing the inspection or test activity including: • The manner, the extent, the location and the timing for the inspection or test and by whom • The documentation and analysis of results • The resolution of functions or conditions not meeting acceptance criteria	
Documentation requirements	Inspection and testing check sheet	Special notes: • Records retained long enough to establish trends • Documentation by exception may be acceptable (for certain tasks and select electric motors)

TABLE 12A-16. Asset Integrity Activities for Switchgear *(Continued)*

Preventive Maintenance		
Example activities and typical frequencies	**Activity**	**Frequency**
	Overhaul of switchgear, including cleaning, inspecting, tightening and adjusting of all components	3 to 6 years, depending on conditions
Technical basis for activity and frequency	• NFPA 70B • Manufacturers' recommendations • Common industry practices	
Sources of acceptance criteria	• Manufacturers'/vendor's recommendations • Industrial insurers' recommendations • Company engineering and/or maintenance standards	
Typical failures of interest	• Failure to provide electrical power to safety-critical devices (e.g., controls) as a result of failure of batteries, connection, transfer switch and/or charging system • Potential ignition source or electrical shock to personnel as a result of a short in device or connection, incorrect overcurrent protection device (e.g., incorrect fuse), or operating device with too high a load (e.g., operates at too high of a temperature)	
Personnel qualifications	• Craft skills and knowledge required by the individual procedures • Training on the specific procedures for the preventive maintenance activities	
Procedure requirements	Written procedures describing the PM activity including: • The manner, the extent, the location and the timing of the activity and by whom • The documentation and analysis of results • The resolution of functions or conditions not meeting acceptance criteria	
Documentation requirements	• Completed/closed work order • Equipment PM record	Special notes: • Records retained long enough to establish trends • Documentation by exception may be acceptable (for certain tasks and select electric motors)

TABLE 12A-16. Asset Integrity Activities for Switchgear *(Continued)*

Repair		
Example activities and typical frequencies	Activity	Frequency
	Removal and installation of circuit breakers	As required
Technical basis for activity and frequency	Common repair activity	
Sources of acceptance criteria	• Electric distribution system specifications • Manufacturers' recommendation • Company engineering and/or maintenance standards	
Typical failures of interest	• Failure to provide electrical power to safety-critical devices (e.g., controls) as a result of incorrect installation (e.g., incorrectly connected or improper sizing of the unit) • Potential ignition source or electrical shock to personnel as a result of improper installation or incorrect overcurrent protection device (e.g., incorrect fuse)	
Personnel qualifications	• Craft skills and knowledge required by the individual procedures • Training on the specific procedures for the repair activities	
Procedure requirements	• Generic written repair procedures for removal and installation of circuit breakers that include references to the Manufacturers' manuals • Procurement and receiving procedures to ensure the use of proper equipment	
Documentation requirements	• Completed/closed work order • Work order or storeroom records of parts/materials used • Return-to-service check sheet	Special notes: • Repair data (such as condition found, parts used, repairs made and condition left) are recorded in the equipment history files, such as condition found, parts used, repairs made and condition left. • Repair history maintained long enough to establish trends

13
AIM PROGRAM IMPLEMENTATION

This chapter discusses issues related to implementing the asset integrity management (AIM) systems and activities previously discussed. Specifically, this chapter contains sections on the following topics:

- AIM program budgeting and resources
- Use of data management systems in AIM programs
- Return on investment of an AIM program.

The budgeting and resources section outlines the resource needs for all phases of AIM program implementation. This is followed by a discussion of computerized maintenance management systems (CMMS) and other data management applications in an AIM program. The last section outlines some of the benefits a facility can expect from an effective AIM program.

13.1 BUDGETING AND RESOURCES

Company and facility management has the responsibility to provide necessary and adequate resources for initial and ongoing implementation of the AIM program. Management often requests an estimate of the resources needed to develop, implement and sustain a successful AIM program. This section provides information that may be useful when estimating these necessary resources.

Establishing an AIM program for either a new facility or an existing facility consists of two major parts. First, initial implementation of the AIM program can be thought of as a major one-time project, including not only program development and baseline establishment but also procedures development and initial personnel training. After initial implementation, a transition is made to the on-going AIM program to maintain the integrity of assets over time.

Many companies use traditional project management tools and techniques to manage AIM program development and implementation. For example, a project timeline depicted as a Gantt chart can be developed to document and communicate goals, targets and milestones for the activities and to monitor activity progress. Project cost monitoring and reporting systems can help track project costs. In addition, periodic project review meetings and status reports help keep efforts focused and provide opportunities to resolve issues as early as possible.

13.1.1 Program Development Resources

Company management may consider providing resources for the following efforts associated with defining and developing an AIM program:

- Defining and documenting the overall management system for the AIM program

- Identifying other AIM program scope issues (e.g., covering program objectives, including nonregulated processes)

- Identifying assets to include in the AIM program

- Developing the inspection, testing and preventive maintenance (ITPM) program and associated schedule

- Identifying and developing written procedures

- Defining the training program, as well as developing and/or obtaining training program materials

- Developing a system for managing asset deficiencies

- Defining quality management activities and writing quality assurance procedures

- Defining any software needs and acquiring appropriate software.

A primary resource for these activities is time—the time spent by personnel who are responsible for program activity, as well as the time given by other personnel who are needed to properly define and develop the program activities (e.g., craftspersons, inspectors, process engineers, operations personnel). These personnel resources can be obtained from in-house resources and/or from outside consultants experienced in AIM program development.

The previous chapters of this book presented specific information associated with many of the topics listed above. Most of those chapters include a section that defines typical roles and responsibilities. In addition, Table 13-1 provides an overview of resources required for typical AIM program development activities. Estimating actual time requirements will depend on many factors, with no two facilities being the same. Variables include the location and size of the facility, prescriptive requirements, and the status of process technology information (drawings, relief calculations, etc.). Additional costs may be associated with specialized personnel training such as inspector training, depending on the approach taken by the facility (e.g., whether or not to develop in-house expertise).

TABLE 13-1. Summary of Resources Required for AIM Program Development Activities

AIM Program Development Activity	Typical Personnel Resources	
	In-House	Outside
Overall management system	• Plant management • AIM coordinator • Maintenance, Engineering, Inspection manager(s) • Maintenance management staff	• Process safety consultant with AIM program and reliability program experience
Other AIM program scope issues	• Plant management • AIM coordinator • Maintenance, Engineering manager(s) • Training coordinator	• Process safety consultant with AIM program and reliability program experience
Asset list	• AIM coordinator • Process engineer(s) • Maintenance management staff	• Process safety consultant with AIM program and reliability program experience
ITPM program	• AIM coordinator • Maintenance, Engineering, Inspection manager(s) • Maintenance management staff • Process engineer(s) • Operations personnel	• Process safety consultant with AIM program and reliability program experience • Inspection company personnel
Procedures	• Maintenance management staff • Craftspersons • Inspection manager • Inspectors • Procedure writers	• Procedure writers
Training	• Training department personnel • Maintenance, Engineering, Inspection manager(s) • Maintenance management staff • Craftspersons • Inspectors • Operations personnel	• Training development consultants
Asset deficiency	• AIM coordinator • Maintenance, Engineering, Inspection manager(s) • Inspectors • Operations personnel	• Process safety consultant with AIM program and reliability program experience
Quality Management program	• Maintenance, Engineering, Inspection manager(s) • Inspectors • Maintenance management staff • Project engineering • Purchasing • Storeroom manager • Procedure writers	• Process safety consultant with AIM program and reliability program experience • Procedure writers
Software	• Maintenance, Engineering, Inspection manager(s) • Inspectors • Maintenance management staff • Purchasing • Storeroom manager	• Maintenance software vendors

Included in Appendix 13A are worksheets that can be used when identifying personnel needed to complete many of the development activities.

Outside resources such as third-party consultants can be used to:

- Provide technical expertise (e.g., knowledge of codes, standards and good engineering practices)

- Facilitate the development of selected activities

- Develop written AIM procedures

- Augment the facility's staff when sufficient personnel are not available.

While outside resources can help with many development activities, successful AIM programs require significant involvement of in-house personnel during the development phase. Often, AIM program development efforts have been unsuccessful simply because they were contracted and did not include enough facility personnel involvement to achieve buy-in by plant personnel and to ensure that the program was practical for the specific culture, organization and long-term resourcing for the facility. While outside resources can be valuable when developing an AIM program (by providing direction, facilitating meetings, providing boilerplate program documents, etc.), AIM programs developed primarily by third parties with little involvement of facility personnel may lack a level of detail and specific process information needed for an effective AIM program. A high level of involvement by facility personnel will help ensure that programs originating from outside resources contain and reflect facility personnel knowledge of the process and the facility culture.

13.1.2 Initial Implementation Resources

During initial implementation, the AIM program moves from paper to people and assets in the field. This phase involves more time from facility personnel than the development phase and is often the most costly phase. Time requirements for initial implementation will vary considerably between facilities, depending on factors such as facility size and amount of missing asset information. Typical initial implementation activities are:

- Gathering and organizing asset information

- Training and qualifying personnel who will perform AIM activities, including obtaining required certifications (e.g., welding, inspector certifications)

- Implementing the ITPM tasks

- Managing ITPM results

- Implementing QA activities

- Managing asset deficiencies

- Obtaining and implementing software to support the AIM program.

Table 13-2 provides a list of common tasks associated with each of these activities. In addition, the following sections provide brief descriptions of these activities and discuss issues to consider when implementing these activities. Some of these initial implementation tasks will continue for the life of the AIM program.

Gathering and Organizing Asset Information. Asset information is gathered and organized early in the AIM program implementation phase. This information is needed during implementation to:

- *Define acceptance criteria* — usually by referencing specific asset file information such as corrosion allowance, dimensional tolerances and acceptable wear. See Sections 2.2.1 and 11.2 for additional information on acceptance criteria.

- *Execute ITPM tasks* — specifically, personnel performing the inspection task reviews selected asset information with regard to the ITPM history, asset details (e.g., specific components requiring inspection), ITPM task details (e.g., condition monitoring locations, inspection techniques) and acceptance criteria.

The asset information is usually located in the facility's equipment files. The information in these files varies, depending on the type of asset and the facility recordkeeping culture and philosophy. Typically, equipment files contain such information as:

- Equipment design and construction data, such as design codes and standards used, design specifications, as-built drawings, materials of construction, dimensions (e.g., wall thickness, impeller diameter) and performance data (e.g., pressure relief valve settings)

- Service history, such as length of time in service, materials handled and changes in service

- ITPM history

- Maintenance history (i.e., failure history)

- Repair history (temporary repairs; repairs based on inspection results; replacement of parts/components including any upgrades from original design)

- Vendor-supplied information, such as installation instructions, dimensional specifications and allowable tolerances (e.g., the diameter of a journal or shaft), bolt materials and torque requirements, gasket and O-ring material requirements, lubricant specifications, maintenance and operating instructions, testing and maintenance recommendations, and performance testing data (e.g., pump performance testing).

TABLE 13-2. Typical Initial Implementation Tasks by Activity

Initial Implementation Activity	Typical Tasks
Gathering and organizing asset information	Compile/organize asset files
	Develop/obtain missing asset information
	Establish/verify acceptance criteria for assets (e.g., minimum wall thickness data, vibration tolerances)
Training and qualifying personnel	Develop a training plan
	Train maintenance craft personnel
	Qualify welders
	Certify inspectors
	Manage training efforts (e.g., maintain training database)[1]
Implementing ITPM tasks	Develop routes
	Develop ITPM schedule for individual asset items
	Enter recurring work orders
	Perform route tasks[1]
	Perform ITPM tasks[1]
	Perform vessel/tank inspections
	Perform piping inspections
	Perform rotating equipment inspections
	Perform instrument/interlock testing
Managing ITPM results	Implement inspection and testing results software
	Review results[1]
	Route results
	Review individual asset ITPM results; perform any needed calculations (e.g., remaining life estimates)
Implementing Quality Management activities	Train personnel (e.g., purchasing, project, storeroom)
	Perform storeroom QA tasks[1]
	Perform project QA tasks[1]
	Review documents (e.g., specifications)
	Inspect major asset items
	Verify installation (i.e., field verification)
Managing asset deficiencies	Pilot test asset deficiency resolution process
	Develop asset deficiency tracking system
	Conduct asset deficiency status review meetings[1]

[1] Indicates that the task is an ongoing task as well as an initial implementation task

This information may also be found in the facility's process technology information. Any necessary information not available at the site will need to be obtained from the asset manufacturers. If the manufacturer is unable to provide the necessary information, facility personnel may need to develop the information or contract with appropriate engineering consultants to develop the information. Table 6-2 in Chapter 6 provides a list of information needed for selected types of assets.

Training and Qualifying Personnel. As outlined in Chapter 8, personnel performing ITPM tasks and other AIM tasks such as asset repairs need to receive training on a variety of topics and may need to obtain certifications before performing certain tasks such as pipe welding in accordance with ASME B31.3. Facilities also need to train personnel on applicable job procedures such as ITPM procedures, repair procedures and QA procedures. For training to be effective, facilities are challenged to (1) identify training methods that are efficient and that keep experienced workers engaged and (2) allocate time for the trainers and trainees to be away from their regular job duties. To help overcome these issues, facilities can employ a variety of training methods (see Section 8.2 for more information on training methods) and can develop a training plan that accounts for personnel scheduling issues. This plan can then be used to estimate the resources required to complete the initial training. Training resources will typically include:

- Overtime to cover the absence of personnel from their regular duties to attend and perform the training

- Costs for outside trainers and/or training materials

- Administrative time to manage the training plan implementation and maintain the training records.

Contractors at the facility will also require training. While this activity may not be as resource-intensive as training facility personnel, resources may need to be allocated to:

- Obtain and review contractor safety and capabilities information

- Train contract employer representatives and/or contract employees

- Review contractor training records

- Ensure that only trained contract employees are used to perform work.

Implementing ITPM Tasks and Managing ITPM Task Results. The most resource-intensive activity during initial implementation of the AIM program is likely the implementation of the ITPM tasks. Implementation of ITPM tasks typically begins with:

- Organizing the frequently performed tasks that can be performed on numerous assets at a single time into maintenance rounds (e.g., lubrication rounds, vibration analysis rounds)

- Developing the inspection and testing schedules for tasks that are performed on individual asset items

- Entering recurring work orders into the CMMS.

The facility staff may consider the following factors when developing a schedule for maintenance rounds and individual ITPM tasks:

- *Manpower loading.* Performing all weekly activities on the same day of the week or all monthly activities on the first of the month is not practical or advisable. Facility staff can devise schedules that spread tasks out over the course of the week, month and year.

- *Availability of qualified manpower* to perform the activities.

- *Operational issues,* such as which assets can or cannot be taken out of service at the same time and the ability to take assets out of service (e.g., storage tanks for internal inspection, relief valve removal and testing, safety instrumented system and interlock functional testing).

The maintenance planning and scheduling group at most facilities provides much of the personnel needed to complete the schedule; however, production management needs to be involved to ensure effective communication between production and maintenance personnel regarding the scope and any associated asset downtime requirements of any maintenance task or activity to be accomplished. If outside contractors are used for certain ITPM tasks such as API 510 inspections of pressure vessels, consideration can be given to involving them in the scheduling process or asking them to develop a proposed schedule and budget.

Once defined, the schedule can be used to identify the required resources and to develop a budget for implementing the tasks. In developing the budget, consider the following issues:

- Internal staffing requirements for routes and individual ITPM tasks

- Need for, and use of, contractors

- Required asset outages

- Turnaround schedules.

In addition to the resource and budget requirements directly associated with the ITPM schedule, the following issues need to be considered when developing a budget for the implementation of ITPM tasks:

- Obtaining special equipment used to perform some tasks (e.g., vibration monitoring equipment, ultrasonic thickness measurement instruments)

- Obtaining and implementing software needed to manage the ITPM program and task results

- Costs associated with preparing assets for inspection (e.g., coring pipe insulation)

- Asset/system changes needed to maintain efficient performance of ITPM tasks (e.g., installation of block valves under pressure instruments).

Resource requirements for managing ITPM task results include time allocation for assigned personnel to review the results and determine what corrective actions are needed. This can include (1) a few hours per week for the maintenance craft supervisor to review the results from weekly routes and (2) a significant number of hours for engineers to review the numerous inspection reports. In addition, managing ITPM results may require purchasing special software to collect and analyze the volumes of data generated. (See Section 13.2 for a discussion of issues related to the need for, and use of, software in an AIM program.)

Implementing Quality Management Activities. The resources required to implement the QA activities will be a function of the QA plan that is established. These resources may include:

- Acquiring code/standard/practice documents and training personnel to use them effectively

- Implementing vendor and contractor QA plans

- Obtaining tools for positive material identification (PMI)

- Providing space for receiving and storage, and furnishing the facilities/assets needed to maintain required storage conditions (e.g., humidity and/or static control)

- Providing personnel training on receiving and storage procedures.

Managing Asset Deficiencies. Initial implementation costs may be associated with managing asset deficiencies. The resources for managing asset deficiencies include the personnel and time needed to (1) develop and document the corrective actions required to bring the assets back to its original specifications/condition and/or implement temporary corrective actions (if appropriate), (2) track asset deficiencies to resolution and (3) document final resolution. When initially implementing an asset deficiency resolution program, resources can be used to:

- Pilot-test the process

- Develop an asset deficiency tracking system on the CMMS or other database

- Implement periodic asset deficiency review meetings to ensure that deficiencies are being resolved in a timely manner.

Because the management of change (MOC) program may be used to manage asset deficiencies, involving personnel who are responsible for the MOC program would be helpful in this effort.

Obtaining and Implementing Software. Most facilities use software to implement and maintain the AIM program. This includes computerized maintenance management programs to help manage (1) the ITPM task schedule and work orders, (2) some ITPM task data such as thickness measurements, (3) the AIM program documents such as procedures, (4) the AIM training activities, and (5) corrective action tracking activities. Section 13.2 provides additional information on the use of software in the AIM program.

During the AIM program implementation phase, facility personnel need to (1) identify software needs, (2) assess the ability of any existing software to satisfy these needs, (3) modify existing software and/or obtain additional or new software if necessary, (4) implement any new software or modifications to the existing software and (5) provide training on any new or modified software systems. Maintenance software vendors and consultants can be a helpful resource during this phase.

13.1.3 Ongoing Efforts

Ongoing efforts for the AIM program involve both the continued execution of the initial implementation tasks listed in Table 13-2 as well as the inclusion of new tasks to be performed. These new tasks focus on:

- Providing ongoing training
- Maintaining and improving existing AIM procedures and developing new procedures as necessary
- Optimizing the ITPM tasks
- Maintaining QA activities
- Managing program changes.

The initial development of the training program typically receives sufficient attention; however, facilities sometimes overlook the ongoing training effort that will be necessary. Allocating resources and budgets for ongoing training is important to the overall success of the AIM program. Ongoing training efforts include refresher training and training on new topics. Many organizations provide annual refresher training that focuses on regulatory-required training, required certification/qualification training and safe work practices. However, many organizations have expanded annual refresher training to include reviews of (1) job tasks and procedures that are of high risk or are performed infrequently and (2) procedures that have proven to be troublesome, such as tasks that are often not properly executed. The ongoing training budget may also recognize training needs for new employees and for reassigned/promoted employees. New training topics for the existing workforce may include new assets or new maintenance/inspection techniques.

Trevor Kletz writes, in *What Went Wrong: Case Histories of Process Plant Disasters* (Reference 13-1), "Procedures are subject to a form of corrosion more rapid than that which affects the steelwork; they vanish without a trace once management stops taking an interest in them…" Without ongoing procedure development and/or modification efforts, the value of the initial procedure development effort will quickly be lost. Ongoing procedure development resources may be needed to update and correct procedures as procedure users discover errors or suggest improved execution steps. Also, facilities need to perform periodic reviews of the procedures, to ensure that they stay current and accurate, as well as personnel performance, to ensure that the procedures are followed. In addition, new procedures will be needed as new tasks are identified and implemented.

Some ITPM task optimization will result from reviewing task results and, if applicable, performing remaining life or similar time-to-retirement calculations. Optimization can result in changing the task activities and/or the task frequency. ITPM task optimization may also involve applying risk management tools to better understand potential failures (or root cause analysis of actual failures) so that appropriate ITPM tasks can be planned. Chapter 15 discusses common risk management tools and includes a section on root cause analysis (RCA). Providing the resources and budget to optimize ITPM tasks is essential to realize some of the benefits of the AIM program, such as:

- Improved safety and environmental performance

- Extended life of assets, such as by identifying needed repairs before damage progresses to a level that makes repair impractical

- Better capital planning, since the need to replace major asset items is better predicted

- Improved asset reliability, including increased on-stream performance and the resulting economic benefits

- Improved ITPM task effectiveness and efficiency.

Resources from various departments and/or organizations within a company are needed to maintain QA activities. To ensure that the QA activities are sustained, these resources need to be identified and responsibilities need to be incorporated into the job duties for the involved job positions. Table 13-3 outlines some of the ongoing resources needed from various organizations

Some of the resources listed in Table 13-3 may be provided from outside the facility, such as corporate purchasing and project engineering. Experience has shown that outside resources, especially for capital projects, typically require more oversight than facility-provided resources. Possible reasons for this include:

- Outside personnel interacting with the facility personnel often change (e.g., different project managers)

- The outside personnel may not be as familiar with the facility AIM program processes and procedures

- The goals, objectives and expectations for personnel from outside the facility may not include issues related to process safety and the AIM program; therefore, these individuals may not be aware that their job functions impact the process safety and AIM programs. For example, corporate project engineers need to understand that the AIM program can affect (and assist with) such issues as contractor selection, equipment fabrication, asset installation and project startup.

To help minimize these issues, facility management may consider (1) involving personnel familiar with the QA requirements of the AIM program early in the project life cycle, (2) communicating the QA requirements of the AIM program to outside personnel and (3) informing and/or training outside personnel on the QA program policies and procedures. Early involvement of AIM program personnel can help ensure that the QA activities are fully and effectively integrated into the project. In addition, facilities can provide facility representatives to support outside project engineering and to help ensure that facility QA requirements are addressed.

Periodically, changes to the AIM program will necessitate the allocation of resources to maintain the program. Some of the typical program changes are:

- *Organizational changes.* If a facility changes its organizational structure, AIM program responsibilities may need to be changed.

- *Process changes.* Process changes can result in changes to the ITPM tasks and the associated task procedures and schedule.

- *Equipment changes.* As assets are added or replaced, the ITPM tasks and asset information may need to be updated.

TABLE 13-3. Examples of Ongoing QA Activities

QA Area	Example QA Activities	Personnel Involved
Project design	• Specification reviews • Design reviews	• Project engineers • Process engineers • Asset engineers • Materials engineers • Operations personnel
Vendor selection and management (both project and maintenance materials)	• Vendor qualification reviews • Vendor audits • Approved vendor list updating • Vendor performance monitoring	• Purchasing personnel • Project engineers • Maintenance personnel • Storeroom personnel
Contractor selection and management (both project and maintenance materials)	• Contractor safety performance and program reviews • Contractor capability reviews • Approved contractor list updating • Contractor safety training sessions for contractor representatives and/or employees	• Purchasing personnel • Project engineers • Maintenance personnel • Safety department personnel
Asset fabrication (off site)	• Vendor asset drawing reviews • Fabricator site visits • Asset inspections and tests	• Project engineers • Outside nondestructive testing (NDT) contractors • Inspectors
Asset fabrication (on site) and installation	• Asset inspections and tests • Walkthroughs, including piping and instrumentation diagram (P&ID) verification • Installation practices (e.g., equipment alignment) • Onsite contractor safety and QA audits • Asset precommissioning activities	• Project engineers • Outside NDT contractors • Inspectors • Operations personnel • Maintenance craftspersons
Spare part and maintenance material procurement, receiving, storage and issuing	• Spare part information updates, including purchasing information • Receiving practices (e.g., matching paperwork, tagging items) • Receipt inspections and tests (e.g., PMI, visual inspection) • Proper storage of items with specific storage conditions; inspection and maintenance of spares as required to maintain viability • Storeroom inventory control practices and audits • Issuing practices and inspections	• Purchasing personnel • Inspectors • Maintenance personnel • Storeroom personnel

Changes to the AIM program, including the ITPM program, need to be included in a facility's management of change program or equivalent. (Some companies have a separate management system for the ITPM program.) An important factor in this is that any change to the ITPM process is recommended by and reviewed by competent persons.

13.2   USE OF DATA MANAGEMENT SYSTEMS

While successfully managing an AIM program using paper-based systems is possible, most facilities find that managing many of the AIM program activities is not practical without digital information technologies. Typically, the most important information system used in AIM is the computerized maintenance management system, or CMMS. The use of a CMMS will be discussed in this section, as well as other AIM-related computer applications. Although the resources described here primarily apply to fixed facilities, similar resources such as pipeline integrity monitoring systems (PIMS) are available for more specialized AIM applications.

Successfully implementing these applications is not always straightforward. The full functionality of some applications is often not understood or employed, and it is often difficult to query and extract information from these systems. Desirable attributes for an AIM-related application include having a user-friendly interface (including for performing queries) and a flexible data structure, as well as having compatibility between applications that are together used to manage asset integrity.

13.2.1   Computerized Maintenance Management

Significant differences in CMMS software capabilities exist; however, almost all CMMS software packages contain the basic functions needed for an AIM program. Two of the most important features are the CMMS work scheduling function and the work order function. The CMMS can track when an ITPM task was last performed, calculate the date that the task is to be repeated, and generate a timely work order.

In addition, the CMMS is typically used to manage the individual ITPM tasks and the asset repair and replacement tasks. For individual ITPM tasks, the generation of a work order is typically the trigger that begins the scheduling process for the task. In addition, the work order can provide or reference the information needed to execute the task, such as asset item, ITPM task description, ITPM task procedure and locations of other pertinent information. Completed work orders can be used to document some ITPM task results:

- The date the task is completed
- The name of the individual performing the task

- A description of the task performed

- An asset identifier such as a tag number or asset number

- The results of the task.

However, many CMMS programs have limited ability to accommodate the volume of data generated during some ITPM tasks (e.g., condition monitoring data, vibration analysis readings), and some programs may be unable to perform necessary calculations (e.g., corrosion rate, remaining life). Therefore, facilities frequently supplement the CMMS with additional software specifically to manage the information from these data-intensive ITPM tasks (see the following subsections for more information on these data management systems).

For repair/replacement tasks, the CMMS can provide such information as the procedure for the task, a reference to other documents needed to perform the task (e.g., manufacturer's manual, permit requirements) and the means to manage asset deficiencies. Work orders or work order logs can be used to communicate asset deficiencies to affected personnel. (Note that the CMMS may need to be supplemented by other forms of communication.) The CMMS may be used to

1. Identify and document corrective actions, including approval of temporary corrective actions

2. Track deficient assets until permanent repairs are implemented

3. Document the resolution and correction of the asset deficiency.

A CMMS is also frequently used to assist facilities with the QA of spare parts and maintenance materials. Most systems have the following capabilities, which can help ensure that only correct parts and materials are used:

- Controlling and generating purchasing information to ensure that the correct spare parts and maintenance materials are ordered.

- Including notes or other information in the purchasing and/or spare parts information that will communicate specific QA requirements to appropriate personnel such as vendors and receiving personnel.

- Integrating spare parts information with work orders to ensure that correct spare parts are ordered from the maintenance storeroom.

- Monitoring storeroom inventory to help ensure that appropriate parts are available.

- Tracking the use of spare parts and maintenance materials.

Additional CMMS features that are helpful for managing an AIM program are failure coding, cost tracking and report generation. CMMS programs frequently have the capability to enter failure codes for work orders. An effective failure coding system can identify asset types (e.g., specific pump type) or specific asset items with repeat failures that require further analysis such as root cause analysis or failure analysis (see Chapter 15 for more information on these analysis techniques). Also, CMMS programs can often record labor and material cost data associated with each work order. This cost information can be used in the performance measurement system and other aspects of AIM program management. Finally, the CMMS needs to be able to provide AIM program management reports such as AIM-covered asset lists, planned/scheduled ITPM task lists, completed ITPM task lists, overdue ITPM task reports and asset deficiency status reports.

13.2.2 Inspection Data Management

Specialized computer applications known as inspection data management systems (IDMS) are used for some ITPM tasks to provide an interface to data collection devices and to manage the quantity of data generated. Nondestructive examination (NDE) techniques, such as eddy current and UT measurement, often use specialized applications to retrieve the data from field data-collection devices and to manage the large amounts of data generated. These applications are typically used to (1) document the data so that reports can be generated, (2) highlight readings outside of acceptable limits and (3) perform relevant calculations.

Specialized applications for instrumentation and vibration analysis provide similar functions for instruments and rotating equipment. Instrument calibration systems typically contain a catalogue of instrument with associated information such as model number, tag number, calibration range and calibration tolerance. These systems (1) communicate with the devices used to calibrate the instruments and (2) help manage the calibration data with features to document data, generate reports, identify out-of-tolerance readings and perform error calculations. In addition, online instrument applications are becoming increasingly common. Online applications can identify malfunctioning instruments (e.g., transmitters, valves) before a failure affects system performance. Similarly, some vibration analysis systems communicate with field vibration analyzers to collect and analyze vibration readings. Usually, the vibration applications can document the data, generate reports and identify readings that might indicate the onset of failures in the rotating equipment. Computer applications are continually evolving as developers add more features such as compressor performance analysis that can be used to support AIM programs.

13.2.3 Training Program Management

Another type of data management system that can assist AIM program management is training database applications. Usually, these systems provide a means to document the training requirements for employees and can manage individual employee training records. These applications often contain features to define the initial training requirements for new employees (or employees transferring to a new position) and refresher training requirements for the current workforce. Such applications can generate training plans, usually on an annual basis, and a training schedule for each employee in the organization. In addition, such applications commonly include features to document the training that each employee receives and maintain that information in individual training records. These records typically document the training topic, the date of training, the duration of the training, the means used to verify that the employee understood the training and whether the employee successfully completed the training. In addition, these applications can generate a variety of reports, such as reports to verify that employees have completed the training required to perform a task or to provide evidence of employee training during audits. Frequently, facilities have other training programs such as operator training and safety training that may provide database resources for the AIM program.

13.2.4 Document Management

Document management systems can be very helpful for many maintenance organizations, especially for those with no experience managing the number of written procedures needed for an effective AIM program. These applications provide document organization and retrieval tools often used to manage AIM procedures, asset file information and manufacturer manuals. AIM programs need to ensure that up-to-date ITPM procedures, repair/replacement procedures, asset file information and manufacturer manuals are available and accessible to the maintenance personnel and inspectors. This can be difficult and cumbersome using a paper-based system. In addition, paper-based systems risk losing or damaging unique paper copies of asset file information and manufacturer manuals. Document management applications allow personnel to maintain procedures, asset file information, inspection reports, and manufacturer manuals in an electronic format that can be retrieved and printed when needed, such as when issuing a work order. In addition, some of these applications can work in conjunction with CMMS programs so that required documents can be printed with the work order.

13.2.5 Risk Management

Chapter 7 discusses several risk management activities that personnel can implement to improve and optimize aspects of the AIM program. Most of the activities focus on improving ITPM tasks. The most common risk management methods used in conjunction with the AIM program are Failure Modes and Effects Analysis (FMEA), risk-based inspection (RBI) and reliability-centered maintenance (RCM).

Several computerized applications are available to help employees perform and document the results of these activities. These applications are used to assist the study leader in structuring an analysis and to provide specific analysis tools such as lists of failure modes and criteria for assessing damage mechanisms. Frequently, such applications can be used to document results and generate study reports. In addition, many of these applications, especially RBI software, include features to update the analysis and the analysis results based on the completed ITPM task results.

13.3 AIM BENEFITS AND RETURN ON INVESTMENT

An AIM program requires considerable resources; most managers are interested in the return on investment (ROI) of those resources. An AIM program can expect benefits in the following areas:

- Asset reliability

- Cost avoidance (including safety, environmental, and financial costs)

- Regulatory compliance and industry association commitments

- Reduced liability and reduced damage to corporate reputation.

The ROI for AIM programs depends on many factors, such as the maintenance systems in place prior to initial AIM program development and implementation, the size of the organization and the number of outside resources needed. Quantifying the overall ROI for an AIM program can be problematic because of the difficulties in measuring some of the AIM program benefits. For example, determining and/or estimating the benefits associated with preventing a fire in a facility, complying with government regulations and/or maintaining market share (e.g., avoiding lost market share as a result of downtime or of negative publicity associated with a catastrophic event) can be very difficult. The following paragraphs briefly describe the benefits and resulting ROI for each of the above-mentioned areas.

Improved Asset Reliability. One of industry's primary objectives of the AIM program is to replace a "breakdown" maintenance philosophy with a more proactive maintenance philosophy (Reference 13-2). The basic purpose of the ITPM program is to define and implement tasks/procedures that will prevent asset failures or detect their onset. For many facilities, this shift in philosophy can result in significant improvements in asset reliability. In addition, an effective AIM program improves the ability of a facility to predict when asset repair or replacement is needed. This increased understanding of asset condition allows the facility to better plan repairs and replacement, which reduces the impact of unexpected asset failures.

In addition, the QA program supports improved asset reliability with activities that help ensure the suitability of the initial design, fabrication and installation of process equipment. While these activities may result in some additional up-front costs, they are vital to ensuring asset reliability over the life of the assets. For example, installing assets with the wrong materials of construction or foregoing installation standards such as for rotating equipment alignment will likely result in premature failures that can lead to both asset repair and lost production costs.

Furthermore, the asset deficiency resolution process helps ensure that (1) asset breakdowns are properly managed and (2) temporary repairs are tracked until they are permanently corrected. (Historically, many temporary repairs have been forgotten until failures occurred that were worse than the initial failures.) In addition, facilities can employ asset deficiency programs to identify chronic failures that can then be subjected to root cause analysis and subsequent correction. These activities are aimed at improving asset reliability, thus reducing unplanned downtime associated with asset failures, as well as operational losses such as lost production.

In addition, AIM training and procedures heighten workforce efficiency and result in more consistent job performance. In general, personnel who are properly trained and have access to up-to-date, correct procedures and other pertinent documentation will perform tasks more consistently and efficiently. This improves downtime planning and eliminates many breakdowns.

With some effort, many of the benefits described in this section can be quantified by comparing asset reliability performance (e.g., process availability, unplanned downtime) and workforce efficiency measures (e.g., repeat work, average repair times for tasks) before and after AIM program implementation. Chapter 14 provides some suggested performance measures that can be used for such a comparison.

Cost Avoidance. Cost avoidance involves the return associated with avoiding the cost of an asset failure. For example, an internal inspection of a storage tank that discovers that the tank bottom is thin helps a facility avoid the costs associated with the following issues:

- Further asset damage, which could increase the repair cost

- Potential incident impacts to personnel and/or related assets

- Potential environmental impacts, including cleanup costs and adverse publicity

- Potential unscheduled downtime and associated production losses.

Measuring the monetary value of these avoided costs can be difficult. However, in the evolving field of cost-benefit analysis, some corporations are creating "equivalent pain" matrices (similar to risk matrices) and other tools designed for making just such measurements.

Regulatory Compliance and Industry Association Commitments. For many organizations, one objective of the AIM program is compliance with relevant regulations and/or industry association commitments such as Responsible Care®. The avoidance of regulatory costs, when quantified, is calculated based on (1) the magnitude of fines issued to date and (2) the harder-to-quantify benefits of industry and company reputation.

Reduced Liability and Reduced Damage to Corporate Reputation. AIM programs focus on maintaining asset integrity so that failures, especially catastrophic failures, do not occur. Therefore, an effective AIM program will result in reduced risk of:

- Negative publicity and/or extended production outages that could result in lost market share

- Employee injuries that could result in litigation

- Offsite injuries and damage that could result in litigation

- Adverse public reaction or perception.

In addition, an AIM program that is effective in preventing asset failures can have a positive impact on workforce morale and can promote good corporate citizenship.

CHAPTER 13 REFERENCES

13-1   Kletz, T., *What Went Wrong? Case Histories of Process Plant Disasters*, 5th *Edition,* Elsevier Science & Technology Books, Burlington, MA, 2009.

13-2   U.S. Occupational Safety and Health Administration, *Process Safety Management of Highly Hazardous Chemicals*, 29 CFR Part 1910, Section 119, Washington, DC, 1992.

APPENDIX 13A. AIM PROGRAM DESIGN ACTIVITY WORKSHEETS

Included in this appendix are suggested worksheets to use when identifying personnel needed to complete many of the AIM program development activities.

AIM Program Design

☐ Establish AIM program scope

       Team Leader _____

       Team _____ _____

                      _____ _____

                      _____ _____

☐ Review existing AIM program scope

       Team Leader _____

       Team _____ _____

                      _____ _____

                      _____ _____

AIM Written Program Development

☐ Create written administrative program

       Team Leader _____

       Team _____ _____

                      _____ _____

                      _____ _____

☐ Review/update written administrative program

    Team Leader _____

    Team _____　　_____

             _____　　_____

             _____　　_____

AIM Assets Program Development and Implementation

☐ Create/update AIM program assets list

    Team Leader _____

    Team _____　　_____

              _____　　_____

             _____　　_____

☐ Identify applicable codes, standards and manufacturers' recommendations

    Team Leader _____

    Team _____　　_____

              _____　　_____

             _____　　_____

☐ Create/update ITPM plan

    Team Leader _____

    Team _____　　_____

              _____　　_____

             _____　　_____

☐ Integrate ITPM plan with CMMS   Responsible party _____

☐ Identify critical repairs

       Team Leader _____

       Team _____   _____

                  _____   _____

                  _____   _____

Quality Assurance Program Development and Implementation

☐ Develop QA plan

       Team Leader _____

☐ QA for design   Responsible person _____

☐ QA for procurement   Responsible person _____

☐ QA for fabrication   Responsible person _____

☐ QA for receiving/stores/issues   Responsible person _____

☐ QA for installation   Responsible person _____

☐ QA for startup   Responsible person _____

☐ QA for asset repair   Responsible person _____

☐ QA for decommissioning/   Responsible person _____
   recommissioning

Asset Deficiency Program Development and Implementation

☐ Create list of types of deficiencies that the program will address

       Team Leader _____

       Team _____   _____

               _____   _____

               _____   _____

☐ Write asset deficiency plan

       Team Leader _____

       Team _____   _____

               _____   _____

               _____   _____

☐ Establish system for managing ITPM results

       Responsible person _____

☐ Establish system for managing deficiencies found during operations

       Responsible person _____

☐ Pilot test asset deficiency plan

       Responsible person _____

       Affected process area(s) _____

☐ Reality check

       Team Leader _____

> Are deficiencies recognized?
> Are affected personnel informed?
> Are precautions being taken to
>   manage the safety of deficient
>   assets in operation?
> Is deficiency tracking up to date?
> Is deficiency resolution documented?

AIM Personnel Management Program

☐ Create list of skills/knowledge requirements for each affected position

Training Team Leader _____

Training Team _____   _____

_____   _____

_____   _____

☐ Identify personnel/departments to perform each task in ITPM and QA plans.
(Task can be done as part of the creation of these plans.)

Responsible person _____

☐ Identify lists of procedures, training and contractor-assigned tasks based on the ITPM and QA plans and the list of critical repairs. Integrate the skill/knowledge requirements into the training plan.

Team Leader _____

Team _____   _____

_____   _____

_____   _____

☐ Define training program parameters (how to qualify trainers, verify training is understood, document training, etc.)

Responsible person(s), if other than Training team _____

☐ Develop procedures and training materials

Team Leader _____

Team _____   _____

_____   _____

_____   _____

AIM Personnel Management Program *(continued)*

☐ Identify contractor procedure/training requirements

      Training Team Leader _____

      Training Team _____   _____

                             _____   _____

                             _____   _____

☐ Identify QA and other administrative procedures to manage contractor activities

      Responsible person _____

☐ Develop contractor-related procedures

      Team Leader _____

      Team _____   _____

                       _____   _____

                       _____   _____

Credit: ABSG Consulting Inc., *Asset Integrity, Course 111*, Process Safety Institute, Houston, Texas, 2004. Included as resource material in *Guidelines for Mechanical Integrity Systems* (New York: AIChE, 2006).

14

METRICS, AUDITS AND CONTINUOUS IMPROVEMENT: LEARNING FROM EXPERIENCE

While much of the asset integrity management efforts at a facility are generally directed at developing and implementing the AIM program, successful facilities also have an objective and desire to learn from experience and continuously improve the program. This often takes the form of the following activities.

- *Establishing Performance Measurement Systems.* Effective performance measures can help organizations evaluate the ongoing performance of the overall AIM program and key AIM program activities such as compliance with the ITPM task schedule. The metrics can include direct measurement of AIM program performance as well as leading indicators (predictors) of program performance.

- *Conducting Periodic Audits and Assessments of Program Activities.* Audits and assessments are typically used to evaluate how the management systems of the AIM program are functioning, such as the ITPM program and the quality management system, and how they satisfy relevant requirements such as process safety regulations. Periodic AIM program audits are generally a mandatory requirement in process safety regulations (e.g., Reference 14-1).

- *Learning Lessons from Equipment Failures, Process Safety Incidents and Near Misses.* Another improvement activity is to systematically evaluate equipment failures, including near misses and excursions outside of integrity operating windows, using structured evaluation processes such as root cause analysis (RCA). Using systematic analyses provides an effective means to identify the contributing and root causes of failures. From the results of these processes, personnel can make recommendations for eliminating underlying causes, then management can develop and implement appropriate actions to eliminate or reduce the likelihood of future failures and excursions. In addition, lessons learned from these analyses can be shared with others who may benefit from the learnings, both within the facility and at other facilities.

These activities are included in Figure 14-1, which shows a simple model illustrating how continuous improvement efforts can contribute to the overall performance objectives of the AIM program. Facilities can consider defining a procedure for ensuring that the results from these continuous improvement activities are implemented and evaluated for effectiveness. Continuous improvement activities can be most effective if (1) they produce corrective actions that are cost-effective and practical (e.g., technically feasible and having technical merit) and (2) the corrective actions are properly implemented. Therefore, each continuous improvement activity can include steps for:

- Conducting a management-level review of recommendations to ensure that the recommendations are effective and practical

- Communicating rejected recommendations (and the basis for the rejection) to the appropriate teams

- Generating appropriate corrective actions based on a review of the recommendations and/or other recommendations resulting from the management review

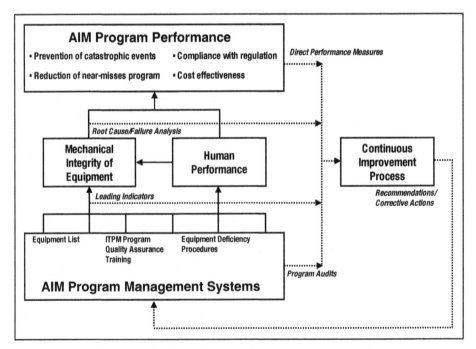

Figure 14-1. AIM program continuous improvement model

- Reviewing and approving corrective actions by management

- Prioritizing corrective actions for implementation, including identifying which corrective actions are to be in place before critical process activities occur, such as placing deficient assets in service or restarting a process

- Periodic (e.g., monthly, quarterly) tracking of the implementation of corrective actions by assigned individuals and management

- Following up to ensure that the corrective actions have been appropriately implemented by assigned individuals and management.

The remaining sections of this chapter provide an overview of:

- AIM program performance measurement and monitoring

- AIM program audits

- AIM-related incident investigations.

Additional information on these topics can be found in the following AIChE Center for Chemical Process Safety (CCPS) publications:

- *Guidelines for Risk Based Process Safety* (Reference 14-2)

- *Guidelines for Process Safety Metrics* (Reference 14-3)

- "Process Safety Leading and Lagging Metrics…You Don't Improve What You Don't Measure" (Reference 14-4)

- *Guidelines for Auditing Process Safety Management Systems, Second Edition* (Reference 14-5)

- *Guidelines for Investigating Chemical Process Incidents, Second Edition* (Reference 14-6).

14.1   PERFORMANCE MEASUREMENT AND MONITORING

Facilities are motivated to implement performance measurement systems for several reasons. Performance measurement enables facilities to:

- *Monitor changeslimprovements.* As the AIM program is improved or changed, the impact of changes/improvements can often be evaluated using performance measures.

- *Make appropriate decisions for supporting the AIM program.* During the implementation and the life of the AIM program, data can be used to make critical decisions. For example, decisions on staffing and budgetary issues for training and procedures can be supported via training and procedure measures.

- ***Track how the AIM program is affecting safety and asset reliability.*** Key benefits of an effective AIM program are improvements in safety performance and increased asset reliability. Safety and asset reliability performance measures (e.g., near misses, asset mean time between failure statistics) are often used to monitor the AIM program. In fact, some organizations include these measures in the daily and weekly key performance indicators.

- ***Identify and publicize achievements.*** Measures can help show whether the allocated resources and the effort expended are truly contributing to the AIM program, which can help individuals and groups understand and take pride in their contribution to the program.

- ***Ensure that the program is maintained.*** Performance that is measured tends to receive closer personal and organizational attention.

One approach for implementing a performance measurement system uses the following steps.

14.1.1 Identify Appropriate Metrics

The metrics selected are tied to the overall goals, objectives and results of the AIM program activities. A comprehensive system will include both lagging indicators and leading indicators (Reference 14-4). The *lagging indicators* directly measure the overall results of the AIM program, usually in terms of the program goals and objectives (e.g., number of loss events, program costs). The *leading indicators* indirectly track the overall AIM program by measuring the program activities such as the percent of ITPM tasks performed on time or the average time assets are operated in a deficient condition. Leading indicators help an organization predict the performance of the AIM program. Because some direct measures of AIM program performance are low-frequency events, such as major releases or fires, leading indicators can be valuable for the decision making process.

> **Lagging Indicator**
>
> Outcome-oriented metric, such as incident rate, downtime, quality defects or other measure of past performance.

> **Leading Indicator**
>
> Process-oriented metric, such as the degree of implementation or conformance to policies and procedures, which supports the PSM program management system and has the capability of predicting performance.

In addition to identifying appropriate metrics, it will need to be determined how to best measure each metric that is chosen to be tracked. In some cases, an ordinal number provides the needed information, such as the total number of maintenance workers. Other cases, such as average years of experience, require that two or more attributes be indexed to provide meaningful information. Still other metrics may need to be tracked as a rate, as in the case of employee turnover. Sometimes, the rate of change may be of greatest interest. Since every situation is different, it will need to be determined how to track and present data to most efficiently monitor the health of AIM programs and systems at its facility.

Table 14-1, compiled from References 14-2, 14-4 and 14-7, is a summary of possible leading-indicator metrics for the elements of an AIM program. When selecting a set of metrics for a given AIM program, it should be kept in mind that more is not always better. It is often more effective to select and use a smaller number of key AIM program effectiveness metrics than to have an overwhelming set of measures to try to track and utilize (Reference 14-7).

TABLE 14-1. AIM Program Leading-Indicator Metrics

Objective	Possible Metric	Comments
Identify equipment and systems that are within the scope of the asset integrity program and assign ITPM tasks	Number of equipment items included in the asset integrity program	Although this number has very little meaning in isolation, it could be used as a basis to compare asset integrity programs, particularly if the company operates similar processes at multiple facilities; the metric is likely to be more meaningful if the data are further subdivided by type of equipment, such as pressure vessels, storage tanks, safety interlocks, pumps, agitators, instrumented control loops, etc.
	Number of ITPM work orders (per month or quarter) that apply to equipment that is no longer present at the facility	A higher-than-expected number may indicate a weak link between the AIM and MOC elements of PSM; if the ITPM plan and preventive maintenance work orders in the CMMS are not updated when equipment is removed from service, it is quite likely that they are not updated when new equipment is installed
Develop and maintain knowledge, skills, procedures and tools	Number of inspectors/ maintenance workers holding each type of required certification	A decline in this metric may be a leading indicator of skill gaps or a higher than acceptable backlog for ITPM tasks
	Number (or percent) of maintenance workers with overdue training	This could also be measured on the basis of number of training modules, maintenance procedures, etc.

TABLE 14-1. *Continued*

Ensure continued fitness for purpose	(Number of inspections of safety-critical items of plant and equipment due during the measurement period and completed on time / Total number of inspections of safety-critical items of plant and equipment due during the measurement period) x 100%	• This metric is one measure of the effectiveness of the process safety management system to ensure that safety-critical plant and equipment is functional • This involves collecting data on the delivery of planned inspection work on safety-critical plant and equipment • To calculate the metric: Define the measurement period for inspection activity, determine the number of inspections of safety-critical plant and equipment planned for the measurement period, determine the number of inspections of safety- critical plant and equipment completed during the measurement period • Inspections not undertaken during the previous measurement period are assumed to be carried forward into the next measurement period
	Number of overdue safety-critical ITPMs	The intent and goal for this metric is zero (see Section 5.4)
	Number (or percent) of overdue ITPM tasks	A high number (or rate) of overdue ITPM tasks may indicate resource constraints or that equipment is not being made available for scheduled maintenance
	Number of emergency/ unplanned repair work orders per month	One of the primary objectives of the asset integrity element is to reduce unplanned/breakdown maintenance work. Although many unplanned failures will not involve equipment included in the scope of the asset integrity element, an increase in this metric may be a leading indicator of an overall slip in the effectiveness of the maintenance program at the facility
	Work order backlog (s); i.e., planned activities not yet past due	Similar to the number of past-due ITPM tasks, a backlog may indicate resource constraints; however, this metric may be a better leading indicator than the number of past-due ITPM tasks. Backlog may be monitored for the inspection group and/or each craft, or it may be monitored for each equipment class
	Work order backlog for each craft; i.e., planned activities not yet past due	Similar to the number of past-due ITPM tasks, a backlog may indicate resource constraints; however, this metric may be a better leading indicator than the number of past-due ITPM tasks
	Total time charged to ITPM tasks each month/quarter	A decline in the amount of time that is charged to these activities may indicate a change in focus for the maintenance department; note that changes could be cyclical by design or could be an intended result (e.g., an effort to rationalize redundant or unnecessary calibration activities ought to result in a decline in time spent on ITPM tasks)

TABLE 14-1. *Continued*

Address equipment failures and deficiencies	(Length of time plant is in production with items of safety-critical plant or equipment in a failed state, as identified by inspection or as a result of breakdown / Length of time plant is in production) x 100%	This is a metric to determine how effectively the safety management system ensures that identified deficiencies of process safety equipment are fixed in a timely manner
	Number of temporary repairs currently in service (deferred maintenance items)	This metric is another leading indicator of risk; it may be a particularly useful measure of efforts to plan maintenance if the metric is limited to repairs that are scheduled to be completed outside of major turnarounds (e.g., when parts arrive) versus repairs that need to be made during an extended outage or turnaround
	Total number of deferred repairs, such as known deficiencies that will be addressed at the next turnaround	Note that this metric will often increase linearly over time until the next maintenance shutdown, when it drops off sharply; however, the rate of increase could be a leading indicator of risk
	Average time to address/correct deficiencies	This can be another leading indicator of risk and may help indicate if a step change has occurred in the ability to quickly repair equipment; however, at a continuous plant, this metric may be heavily influenced by a few deficiencies that are scheduled to be repaired at the next turnaround; facilities may need to exclude "turnaround jobs" from this metric to provide a meaningful trend line
Analyze data	Number (or percent) of ITPM tasks that uncover a failure, require remedial action or identify potential issues tending toward deficiencies	Clearly, one objective of the asset integrity element is to discover and correct hidden failures before they lead to catastrophic incidents; however, an increase in this metric may indicate that risk associated with equipment failure is gradually increasing or equipment degradation is starting to occur
	Equipment reliability (or availability)	Similar to the previous metric, a decrease in reliability (or availability) may indicate that risk associated with equipment failure is gradually increasing

Metrics will be different depending on the level of the organization where they will be used. Senior management will focus more on lagging indicators and a select set of leading indicators and key performance indicators (KPIs), whereas the maintenance department will focus on metrics related to measures such as resource usage, backlog and recurring failures.

Developing one or more "process maps" of the AIM program can help ensure that the correct measures are used to evaluate the program. The process map typically starts with the program objectives (e.g., prevent catastrophic events) at the top and then works backwards through supporting program goals (e.g., preventing releases of hazardous materials) to AIM program activities. Once the objectives, goals, and activities are identified, reasonable and practical measures can be developed for each of these items. While one or several measures could be identified for each item of the process map, an organization can realistically manage only a limited number of measurements. Figure 14-2 on pages 418 and 419 illustrates this process mapping concept and provides suggested measures.

14.1.2 Collect and Analyze Performance Data

In addition to identifying AIM metrics, facilities will need to determine the appropriate frequency for collecting and monitoring the performance data used to generate the metrics. Once facilities begin to collect the measurement data, the data needs to be converted into information that is used on a continual basis to assess and improve the AIM program. Processes for analyzing and using this information need to be defined and implemented, followed by management commitment to using the information to make decisions supporting the AIM program.

Many facilities have developed and implemented computer-based systems for collecting and analyzing the data related to performance measures. Some of the data for the AIM program, such as an overdue report for an ITPM task work order, can be generated from the CMMS or other databases associated with the CMMS. Other data can be generated directly from the DCS, such as the number of excursions outside the integrity operating window (Reference 14-7). The data analysis includes any required calculations or data manipulation (e.g., filtering out of data anomalies) and tracking the results to look for trends, especially negative trends such as increases in overdue ITPM task work orders. Facilities often use procedures or guidelines to ensure the consistency and quality of the data collected and the reporting of the analysis conclusions.

AIM data can be analyzed both for programmatic issues as well as for equipment-specific issues (i.e., reliability analysis). It is possible to do higher-level analysis than just at the local plant level if data is collected company-wide in a consistent manner; e.g., using specific audit questions or sampling the same data.

Pertinent to the topic of this section, the following "Six Laws of Measurement" are presented in the book *Risk-based Management: A Reliability-centered Approach* (Reference 14-8):

1. ***Anything Can Be Measured.*** Computers have made it possible to measure almost everything.

2. ***Just Because Something Can Be Measured Does Not Mean it Should Be.*** As a result of Law 1, determining what should be measured becomes more important. Facilities also need to clearly define how measurement results will be used (i.e., types of decisions to be influenced by the results).

3. ***Every Measurement Process Contains Errors.*** Even the most accurate measurement contains some error, and that error can adversely influence the result and thus the decision. Management needs to attempt to understand the magnitude of the error and the impact that errors can have on the results.

4. ***Every Measurement Carries the Potential for Changing the System.*** Measurement processes, especially those involving human intervention, can bias the measurement data and results. For example, simply measuring the air pressure in a tire results in a lowering of the pressure.

5. ***The Human Is an Integral Part of the Measurement Process.*** Humans involved in the measurement process affect and influence the measurements.

6. ***(Jones' Law) You Are What You Measure.*** Each measurement ought to be part of the overall management strategy for the AIM program. The following guideline can be applied: Each measurement needs to be directly related to achieving the mission of the organization (e.g., the AIM program mission).

14.1.3 Monitor the Performance Measures and KPIs

Performance measurement systems can be set up to communicate necessary information to the personnel assigned to correct negative trends to improve the AIM program. In addition, performance measures can be used to communicate the status of the AIM program, and the improvements/effects resulting from the program, to the entire organization.

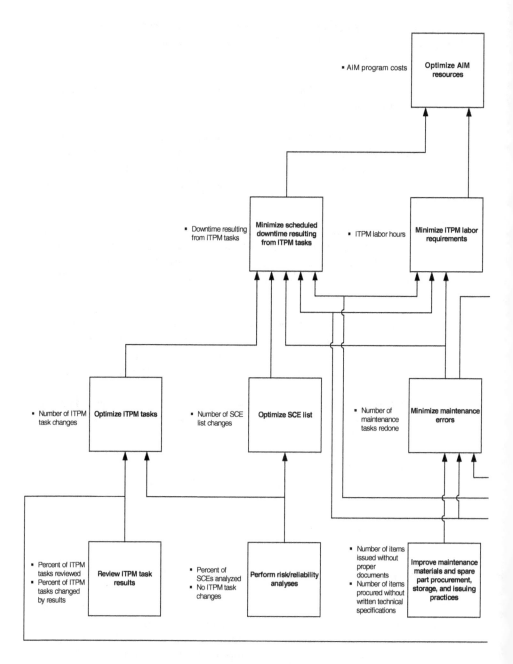

Figure 14-2. Example AIM process map with suggested performance measures
(Credit: ABSG Consulting)

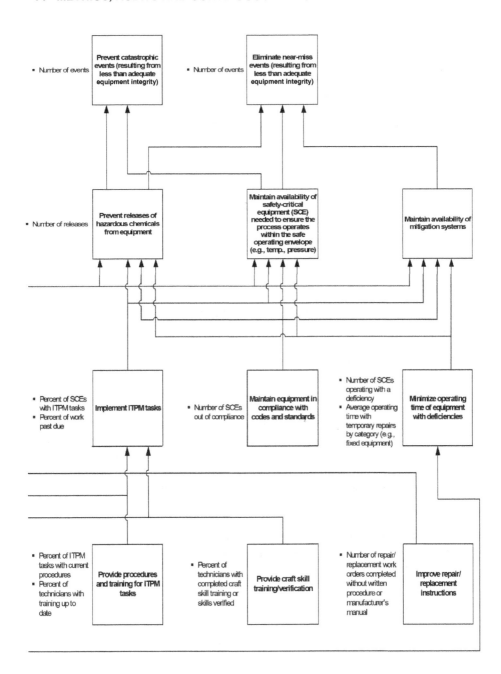

Figure 14-2. *Continued*

Performance measures and the evaluation of the results may be periodically summarized and distributed to managers and decision makers within the organization. These reporting and communication activities help the AIM program receive the continual attention that it needs.

Many facilities collect and monitor AIM program measures on a daily or weekly basis. Sometimes, these frequent measures are included in the facility's measurement "dashboard" (i.e., continuous/daily/weekly publication of the facility's key performance indicators, or KPIs). Another approach is to display a simple chart showing which facilities or units have fully implemented specific programs or practices. Sometimes simply displaying this information, and reviewing it in management team meetings, will spur management interest and support (Reference 14-2).

14.2   AIM PROGRAM AND IMPLEMENTATION AUDITS

Program and implementation auditing is one of the basic tools for continuous improvement of an AIM program. A good audit is a proactive means to uncover program and implementation deficiencies that can be corrected before an incident occurs.

In a risk-based process safety (RBPS) program, audits are intended to evaluate whether management systems are performing as intended. Such management system audits are likely to deliver greater benefits in terms of overall effectiveness than audits that focus only on whether specific compliance requirements are met (Reference 14-5).

Audits complement other RBPS control and monitoring activities in elements such as management review, metrics and inspection work activities that are part of the asset integrity and conduct of operations elements. The auditing element of RBPS comprises a system for scheduling, staffing, effectively performing and documenting periodic evaluations of all RBPS elements, as well as providing systems for managing the resolution of findings and corrective actions generated by the audits.

Various aspects of an AIM program and its implementation can be readily audited. Some audits look at the overall program, while others may target select areas of implementation or of suspected weakness. Some of the more common program features to examine are:

- *AIM program asset selection.* Are all assets included in the AIM program that, if they fail, could result in consequences of concern? Are all assets used to prevent hazardous/unwanted conditions, or to detect and mitigate loss events, included in the AIM program?

- *ITPM tasks and schedules.* Have ITPM and anticipated repair/rebuild activities been appropriately determined and their frequencies established? Are the ITPM tasks and schedules consistent with applicable codes, standards and practices? Are ITPM tasks performed thoroughly and on schedule? Are ITPM tasks properly documented?

- *Written procedures for AIM program tasks.* Do written procedures exist for all AIM program tasks? Do the procedures contain adequate information, and are they kept current? Is work performed in accordance with the procedures? Has the facility implemented a means or program, such as periodic supervisory audits, to ensure that procedures are accurate and personnel are performing tasks in accordance with the procedures?

- *AIM training.* Are personnel trained in an overview of the process and its associated hazards, and in the procedures applicable to their job tasks, to ensure that the tasks can be performed correctly, safely and consistently? Do craftspersons and inspectors have needed qualifications/certifications?

- *Correcting asset deficiencies.* Is a program/process in place to ensure that asset deficiencies are identified and communicated? Are ITPM results reviewed for potential deficiencies? Are continued operation and associated temporary corrective measures such as temporary repairs reviewed and approved by appropriate personnel? Are temporary corrective actions communicated? Are temporary actions tracked to ensure that they are corrected in a timely manner?

- *QA for asset fabrication and installation and spare parts/equipment.* Is a program/process in place to ensure that assets are fabricated and installed in accordance with specifications? Are materials of construction verified for equipment, spare parts and maintenance materials? Has a contractor selection and auditing process been established? Are off-the-shelf spare parts, equipment items and other materials/supplies (e.g., bolts, fittings, gaskets, O-rings, welding rods) correctly ordered, received, placed into inventory and retrieved from inventory before use or installation?

Comparing written records and policies with field observations and personnel interviews provides an auditor with a more accurate picture of the complete AIM program. More information about auditing techniques is presented in CCPS' *Guidelines for Auditing Process Safety Management Systems* (Reference 14-5).

Preparation for an AIM program audit usually includes deciding on an audit protocol and any particular focus areas. Input to these decisions may include information on significant equipment failures and trends, as well as incidents where managing asset integrity was a root cause or contributing factor.

The audit then typically begins with a review of written programs/procedures and interviews with personnel involved, in order to gain an understanding of the intended functions of the management system(s) associated with each audit topic. The auditors then determine if the management systems actually function as intended and if other requirements (e.g., jurisdictional requirements) are met. Table 14-2 summarizes one approach for evaluating AIM management system functions.

TABLE 14-2. AIM Audit Approach

Audit Topic	Audit Approach Summary
AIM program assets	• Use facility tours and/or examination of P&IDs to select a representative sample of the different types of assets (e.g., pressure vessels, storage tanks, piping segments, pumps, PSVs, rupture disks, alarms and interlocks, release detectors, emergency shutdown systems, deluge systems) that are intended/required to be in the AIM program, and confirm that each is actually in the AIM program (i.e., AIM program tasks are scheduled and records show that the tasks are being performed)
	• Verify the AIM program includes all intended equipment items, such as (1) assets that under normal operating conditions contain a hazardous chemical, (2) assets that are designed to relieve pressure or vent hazardous chemicals, (3) emergency shutdown systems, (4) safety devices/systems such as detection systems, suppression systems, alarms and interlocks, and (5) critical utility systems
ITPM tasks and schedules	• Review the ITPM tasks and schedules for the selected assets to determine (1) if AIM program tasks are identified and schedules exist for these tasks and (2) whether the tasks and schedules are appropriate
	• Verify ITPM tasks and schedules are (1) based on codes, standards and/or manufacturer recommendations and (2) more stringent and/or performed more frequently if dictated by facility experience (e.g., PSVs are on a more frequent ITPM schedule if they have failed prior tests/inspections)
	• Review the completed AIM program task records for the selected assets. Generate/review reports listing a backlog or past-due AIM program tasks
	• Interview maintenance personnel to determine if (1) AIM program tasks are being performed on schedule (as defined or understood within the facility), (2) tasks are being performed thoroughly (e.g., all condition monitoring locations are addressed in each ultrasonic test), (3) the results of the AIM program tasks are documented and the records contain all intended/ required information, and (4) the results of AIM program tasks are reviewed by a qualified person to determine if an asset deficiency exists and, if so, recommend corrective actions
Written procedures for AIM program tasks	• Review the written AIM program task procedures for the selected assets and interview maintenance personnel as necessary to determine whether (1) written procedures exist for all tasks that could affect asset integrity, (2) the content and level of detail in the procedures are sufficient based on the complexity and criticality of the tasks and the minimum skill level of maintenance personnel, (3) the procedures are being kept current, and (4) the procedures are always accessible to personnel performing the tasks
	• To the extent that any written procedures provided in manufacturer/vendor manuals do not meet internal guidelines (e.g., address issues unique to the facility, such as PPE and special tool requirements) and/or external requirements (e.g., regulatory requirements), such procedures need to be supplemented to satisfy the need for written procedures. Note also that AIM program procedures are typically kept current through periodic review by maintenance personnel and revision as necessary, and/or by including the procedures in the facility's management of change process

TABLE 14-2. *Continued*

Audit Topic	Audit Approach Summary
AIM training	• Review a representative sample of AIM training records and interview a representative sample of maintenance personnel to determine (1) which procedures and additional training topics specifically apply to which personnel (e.g., an overview of the processes and their hazards); (2) whether each craftsperson has been trained as intended/required on the procedures and topics that apply to their jobs before being required/allowed to perform a task; (3) whether training is also provided on an ongoing basis as new tools, equipment, techniques and other changes are introduced; (4) whether personnel have been qualified and certified as required by applicable codes, standards and regulations; and (5) whether training is documented; note that training needs to be documented for AIM activities defined by codes and standards requiring special certifications, such as American Petroleum Institute inspection standards (e.g., API 510, API 653) and welding standards (e.g., ASME B31.3)
	• Note that training documentation may not be required by the AIM program; however, if training is not documented, determine how the facility (1) internally tracks training to ensure that personnel receive needed training, (2) demonstrates to a compliance officer or auditor that the training has been provided and (3) provides evidence that personnel understood the training
Correcting asset deficiencies	• Review any corrective action records such as completed work orders and interview maintenance personnel and/or operators to determine if corrective actions are addressed as necessary; determine whether (1) corrective actions are consistently implemented before further use of the equipment, (2) adequate interim measures are consistently implemented to ensure safe operation when corrective actions are not taken before further use of the asset and (3) asset records document how and when identified deficiencies were corrected (and preferably any interim measures taken to ensure safe operation) and that any interim measures are tracked through permanent repair
QA for equipment/ spare parts	• Review documentation such as OEM equipment lists, purchase orders, shipping papers, metallurgical analyses records, and interview personnel who order, receive, place into inventory, retrieve from inventory and issue spare parts/equipment or materials/supplies (e.g., bolts, fittings, gaskets, O-rings, pump shafts, seals, welding rods), to determine whether (1) the correct items are consistently ordered, (2) any items received that are not what was ordered are consistently identified during the receiving process and returned to the supplier (or appropriately reviewed before being accepted), (3) received items are consistently placed correctly into inventory, (4) stored items are tested or inspected as required to ensure they remain functional, and (5) the correct items are consistently retrieved from inventory for installation
	• Many facilities (1) order OEM spare parts/equipment/supplies whenever they are available, (2) use an MOC process when ordering other-than-OEM parts (even when the specifications appear to be the same), (3) require that persons ordering items personally examine them when received to ensure that correct items are received and (4) allow only designated personnel to place items into and retrieve items from inventory; some facilities also require suppliers to provide a metallurgical analysis or are testing metallurgy themselves when metallurgy is important

TABLE 14-2. *Continued*

Audit Topic	Audit Approach Summary
QA for equipment fabrication/ installation	• Interview personnel who order and/or oversee fabrication and installation of specialty equipment and examine documentation as available such as written QA plans and installation checklists to determine whether the facility takes appropriate measures to ensure that equipment is fabricated and installed in a way that is suitable for the process application
	• Frequently, QA for equipment fabrication and installation involves (1) developing equipment fabrication specifications; (2) requiring various inspections and tests to be performed during fabrication to ensure that the equipment specifications are met; and (3) developing installation plans, schedules, and checklists to ensure that field installation issues such as bolt torquing, welding rod selection, gasket and packing materials and lubrication are thoroughly and correctly addressed
AIM documenta- tion	• Determine whether all internally or externally required documentation of AIM activities is created and retained

	• Some AIM programs require only the following documentation:
	  • Written ITPM task procedures
	  • Required ITPM task lists and their schedules
	  • ITPM task records showing the results of the tasks performed

	• However, in most effective AIM programs, the following additional documentation is also provided:
	  • Roles and responsibilities assignments for AIM program activities
	  • AIM program equipment list(s)
	  • AIM training records
	  • Records showing that deficiencies have been corrected
	  • Records, which are frequently in project files, of QA activities for equipment fabrication and installation

	• Many facilities also describe the overall program in a written AIM program document

When performing an audit of an AIM program, the following issues are worth considering.

- *Level of detail.* Company safety audits will generally include the AIM program; however, typical safety audit teams may not have the resources available or the expertise to thoroughly review AIM. Some companies schedule specific AIM program audits or even audits of specific AIM practices (e.g., QA) to better focus on AIM improvement.

- *Objectivity.* Some auditing can be done within a department to periodically verify program adherence (e.g., ITPM schedule adherence). However, periodic reviews by personnel from outside departments, sister plants, the parent company and/or persons external to the company add perspective and can often elicit helpful suggestions for program enhancements.

- *Documentation.* Although not all program audits require documentation (unless the audit is for compliance with a jurisdictional regulation), most audits are documented with a report. Formal reporting helps ensure that any necessary follow-up actions are completed and provides an audit history for charting AIM program progress. In addition to documenting deficiencies, management systems and practices that fully meet expectations and requirements can also be documented.

- *Audit follow-up.* When an audit is completed, facility management needs to promptly respond to audit findings and have required changes implemented in a timely manner. Management responsibilities include (1) deciding on appropriate action to address each audit finding, (2) assigning follow-up responsibilities for every audit finding, (3) providing resources needed to complete each action, (4) ensuring that a system is in place to track each action item to closure, (5) holding assigned personnel accountable for implementing their assigned actions, and (6) ensuring the closure of every audit finding is documented. Management action is a key aspect in ensuring that the changes are implemented. Lack of diligence by management can result in lost opportunities for improving the AIM program.

Learning from Inspections at Other Facilities. One additional means of identifying possible AIM program weaknesses is to look through a listing of items that have been cited by regulators as program deficiencies, then determine whether any of the items are relevant to the facility at hand. Included in Appendix 14A is a listing of AIM-related inspection results, from U.S. OSHA's Refinery National Emphasis Program (NEP), grouped by specific regulatory requirements in the mechanical integrity element of the OSHA PSM Standard.

Management Reviews. The overall design and conduct of management reviews is described in Chapter 22 of CCPS' *Guidelines for Risk Based Process Safety* (Reference 14-2). However, many specific questions/discussion topics exist that management may want to check periodically to ensure that the management system for the asset integrity element is working properly.

A proper starting place is for management to seek to understand whether the system being reviewed is producing the desired results. If the organization's level of asset integrity is less than satisfactory, or it is not improving as a result of management system changes, then management can identify possible corrective actions and pursue them. Possibly, the organization is not working on the correct activities, or the organization is not doing the necessary activities well. Even if the results are satisfactory, management review can help determine if resources are being used effectively, asking questions such as whether there are tasks that could be done more efficiently or tasks that do not need to be done at all. Management can combine metrics listed in the previous section with personal observations, direct questioning, audit results and feedback on various topics to help answer these questions. Activities and topics for discussion include the following:

- Compare the plan that is used to assign ITPM tasks at the facility to corporate standards. Does the plan contain any blanket exceptions? If so, review the justification(s) for these exceptions to ensure they are still valid.

- Review individual exceptions to the ITPM plan that have been authorized since the last management review. (Note, this item ought only address tasks not being performed, not individual ITPM tasks that are currently past due or were performed late.) As a group, are these exceptions justified, or does the facility have an underlying tendency to not perform certain ITPM tasks (or not perform a variety of ITPM tasks in a certain unit)?

- Review inspection or test results that have been obtained using new/trial methods. Did these methods deliver the expected benefits and can their use be extended to other areas/equipment at the facility?

- Review the special certifications that maintenance department personnel hold. Has this changed since the previous management review? If any personnel holding critical certifications are planning to retire soon, review plans to train their replacements.

- Determine if failure to comply with the ITPM work schedule is more often caused by random events or systemic issues. For each systemic issue that is identified, determine if it was reviewed in advance to assess the risk.

- Determine the dominant causes of past-due ITPM tasks that appear to be random.
  - Is equipment not available?
  - Are intervals for ITPM tasks intentionally set low because a widely held belief exists that tasks are not normally completed on time (i.e., the "snooze button" approach to scheduling)?
  - Was the maintenance or inspection group unprepared to conduct the activity when equipment was available?

- Review which inspection activities are routinely contracted out and which are performed by maintenance department employees. Does this make sense from a risk management and business perspective?

- Review metrics related to the fraction of maintenance work that is planned. If the work is planned, the OEM parts and shop manuals are much more likely to be available, and the work can be scheduled when a fully staffed and trained work crew is available. Furthermore, an increase in unplanned maintenance work may reflect an underlying increase in the rate of unplanned breakdowns.

- Review new maintenance programs and work methods that are designed to reduce maintenance costs. What, if any, impact has the cost cutting had on equipment reliability, the fraction of ITPM tasks completed as scheduled, or other key metrics for this element?

- Examine Pareto charts showing chronic equipment failures and determine if the facility has a practice to report and investigate these failures as chronic (possibly near miss) incidents.

Reliable equipment, coupled with reliable human performance, is key to managing risk. In addition, both are necessary conditions for reliable operation. The management review for this element often starts with questions related to metrics such as equipment availability or performance. However, the management review needs to delve deeper, examining the quality of work activities that underpin the asset integrity element. Just as the lack of a catastrophic process safety incident in the past ten years is not necessarily a reliable predictor of the likelihood of a catastrophic incident in the next year, the lack of catastrophic equipment failure does not indicate that the asset integrity element is fully functional. Piping may be about to fail as a result of corrosion, nozzles on vessels may be cracked, or safety systems that mitigate the consequences of an incident may have already failed. Thus, the management review for this element needs to examine the details, asking hard questions such as, "How many ITPM tasks have we missed?" and "What temporary repairs or deficient equipment are currently in service?" Also, maintenance and contract personnel can make errors that compromise the integrity of equipment. Asking these and other tough but fundamental questions is critical to understanding the true health of the asset integrity element.

14.3   CONTINUOUS IMPROVEMENT

Management Review and Continuous Improvement is one of the essential elements of risk-based process safety management, as part of learning from experience (Reference 14-2). One key continuous improvement activity is to systematically evaluate equipment failures, process safety incidents and near misses, including excursions outside of integrity operating windows, using structured evaluation processes such as root cause analysis. Using systematic analyses provides an effective means to identify the contributing and root causes of failures. From the results of these processes, personnel can make recommendations for eliminating underlying causes, then management can develop and implement appropriate actions to eliminate or reduce the likelihood of future failures and excursions. In addition, lessons learned from these analyses can be shared with others who may benefit from the learnings, both within the facility and at other facilities.

14.3.1 Incident Investigations

Incident investigation is a process for reporting, tracking and investigating incidents that includes (1) a formal process for investigating incidents, including staffing, performing, documenting, and tracking investigations of process safety incidents and (2) the trending of incidents and incident investigation data to identify recurring incidents. This process also manages the resolution and documentation of recommendations generated by the investigations, and includes a two-way sharing of learnings from failures (program level, equipment level, internal and external industry lessons).

At some facilities, an incident investigation is used to determine what employee (or contractor) caused the incident. This approach results in ineffective recommendations being implemented. A more effective approach is to develop recommendations that address the systemic, management-system failures that led to the incidents. The incident investigation element is not a process to assign blame but a process to develop effective recommendations to address the underlying, system-related causes of incidents. Some hazardous material releases, for example, may be due to failures in the managing of the ongoing integrity of equipment, whereas others may be due to design or specification errors that would have very different underlying causes.

A good starting point for incident investigations is determining whether or not an "incident" has occurred. Reference 14-4 gives consensus definitions of *process safety incident* and *process safety event* as well as how to calculate lagging metrics (incident rates) based on those definitions.

Some companies collect and report information on process excursions that did not actually lead to a release, fire or explosion but had the potential to do so. A good practice is to collect information when standby safety systems function (or fail to function) on demand, as an indication of the frequency at which such systems are being challenged. An organization could also collect information from the basic process control system (BPCS) on how often process deviations occur, or what percentage of the time the system is operating out of limits, or the number of instruments operating out of range.

14.3.2 Asset Failure and Root Cause Analyses

Facilities can incorporate lessons learned into the AIM program by using a structured process to analyze asset failures and associated management system failures. This process can include applying two different analysis techniques: failure analysis and root cause analysis (RCA). Details on asset failure analysis and root cause analysis can be found in Section 15.5 of this document.

CHAPTER 14 REFERENCES

14-1   U.S. Occupational Safety and Health Administration, *Process Safety Management of Highly Hazardous Chemicals*, 29 CFR Part 1910, Section 119, Washington, DC, 1992.

14-2   Center for Chemical Process Safety, *Guidelines for Risk Based Process Safety*, American Institute of Chemical Engineers, New York, NY, 2007.

14-3   Center for Chemical Process Safety, *Guidelines for Process Safety Metrics*, American Institute of Chemical Engineers, New York, NY, 2010.

14-4   Center for Chemical Process Safety, "Process Safety Leading and Lagging Metrics …You Don't Improve What You Don't Measure," American Institute of Chemical Engineers, New York, NY, www.aiche.org/ccps.

14-5   Center for Chemical Process Safety, *Guidelines for Auditing Process Safety Management Systems, Second Edition*, American Institute of Chemical Engineers, New York, NY, 2011.

14-6   Center for Chemical Process Safety, *Guidelines for Investigating Chemical Process Incidents, Second Edition*, American Institute of Chemical Engineers, New York, NY, 2003.

14-7   Broadribb, M.P., B. Boyle and S.J. Tanzi, "Cheddar or Swiss? How Strong Are Your Barriers?" *Process Safety Progress*, *28*(4), December 2009, 367-372.

14-8   Jones, R., *Risk-based Management, A Reliability-centered Approach*, Gulf Coast Publishing, Houston, TX, 1995.

14-9   U.S. Occupational Safety and Health Administration, Refinery NEP, Citation Summaries, https://www.osha.gov/dep/neps/inspection_results_attachment_b.html, accessed 10/1/2015.

Additional Chapter 14 Resources

Perry, Robert G. and J. Steven Arendt, *If You Can't Measure It, You Can't Control It; ProSmart Process Safety Management*, AIChE CCPS International Conference and Workshop, Toronto, Ontario, 2001.

API Recommended Practice 585, *Pressure Equipment Integrity Incident Investigation*, American Petroleum Institute, Washington, DC.

APPENDIX 14A. AIM-RELATED REGULATORY CITATIONS

One means of identifying possible AIM program weaknesses is to look through a listing of items that have been cited by regulators as program deficiencies. Included in this appendix is a listing of AIM-related inspection results from U.S. OSHA's Refinery National Emphasis Program (NEP). Each relevant requirement from the OSHA Process Safety Management (PSM) Standard is given, followed by the pertinent results summary paraphrased from citations (Reference 14-9).

TABLE 14-3. AIM-Related U.S. OSHA Refinery NEP Inspection Results

---

29 CFR 1910.119(j)(2) Written procedures. The employer shall establish and implement written procedures to maintain the on-going integrity of process equipment.

The employer did not develop an inspection procedure to inspect for corrosion under insulation on pressure vessels.

The employer had not developed a Mechanical Integrity program for corrosion-under-insulation inspections for critical piping containing large quantities of flammable materials.

The employer did not establish and implement written procedures to maintain the ongoing mechanical integrity of process equipment, including but not limited to procedures addressing:
- Inspecting pressure vessels with integrally bonded liners.
- Corrosion-under-insulation inspections for pressure vessels and piping.
- Determining the safe operation of equipment such as pressure vessels and piping after a temporary repair.
- Monitoring recommending follow-up inspections on equipment such as pressure vessels or piping that are operating with a deficiency.
- Addressing anomalous readings pertaining to metal thickness in pressure vessels and piping.
- Updating the mechanical integrity database within a short time frame after a pressure vessel or piping inspection to determine if the equipment has exceeded the retirement thickness or if the inspection interval needs to be reduced.
- Inspecting non-metallic linings in pressure vessels.
- Inspecting injection points for corrosion inhibitors, emulsifiers, anti-foam, methanol, etc. into the refinery piping.

The employer did not establish and implement written procedures to maintain the ongoing mechanical integrity of process equipment, including:
- Inspections for non-metallic linings of pressure vessels.
- Inspections for integrally bonded linings such as strip plating or plate lining of pressure vessels.
- Procedure requiring periodic internal as well as external (on-stream) inspection of pressure vessels with integrally bonded liners.
- Thickness measurement frequency for pressure vessels.
- Thickness measurement strategy for identifying TMLs on pressure vessels and for determining the representative number of thickness measurements performed for internal and on-stream inspections.
- Thickness measurement strategy for identifying TMLs on piping and for determining the representative number of thickness measurements on each piping circuit.
- Procedure for anomalous data related to thickness growth recorded in piping.
- Identifying person(s) permitted to conduct re-rating of pressure vessels.
- Information on alterations of vessels including welding qualifications and certifications for adding a nozzle to a heat exchanger.

TABLE 14-3. *Continued*

The employer did not write procedures to maintain the on-going mechanical integrity of process equipment in the refinery such as, but not limited to:

- Inspection procedures for non-metallic linings of pressure vessels.
- Inspection procedures for integrally bonded liners such as strip or plate lining of pressure vessels.
- Procedure requiring the next scheduled inspection after on-steam inspection to be an internal inspection for pressure vessels with integrally bonded liners.
- Thickness measurement frequency for pressure vessels.
- Condition monitoring strategy for pressure vessels and for determining the representative number of thickness measurements performed to satisfy requirement for internal and on-steam inspections.
- Thickness measurement strategy for the frequency of measuring each thickness measurement location (TML) on piping circuits.
- Procedure for anomalous data related to thickness growths recorded in piping with criteria as to what would be considered an increase from a previous thickness measurement.

The employer did not establish and implement written procedures to maintain the ongoing mechanical integrity of process equipment:

- The employer did not have procedures that identified and documented critical equipment to ensure that proper spare parts and/or spare equipment were available.
- The employer did not implement a written mechanical integrity inspection program that included relief devices.
- The employer had not established and implemented a written program procedure for establishing thickness measurement locations (TML) for injection points and mixing points on piping.
- The employer had not established and implemented a written procedure for managing permanent repair and replacement information related to the mechanical integrity of piping circuits in the risk-based mechanical integrity program. Work processes have not been established and implemented to capture, document and maintain material specifications, installation-related information (welding processes, personnel documentation, qualification, certifications, etc.) and physical location for the repair.
- The employer did not implement local and corporate policy regarding identified equipment deficiencies for fixed equipment and had not established and maintained written procedures, work practices and job duties to track temporary repairs, track completed repairs, modify inspection plans and follow modified inspection plans.
- The employer had not established and implemented a written procedure for the analysis of incoming data from inspection and testing reports for fixed equipment. Inspection and testing reports must be analyzed to identify and disposition rejectable indications, such as thickness monitoring locations with data below minimum thickness levels.
- The employer had not established and implemented a written procedure for managing the temporary repair of piping circuits, such as the use of the pipe clamps for leak control, which is common at the refinery.
- The employer had not established a written program for implementing and managing TMLs for piping circuits and pressure vessels.
- The employer had not implemented a system for TML identification, designation, and reference on isometric sketches for use in field.

The employer did not ensure that mechanical integrity inspection procedures for relief devices were implemented at the facility. Multiple instances of relief valves were found that were past their individual inspection intervals for inspection and testing in the Hydrocracker and Alkylation Units.

TABLE 14-3. *Continued*

Written mechanical integrity program procedures were not developed for process instrumentation and controls including, but not limited to, monitoring devices and sensors, alarms, interlocks, emergency shutdown systems (i.e. safety instrumented systems, safety shutdown systems, protective instruments systems, and safety interlock system), such as:
- Emergency steam to the riser control valve and associated equipment.
- The spent catalyst slide valve.
- Non-destructive evaluation (NDE) written procedures for pressure vessel and piping inspections.

The employer did not have written procedures for inspections and tests of critical instrumentation in accordance with RAGAGEP in the De-Coker and Alkylation units, such as, but not limited to, inspections and tests procedures for critical instrumentation that did not follow RAGAGEP. ISA S84.01 was referenced in the MI program, but was not being followed.

---

29 CFR 1910.119(j)(3) Training for process maintenance activities. The employer shall train each employee involved in maintaining the on-going integrity of process equipment in an overview of that process and its hazards and in the procedures applicable to the employee's job tasks to assure that the employee can perform the job tasks in a safe manner.

The employer did not train inspectors involved in maintaining the ongoing integrity of the process piping and pressure vessels in the employer's mechanical integrity program and procedures applicable to the employees' job tasks.

The employer did not train vibration analysts involved in maintaining the ongoing integrity of rotating and reciprocating equipment in the employer's mechanical integrity program and procedures applicable to the employees' job tasks.

Each employee, including operators, supervisors, coordinators for contractors, and contractor's employees involved in a line break were not adequately trained on specific lockout/tagout procedures to ensure that each employee performed their job tasks in a safe manner.

---

29 CFR 1910.119(j)(4)(i) Inspections and tests shall be performed on process equipment.

A hot relief high temperature alarm on a disengaging drum was mistakenly removed from the critical instruments program. The critical instrument had not been inspected since 2003.

The employer does not ensure that thickness monitoring for process equipment including, a piping circuit, is conducted in accordance with the MI inspection program.

Crude/vacuum unit pressure vessels had not received an inspection at the frequency required by the employer's mechanical integrity inspection program.

The employer failed to ensure that inspection testing was conducted on a heat exchanger on a five year inspection interval, as indicated by the vessel inspection plan.

The employer failed to ensure that ultrasonic testing was conducted on a heat exchanger nozzle.

Inspections and tests were not performed on process equipment to maintain its mechanical integrity:
- The employer did not ensure that relief devices and systems including the blowdown drums and vent stacks were included in a mechanical integrity inspection program. There was no inspection plan for the drum/stack and there had not been any non-destructive testing since 1992.
- The employer had not performed radiographic or ultrasonic testing for internal corrosion inspections on piping circuits in corrosive service.

The employer does not ensure comprehensive thickness readings were being conducted for an FCCU reactor.

The employer did not perform tests in the De-Coker and Alkylation units to maintain their mechanical integrity such as but not limited to control valves and process critical instruments, including critical alarms, critical indicators, and critical controls.

---

29 CFR 1910.119(j)(4)(ii) Inspection and testing procedures shall follow recognized and generally accepted good engineering practices.

Inspections ports on insulated piping were not adequately plugged after inspections were performed. Excessive corrosion could occur due to the open inspection ports.

TABLE 14-3. *Continued*

The employer did not follow RAGAGEP such as API 510 and API 580, when they established a 10-year inspection and testing interval for twenty safety relief valves.
Inspections and testing procedures did not follow RAGAGEP: • Safety valve failed to pop at its set pressure due to fouling. The valve was repaired and returned to service with the same 6 year inspection frequency. API 510 6.6.2.3 requires that the inspection interval shall be reduced when a pressure relieving device is heavily fouled.
The employer did not follow RAGAGEP by not ensuring an adequate number of TMLs and not conducting a representative number of thickness measurements for internal and on-stream inspections.
The employer did not follow RAGAGEP by not ensuring an adequate number of piping TMLs and not conducting a representative number of thickness measurements.
The employer did not follow RAGAGEP when it failed to resolve anomalous data for pressure vessels and piping, such as increasing thickness measurements over previous readings.
The employer did not follow RAGAGEP when it failed to compile necessary and complete piping inspection records and other inspection records related to thickness measurements for piping circuits.
In the (FCCU), the employer did not follow RAGAGEP when it failed in conduct thorough pressure vessel inspections.
In the Alkylation and FCCU, the employer did not follow RAGAGEP when it failed to conduct thorough piping inspection by not testing at designed thickness measurement locations (TML) each time.
In the Alkylation and FCCU, the employer did not follow RAGAGEP when it failed to resolve anomalous data for piping.
Throughout the refinery, the employer did not follow RAGAGEP when it failed to inspect and test emergency shutdown systems and controls, such as but not limited to monitoring devices, sensors, alarms and interlocks. A program was not in place for Identification, testing and documenting the results of the tests for safety systems.
In the Alkylation unit, previous condition monitoring required inspection of the recycle cooler in 2005, which did not occur.
The employer did not follow RAGAGEP when it failed to inspect a pressure vessel at the designated thickness monitoring locations (TML) and in the number of thickness locations in the inspection plan.
The employer did not follow RAGAGEP for piping inspections when it failed to compile necessary and complete piping inspection records related to repairs and replacements for 3 piping circuit.
29 CFR 1910.119(j)(4)(iii) The frequency of inspections and tests of process equipment shall be consistent with applicable manufacturers' recommendations and good engineering practices, and more frequently if determined to be necessary by prior operating experience.
Pump vibration analysis was not performed according to the frequency established by the employer's mechanical integrity program procedure.
Thickness measurements were not performed according to the frequency established by the employer's mechanical integrity program procedure.
The frequency of inspections and tests of process equipment was not conducted as specified by the employer's master plan to maintain mechanical integrity; such as but not limited to: a cathodic protection system, and process analyzers.
The employer did not ensure that relief devices were inspected on a frequency consistent with the company's mechanical integrity program and with RAGAGEP (liquid propane, butane service).
The employer had not performed a piping inspection at the interval required by the risk-based mechanical integrity (RBMI) database.
The employer did not meet its set inspection frequencies and determine and adjust its frequency of inspections and tests based on findings from prior operating experience.

TABLE 14-3. *Continued*

Relief valves inspected beyond their set date based on a four year cycle: • The FCC Complex Blowdown Drum per the frequency set by operating experience. • The Crude Units Blowdown Drum per the frequency set by operating experience.
The employer did not ensure that inspections and test of process equipment occurred with the necessary frequency such as but not limited to: • The employer did not ensure that inspections and testing were being conducted for critical valves and pressure safety valves (PSV) at the Boiler House and at the De-Coker unit. • The employer did not ensure that the frequency of inspection and the testing of 2 Alkylation unit piping circuits were being maintained. • The employer did not ensure the frequency of inspection and tests for 2 piping circuits De-Coker unit.
29 CFR 1910.119(j)(4)(iv) The employer shall document each inspection and test that has been performed on process equipment. The documentation shall identify the date of the inspection or test, the name of the person who performed the inspection or test, the serial number or other identifier of the equipment on which the inspection or test was performed, a description of the inspection or test performed, and the results of the inspection or test.
The employer did not adequately document the inspection and test of interlocks in the FCCU in that the records did not identify the person who performed the test, the serial number or other identifier of the equipment.
Critical Piping circuit Item #1 did not have any inspection data established since the line was installed in 1972.
The employer did not document an internal inspection for two HF acid storage drums.
The employer did not document each inspection and test that had been performed on process equipment to maintain its mechanical integrity. • For profile radiographic testing performing in conjunction with an amine leak, the employer did not adequately document the information in the RBMI database. • The employer did not adequately document guided ultrasonic longwave (GUL) testing performed on piping circuit during the 2004 Crude unit turnaround. There are no inspection results, the testing data has not been captured and entered into the risk based mechanical integrity database and there is no information in the asset file for this piping circuit. • Historical piping inspection results and descriptions are not available prior to 2001, which does not provide the information needed to estimate accurate corrosion rates. • Profile radiographic testing performed on piping circuit was not documented.
The employer did not ensure that thickness readings were documented for an FCCU.
The employer did not ensure that inspection and testing documentation for function tests of emergency shutdown systems and safety-critical controls were maintained.
The employer did not ensure that inspections and tests were adequately documented. Inspections and tests to determine if a leak had been repaired were not adequately documented or not documented.
29 CFR 1910.119(j)(5) Equipment deficiencies. The employer shall correct deficiencies in equipment that are outside acceptable limits (defined by the process safety information in paragraph (d) of this section) before further use or in a safe and timely manner when necessary means are taken to assure safe operation.
The employer continued to use an undersized Blowdown drum for high volume relief of hot liquid/vapor hydrocarbons.
The employer also continued to use the blowdown drum after isolating the quench water system in late 2004 and did not use other means to assure safe operation.

TABLE 14-3. *Continued*

The Employer did not correct deficiencies in the equipment that were outside acceptable limits including: • Gage glasses were not maintained in a clean/readable condition. • Exchanger supports were visibly deteriorated with concrete spalled off exposing reinforcing steel. • Fireproofing insulation removed from conduits powering EIVs in LUU4 was never replaced, and continued to fail critical instrument inspections.
The employer did not correct deficiencies in equipment that were outside acceptable limits, including: • Piping with thickness reading below the recommended minimum thickness (3 instances).
The employer did not correct deficiencies that were outside acceptable limits before further use or in a safe and timely manner: • Poly Charge Drum had heavy scale corrosion between the shell and support saddles with no determination as to the extent of the corrosion. • Toxic and flammable gas detectors located in duct work and the control room for H2S, and LEL were not in working order. • The relief valve for a depropanizer reboiler was set at 270 psi for protection of the tube side that has a maximum allowable working pressure of 150 psi. There was also inadequate overpressure protection in the event of a tube leak or rupture scenario. • The relief devices providing overpressure protection for a heat exchanger had set pressures of 540 and 600 psi, while the maximum allowable working pressure for the exchanger was 480 psi.
The employer did not correct deficiencies that were outside acceptable limits before further use or in a safe and timely manner, including: • Chains missing on chain operated valves. • TI not functional. • PI pegged at max. • PI damaged. • Sulfur mist control system not functional. • Corroded piping not repaired.
The employer failed to correct deficiencies in equipment to ensure safety in operation and maintenance before further use or in a safe and timely manner: • Three pressurized enclosures for electrical equipment did not have an alarm or did not have an alarm monitored at a constantly attended location. • Broken bolts in a flange connection on the Splitter Reboiler Exchanger were not repaired.
The employer did not correct deficiencies in equipment that were outside acceptable limits before further use or in a safe and timely manner: • A relief valve was evaluated to be undersized in regards to the external fire case scenario. • A relief valve was found to be discharging at an elevation below the flare header, which could allow condensation to damage the relief device and related piping. • A relief valve was evaluated to be undersized for several relief scenarios. • A relief valve was evaluated to be undersized for protection against the loss of cooling and power failure relief scenarios. • The Hydrocarbon Relief K.O. Drum was evaluated to not be able to provide adequate retention time to prevent liquid carryover to the flare.
The employer did not correct deficiencies in equipment that was outside of acceptable limits, in that, the employer had not resolved ultrasonic thickness data below minimums for the hydrocracker pretreater and profile radiographic thickness measurements below minimums for four circuits.

TABLE 14-3. *Continued*

The employer did not correct deficiencies in equipment that was outside of acceptable limits, in that "illegal type bushings" were not identified and replaced when the facility was made aware of insufficient metal thickness after both OD and ID threading of the reducer through corporate instruction and findings of an incident investigation.

The employer did not correct deficiencies in its process equipment prior to continued operation, including:

- The Crude Unit Blowdown Drum was documented with uncorrected mechanical deficiencies.
- The FCC Unit complex Blowdown Drum was documented with uncorrected mechanical deficiencies.
- The FCC Unit complex Blowdown Drum flare header piping sections were documented with uncorrected mechanical deficiencies (missing support shoes).
- In the Crude Units and the FCC Unit complex, relief system piping including flare header and blowdown header piping was documented with uncorrected mechanical deficiencies (exterior corrosion and degraded protective painting).
- The old/secondary Flare stack was documented in poor mechanical condition (structural integrity concern).

The employer did not ensure that a heat exchanger was provided with proper safety relief for the dead heading condition.

The employer did not ensure that a heater was provided with proper relief for a blocked discharge condition.

The employer did not ensure that the HDPE underground piping for the HF mitigation system was designed properly, in that the designed pressure for the underground piping was less that the pump discharge pressure.

---

29 CFR 1910.119(j)(6)(ii) Appropriate checks and inspections shall be performed to assure that equipment is installed properly and consistent with design specifications and the manufacturer's instructions.

The employer did not check to ensure that the blowdown quench water systems could provide the design quantity of approximately 12,00.gpm [sic] of quench water when required.

Employer did not perform timely checks and inspections; specifically, positive material identification tests on adjacent piping circuits in similar service were not performed following failure of a piping system component.

---

29 CFR 1910.119(j)(6)(iii) The employer shall assure that maintenance materials, spare parts and equipment are suitable for the process application for which they will be used.

The employer did not assure that replacement valves were suitable for applications including use of wrong materials for replacement valves on heaters.

The employer failed to ensure that the correct gaskets were used on various pieces of equipment, including vessels and compressors.

15

OTHER ASSET MANAGEMENT TOOLS

This chapter provides an overview of some analytical techniques that can be used to assist in the managing of asset integrity. Following two sections that give a side-by-side comparison of risk-based analytical techniques and examine the incorporation of risk concepts into AIM decision making, this chapter briefly discusses the application of five specific techniques and tools:

- Reliability-Centered Maintenance (RCM)

- Layer of Protection Analysis (LOPA) and similar analysis approaches

- Fault Tree Analysis (FTA) and Markov analyses

- Equipment failure analysis

- Root Cause Analysis (RCA).

Other analytical techniques were introduced in Chapter 7 of this document; namely, Risk-Based Inspection (RBI) and Failure Modes and Effects Analysis (FMEA).

15.1 INTRODUCTION TO COMMON RISK-BASED ANALYTICAL TECHNIQUES USED IN AIM PROGRAMS

In recent years, facilities have applied risk-based analytical techniques in an effort to develop more performance-based AIM programs. Some inspection standards, such as American Petroleum Institute (API) 510 and API 570 (References 15-1 and 15-2), now include provisions for determining inspection requirements based on risk. Also, API and other organizations have developed standards and recommended practices that encourage the use of risk-based techniques to define inspection and testing requirements, including:

- API Recommended Practice 580, *Risk-Based Inspection*

- API RP 581, *Risk-Based Inspection Technology*

- IEC 61511, *Functional Safety: Safety Instrumented Systems for the Process Industry Sector - Part 1: Framework, Definitions, System, Hardware and Software Requirements*. Also available as ANSI/ISA-84.00.01 Part 1 (IEC 61511-1 Mod).

Many organizations find that using risk analysis techniques to aid in the management of process safety provides several benefits, such as:

- Assurance that a structured, systematic and technically defensible approach is used to make decisions

- Improved knowledge of the system operation and the cause/effect relationships that result from specific asset failures

- Confidence that AIM resources are being focused on the most important failures by explicitly assessing the risk and then assigning resources to those areas for which they will be the most effective.

This section summarizes some of the important attributes of risk-based analytical techniques.

Various analytical techniques can assist decision making in the AIM program. However, facilities need to recognize when these techniques are best applied, what decisions can be made with the results, and other related issues using the techniques such as timing, advantages and resources. In general, the following are typical uses of analytical techniques in an AIM program:

- *FMEA/FMECA* to identify and prioritize potential equipment failure modes that need to be addressed by ITPM tasks

- *RCM* to optimize proactive maintenance tasks such as predictive maintenance, preventive maintenance, and failure-finding tasks; usually applied to functional failures of active components (e.g., pumps failing off, erratic control)

- *RBI* to optimize inspection tasks and frequencies for fixed equipment (e.g., vessels, tanks, and piping) and pressure relief devices

- *LOPA* and similar analysis approaches to define performance requirements for independent protection layers (IPLs), including safety instrumented functions (SIFs)

- *Fault Tree Analysis and Markov Analyses* to verify that SIF and IPL designs meet performance requirements (i.e., targeted PFD).

Quantitative risk analysis might also be used in particular situations for identifying vulnerabilities of assets due to flame impingement, thermal radiation/excessive heat, blast overpressure, missile impact, etc. Table 15-1 summarizes different key attributes for each of the techniques listed above. Sections 15-3 to 15-5 discuss the methods not already presented in Chapter 7.

TABLE 15-1. Summary of Analytical Techniques

	Brief Description
FMEA/ FMECA	• Inductive reasoning approach that evaluates how the equipment can fail and the effect that these failures have on process or system performance, and ensures that appropriate safeguards against the failure(s) are in place • FMECA is an FMEA that assesses the criticality of the failure modes and resulting effects using qualitative, semi-quantitative, or quantitative risk measures
RCM	Comprehensive review and analysis of systems and their components using (1) an FMEA/FMECA to identify potential equipment failures and their impact on system/process performance and (2) decision tree (or similar tools) to determine appropriate failure management strategies (e.g., ITPM tasks)
RBI	• Risk assessment and risk management process that assesses the likelihood and consequences of a loss of containment in process equipment used as an ongoing part of the AIM program • It integrates the traditional RAGAGEP standards with flexibility to focus and optimize the activities on risk reduction by identifying higher-risk equipment and failure mechanisms
LOPA and alternate approaches	• LOPA – an order-of-magnitude analysis of the risk of a scenario; each scenario has a consequence with its associated severity and one initiating event with its associated frequency; IPLs are evaluated for applicable risk reduction; additional layers, such as SIFs, can be added to meet a risk target • Alternate approaches – same analysis objectives as LOPA, but typically using fully quantitative analysis approaches, such as Event Tree Analysis
Fault Tree and Markov analysis	Quantitative analysis tools used to estimate the unavailability (probability of failure on demand) of protection layers, including SIFs
	Types of Equipment
FMEA/ FMECA	Mechanical (e.g., pumps, compressors) and electrical equipment
RCM	All types of equipment, but is typically best applied to mechanical (e.g., pumps, compressors) and electrical equipment, and instrument systems
RBI	Pressure vessels, storage tanks and piping systems; in addition, recent use on pressure relief systems
LOPA and alternate approaches	Process control safety functions, safety shutdown systems, operator responses to alarms, relief devices and other IPLs
Fault Tree and Markov analysis	Process control safety functions, safety shutdown systems, operator responses to alarms, relief devices and other IPLs

TABLE 15-1. *Continued*

	Suggested Application
FMEA/ FMECA	Critical and/or complex systems in which failure mode cause and/or effect on process/system performance is not known or well understood
RCM	Critical and/or complex systems in which failure mode cause and/or effect on process/system performance is not known or well understood
RBI	In-service inspection of pressure vessels, storage tanks, piping systems and pressure relief devices for loss of containment issues
LOPA and alternate approaches	Processes with higher risk incident scenarios (e.g., fire, explosions) that require more in-depth evaluation than provided by process hazard analyses (PHAs) to determine if sufficient layers of protection are provided to meet risk criteria; in addition, to determine the required PFD for safety systems, especially SIFs
Fault Tree and Markov analysis	Evaluation/verification that SIF and IPL designs meet the PFD requirements (typically defined via LOPA or one of the alternate approaches)

	Use of the Results in the ITPM Program
FMEA/ FMECA	• Identifies system/equipment failure modes and specific failure causes to be addressed by ITPM tasks • Provides risk/criticality rankings that can be used to establish ITPM task frequencies and to prioritize ITPM tasks
RCM	• Identifies appropriate ITPM tasks and frequencies needed to address system/equipment failure modes and specific failure causes • Provides risk/criticality rankings that can be used to establish ITPM task frequencies and to prioritize ITPM tasks
RBI	• Identifies inspection strategy for equipment; note that activities are based heavily on the API inspection standards, with added flexibility on the extent and frequency of inspections based on risk • Uses inspection results to update the extent and frequency of inspections in order to manage the likelihood of failure
LOPA and alternate approaches	• Does not typically define ITPM tasks and frequencies directly or in detail. • Can be used to identify critical safety systems needing to be addressed by ITPM tasks
Fault Tree and Markov analysis	Defines the testing frequency required for SIFs (and potentially other IPLs) to achieve the required PFD

	Use of the Results in Other AIM Program Activities
FMEA/ FMECA	• Can be design-level quality assurance (QA) activity, as part of the design review to identify potential failures • Can also identify potential maintenance errors resulting in system/equipment failures to be addressed by training or procedures • Can be used to develop an equipment troubleshooting guide
RCM	• Can be design-level QA activity, as part of the design review to identify potential failures and develop strategies for managing the failures (e.g., redesign, start-up considerations)

TABLE 15-1. *Continued*

RCM *Continued*	• Can also identify potential maintenance errors resulting in system/equipment failures to be addressed by training or procedures • Can be used to develop an equipment troubleshooting guide
RBI	Typically, not applied to other AIM program activities
LOPA and alternate approaches	Used during the design phase to establish the risk reduction of IPLs, including the SIL of an SIF
Fault Tree and Markov analysis	Used during the design phase to define specific design requirements (e.g., level of redundancy, dangerous failure rate targets) for SIFs

Timing	
FMEA/ FMECA	Initial development of the ITPM program when appropriate tasks are not apparent or results of an existing ITPM program are not adequate
RCM	• Initial development of the ITPM program when appropriate tasks are not apparent or results of an existing ITPM program are not adequate • Initial design of system/equipment to identify opportunities to improve system/equipment reliability and integrity; this timing probably provides the greatest value, but is seldom done by facilities
RBI	• Initial development of the ITPM program • Optimization of initial inspection efforts • Ongoing program to better focus and optimize regular inspections
LOPA and alternate approaches	• Initial design of a process and its safety systems • Applicable in existing process units to check or verify the integrity of the existing systems
Fault Tree and Markov analysis	• Initial design of SIFs and/or IPLs • Applicable in existing process units to check or verify SIF/IPL integrity of the existing systems

Effort/Resource Requirements	
FMEA/ FMECA	Variable – can be as simple as using generic FMEA/FMECA results (i.e., templates) for standard equipment types, to more resource-intensive development of FMEA/FMECA results requiring input from reliability personnel, maintenance personnel, operations personnel, process engineers and other engineering specialists if warranted
RCM	Resource-intensive; however, can be reduced by the application of FMEA/FMECA templates and generic ITPM plan; these resource reduction efforts can decrease some RCM benefits (e.g., specific equipment failure management strategies)
RBI	Greater initial effort than a conventional program, but typically with rapid payback
LOPA and alternate approaches	• LOPA requires a moderate level of resources; the analysis requires (or begins) a PHA to identify the incident scenarios, and then a team to evaluate the sufficiency of the IPLs • Alternate analysis approaches (e.g., Event Tree Analysis) require more resources and can be resource intensive depending on the number and complexity of the IPLs, the availability and applicability of failure data, and other modeling needed to determine PFD of the IPLs

TABLE 15-1. *Continued*

Fault Tree and Markov analysis	• Moderate to high, depending on number and complexity of SIFs and IPLS evaluated, availability and applicability of failure data, and analysis technique used to model the PFD of SIFs/IPLs
	• ISA has developed simplified equations that can be used to determine the SIL/IPL PFD; these equations can reduce the resources required to determine SIFs/IPL PFDs

Advantages	
FMEA/ FMECA	• Thorough, logical approach for identifying and evaluating system/equipment failures and their importance (based on impact on system/process performance)
	• Objective method for prioritizing equipment failures through the use of risk or criticality rankings
RCM	• Thorough, logical approach for developing ITPM task plans (i.e., provides a rationale or basis for ITPM tasks for which no applicable RAGAGEPs are available)
	• Effective in evaluating system/equipment designs and developing appropriate strategies for managing potential failures
RBI	• Better focus on controlling risk instead of just generating inspection results
	• Lower costs
	• Less likely to lose sight of the high-risk items in the quantity of low-risk data
LOPA and alternate approaches	• Compared to other techniques, LOPA is a simple approach for accomplishing the objective
	• Objectively evaluates the sufficiency of IPLs (i.e., less team judgment than more qualitative methods)
Fault Tree and Markov analysis	• More complex and rigorous approaches; both methods are good where common mode failures exist; Markov analysis is of particular value when a time dependency for detecting, repairing, or restoring a degraded system to full functionality exists
	• Objective evaluation of the sufficiency of IPLs (i.e., less team judgment than more qualitative methods)

Disadvantages	
FMEA/ FMECA	• Can be resource intensive
	• Only identifies single-initiated-event failures
	• Results depend somewhat on analysis team's knowledge of the system/equipment
RCM	• Tends to be resource intensive
	• Only identifies single-initiated-event failures
	• Results depend somewhat on analysis team's knowledge of the system/equipment
RBI	Requires greater technical knowledge/training to successfully set up and maintain than a conventional program
LOPA and alternate approaches	May yield overly conservative results for systems with potential common mode failures (e.g., interlocks within the same process control system)
Fault Tree and Markov analysis	• Most resource intensive
	• Requires highest level of training

15.2 INCORPORATING RISK INTO AIM DECISIONS

A primary objective of an AIM program is to reduce unreliable performance as a result of asset failures (Reference 15-3). An AIM program can accomplish this by identifying and prescribing tasks to prevent asset failures, detect the onset of asset failures, and/or discover hidden asset failures before they impact system performance, safety and/or the environment. Risk-based analytical techniques can be used to assess unreliable performance by identifying potential loss exposures and then estimating their risks. Based on this analysis, facility staff can determine the most effective AIM tasks for reducing those loss events.

Risk is the combination of the severity of consequences and the frequency (likelihood) of occurrence of an undesired event. Consequences may include safety, health, economic, environmental and/or other types of loss and harm impacts. The likelihood aspect addresses the frequency of a single occurrence of the undesired event (e.g., asset failure) or the average frequency of occurrence of the undesired event.

For example, an equipment item that fails once a year causing $100,000 in lost production each time it fails has the same risk (annualized loss rate) as another equipment item that fails 10 times a year causing $10,000 in lost production each time it fails. Every year, the failures of each item results in $100,000 in lost production. The economic risks associated with operating these two equipment items are equivalent. Using this example, the risk of operating the two equipment items is measured as an annual cost (dollars per year). This example expresses risk in terms of dollars to illustrate the calculation of risk. Other adverse effects (e.g., safety, environmental) can be measured with units such as injuries per year, affected area of fires per year, or mass of a chemical released per year.

Understanding system risk can help a facility develop a performance-based AIM strategy. Identifying higher-risk areas provides better opportunities for risk reduction. Priorities can be established by comparing systems, system functions and component failure modes. Also, risk characterization tools can be used during AIM task development to predict the effectiveness of the tasks being assigned. When assigning tasks, the objective is to reduce the risk of a failure to a tolerable risk threshold, which can be considered as "balancing the risk." Risk is considered to be in a balanced state when the assigned maintenance tasks are effective in reducing the risk to a tolerable level.

A tolerable-risk threshold is known as a *risk tolerance criterion*—the rate of loss that decision makers are willing to tolerate for a given consequence. This tolerable rate of loss will vary with many factors, including the type of loss, the magnitude of loss, and the individuals or groups affected by the loss. Risks that exceed the risk tolerance criterion require actions to lower the risk.

A common risk characterization tool used for AIM decisions (and other risk decisions) is a risk matrix that has frequency of occurrence on one axis and severity of consequences (impacts) on the other axis. Figure 15-1 is an example risk matrix with severity of consequences on the horizontal axis and frequency on the vertical axis. The severity and frequency ratings are typically defined as categories (i.e., range of impacts and frequency). These ratings can be defined qualitatively (e.g., insignificant, catastrophic, extremely remote, frequent) or quantitatively (e.g., 1 to 10 times per year, $1M to $10M loss). Each cell in the risk matrix represents a risk characterization that is defined by the intersection of the severity and frequency ratings.

In addition, risk tolerance criteria can be included on the risk matrix by defining a risk level for each cell. The results are groups of cells with the same or similar risk levels. These risk levels are then used along with risk tolerance criteria to indicate which risk levels require reduction and which are tolerable. The risk tolerance criteria are often indicated by lines that divide the risk matrix into two or more risk regions. The meanings of these risk regions and the terminology used varies between countries and between different companies. In some usages, any risk found to be outside the "Tolerable or Broadly Acceptable" region would require further risk reduction. In others, those risks following in an intermediate "ALARP Region" (see following paragraph) are treated differently. Determining where risk boundaries are set is discussed in depth in Reference 15-4.

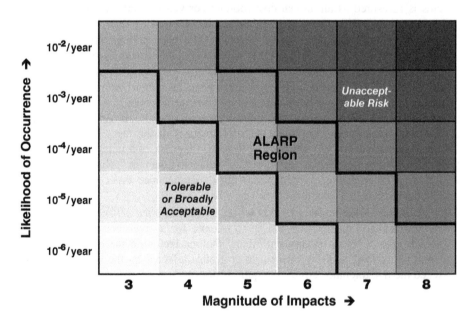

Figure 15-1. Example risk matrix with ALARP region *(see text)*

Another risk tolerance measure is known as "as low as reasonably practicable" (ALARP). An ALARP risk matrix typically contains three regions: broadly acceptable risk, unacceptable risk, and ALARP. The example risk matrix of Figure 15-1 shows these three regions. Risks that fall in the unacceptable region require actions to lower the risk into the ALARP or broadly acceptable risk regions. Risks that fall in the broadly acceptable risk region do not require any further risk reduction. Risks that fall in the ALARP region need to be further evaluated to determine if the risk could be reduced by implementing reasonable risk-reduction actions.

To use a risk matrix, an analysis team first selects a loss event of interest (e.g., fire/explosion caused by leak of flammable material). It then determines the severity and frequency categories that best characterize the event and identifies the risk level on the matrix. The team then bases the need for action on the risk level.

15.3 RELIABILITY-CENTERED MAINTENANCE

Unlike the four approaches outlined in Chapter 7 (code/standard, regulatory, company-specific, and RBI approaches) that are used to develop AIM inspection and test plans, Reliability-Centered Maintenance (RCM) is an established technique that is used for defining maintenance tasks, including ITPM tasks. RCM deals only with specific preventive maintenance and inspection tasks for an asset and may not include sufficient detail to be called a true inspection plan that meets code/standard or regulatory requirements.

Originally developed for the aviation industry and later expanded to other industries, RCM can be used in an AIM program to evaluate critical and complex systems to determine (1) which potential failures pose the highest risk and (2) what ITPM tasks and task frequencies are needed to address the potential failures; i.e., what is best strategy for managing the potential failures. It is a useful tool to determine which tasks are important to meet the operating context of an asset (i.e., how the asset will be employed and what its intended function will be). The RCM method is most commonly applied to an asset with a goal of increasing equipment availability (uptime) rather than meeting code/standard, regulatory or corporate requirements from an AIM perspective.

RCM provides a systematic approach for identifying potential failures that can affect process or system performance. The characteristics of the potential failures are then evaluated to determine (1) appropriate maintenance tasks or (2) potential design or operational changes (referred to as *one-time changes* in RCM). The RCM approach is based on systematically answering the following seven questions (Reference 15-5):

1. *Function and performance.* What process/system functions need to be preserved? (For process safety, these functions will typically relate to containing and controlling hazardous materials and energies and ensuring functionality of critical assets.)

2. ***Functional failure.*** How can the process or system fail to fulfill these functions (i.e., what functional failures can occur)?

3. ***Failure mode.*** What specific asset failures can cause each functional failure?

4. ***Failure effects.*** What happens when the failure occurs (i.e., what is the effect on system performance)?

5. ***Failure consequences.*** Why does the failure matter (i.e., what is the potential effect severity)?

6. ***Preventive tasks and task intervals.*** What proactive maintenance such as ITPM tasks need to be done to predict or prevent the failure? (This step is the primary link to the development of ITPM inspection plans.)

7. ***Default actions.*** What needs to be done, such as design or operational changes, if appropriate proactive maintenance tasks cannot be found or are not effective?

An RCM analysis typically uses two analysis tools, Failure Modes and Effects Analysis (FMEA) and a decision tree, to answer these questions.

The ***FMEA*** is used to help answer questions 1 through 5. Table 15-2 gives an example FMEA from an RCM analysis. Details of the FMEA methodology were previously presented in Section 4.7 of this document.

The ***decision tree*** is used to answer questions 6 and 7. Figure 15-2 provides an example RCM decision tree.

TABLE 15-2. Sample RCM FMEA Worksheet

Asset: Pump 1A, including the gearbox and the motor										
			Effects			Risk Characterization[1]				
Failure Mode	Failure Characteristic	Hidden/ Evident	Local	Functional Failure	End	C	UL	UR	ML	MR
External leak	Wear out	Evident	Release of hazardous material	Loss of containment	Potential severe injury to employees	Major	Occasional	Medium	Remote	Medium
Fails off	Random	Evident	Brief loss of flow until the spare pump is started	Transfer time too long	Brief interruption in production	Minor	Frequent	Medium	Occasional	Low
Degraded head	Wear out	Evident	Reduced flow of material	Transfer time too long	Production rate reduced	Moderate	Occasional	Medium	Remote	Low

[1]Risk characterization abbreviations: UL = unmitigated likelihood   ML = mitigated likelihood
          C = consequence (severity)   UR = unmitigated risk   MR = mitigated risk

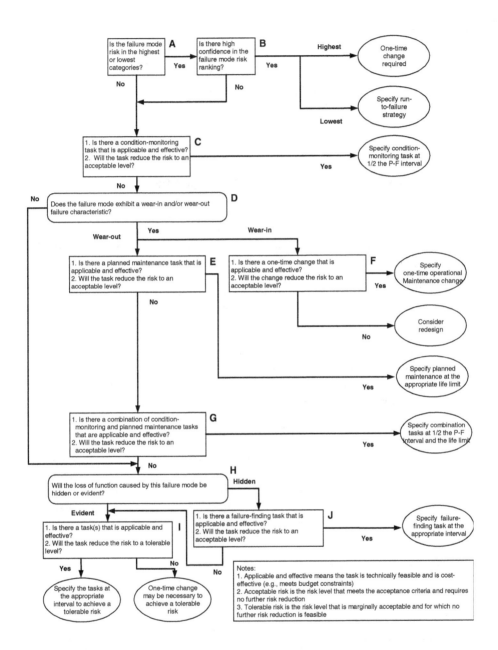

Figure 15-2. Example RCM decision tree

RCM Resources. SAE International has published an RCM standard, SAE JA1011, entitled *Evaluation Criteria for Reliability-Centered Maintenance (RCM) Processes* (Reference 15-6). It is intended to set out the criteria with which any work process must comply in order to be truly called "RCM". A companion guide is also available as SAE JA1012, *A Guide to the Reliability-Centered Maintenance (RCM) Standard* (Reference 15-7). The most prominent of several books that have been written on RCM is *Reliability-Centered Maintenance, Second Edition* by Moubray (Reference 15-8).

15.4 PROTECTION LAYER ANALYSIS TECHNIQUES

As defined in IEC 61511, *Functional Safety: Safety Instrumented Systems for the Process Industry Sector - Part 1: Framework, Definitions, System, Hardware and Software Requirements*, a safety instrumented system (SIS) is a system composed of sensors, logic solvers, and final control elements for the purpose of taking the process to a safe state when predetermined conditions are violated. (Other terms used for SIS include emergency shutdown system, safety shutdown system, and safety interlock system.) IEC 61511 also defines a safety life cycle that includes the application of two analyses:

1. ***Layer of Protection Analysis (LOPA)*** or an alternate tool to determine if additional independent protection layers (IPLs), which may include one or more safety instrumented functions (SIFs), are needed to achieve tolerable risk levels and, if tolerable risk is not achieved, to define the required risk reduction. Other techniques including Hazard and Operability (HAZOP) Studies can be used for this purpose.

2. ***Reliability analysis*** (unavailability analysis) to determine the protection layer configuration (including redundancy, if required) and functional test frequency needed to achieve the required risk reduction for SIFs in terms of the overall probability of failure on demand (PFD), based on specified system components and their failure rates. Tools include Fault Tree Analysis (FTA), simplified equations, and Markov analysis.

The following paragraphs provide an overview of these two types of analyses.

Layer of Protection Analysis (LOPA) is a simplified risk analysis approach that can be used to (1) identify the need for additional IPLs, including one or more SIFs, and (2) determine the performance required by each protection layer to achieve a tolerable level of risk. LOPA is an analysis tool that builds on the information developed during a hazard study such as a process hazard analysis. Specifically, the hazard study is used to identify (1) the incident scenarios to be evaluated and (2) the non-SIF safeguards that are in place. (Note: A non-SIF safeguard is one that is not part of a defined SIS. Examples include safety functions implemented in the basic process control system, mechanical safety devices, operator alerting and response, emergency relief systems, and containment dikes.) LOPA is performed using order-of-magnitude estimates of the cause frequency, consequence severity and likelihood of failure of the non-SIF safeguards to determine if an SIF is needed to achieve a tolerable risk level. Qualitative rules (e.g., a risk matrix) or a combination of approaches can also be used to perform a LOPA. A LOPA generally includes the following steps (Reference 15-9):

1. Identify and select incident scenarios (typically, as part of a PHA).

2. For each scenario to be evaluated, identify the initiating event and estimate the initiating event frequency or use a value from a rule set.

3. Identify the consequence and estimate its severity (if not already done as part of the PHA).

4. Identify the IPLs and estimate the probability of failure on demand (PFD) for each IPL or use PFD values from a rule set.

5. Calculate the risk of the scenario by combining the initiating event frequency, the PFD for IPLs, and the consequence severity. (Enabling condition and conditional modifier probabilities might also be employed.)

6. Evaluate the risk to determine if an SIF is needed and/or to determine the performance required (e.g., PFD needed by the SIF[s]) to achieve a tolerable level of risk.

Figure 15-3 shows an example of a completed LOPA worksheet. The LOPA results can be used in the ITPM task planning process to:

- Identify the important non-SIF safeguards that need to be maintained by the AIM program.

- Define the performance requirements for the SIFs, including the type and frequency of testing needed to ensure that the SIF achieves the desired performance.

The results of a LOPA are then used to design the SIF (e.g., identify the redundancy needed, control system architecture, component failure rate). Specifically, the LOPA identifies (1) the safety function (i.e., what the SIF has to do) and (2) the required risk reduction (e.g., PFD required). The risk reduction is defined in terms of a safety integrity level (SIL 1, SIL 2 or SIL 3; see Table 15-3 and Reference 15-10) or a non-SIL-rated interlock when the required risk reduction is very low.

Scenario: Compressor knockout drum V-9 level control failure causes a high level in V-9, which overflows and results in liquid flow to suction of compressor C-1, followed by catastrophic compressor failure, release of highly energetic compressor parts, and potential for severe injury up to and including fatality

Tolerable frequency for this scenario *(example value)*	$\leq 10^{-6}$/yr	Severity Category 7 event
Initiating event	10^{-1}/yr	Control failure (LC-5) causes V-9 control valve LV-5 to close
Enabling condition	N/A	Compressor is in continuous operation most of the year
Human response IPL	0.1	Operator response to critical high level alarm LAH-10 on V-9; alarm is independent of the level control failure initiating the scenario; operator has 15 min to shut off compressor via ESD; operator action included in SOP and verified training
High-high level SIF	0.1	SIL 1 safety shutdown system trips power off C-1 when high-high level is detected by independent sensor LSHH-11
High compressor vibration SIF	0.1	SIL 1 safety shutdown system trips off C-1 via second independent relay when high vibration threshold level is exceeded by vibration sensor
Probability of personnel presence	0.1	Personnel excluded from compressor building when compressor is in operation; effective administrative control in place
Calculated scenario risk	10^{-5}/yr	**Tolerable risk level NOT MET for this scenario**
Risk reduction required	$\geq 10x$	This could be achieved by making one of the two SIFs = SIL 2

Figure 15-3. Sample LOPA worksheet (Reference 15-11)

TABLE 15-3. IEC 61511 Safety Integrity Levels

Safety Integrity Level	Average Probability of Failure on Demand (PFD_{AVG})
SIL 1	$\geq 10^{-2}$, $< 10^{-1}$
SIL 2	$\geq 10^{-3}$, $< 10^{-2}$
SIL 3	$\geq 10^{-4}$, $< 10^{-3}$

The SIL and PFD$_{AVG}$ are used to design the SIF. The SIS design provides necessary fault tolerance. The SIS must be tested regularly and rigorously to eliminate systemic error and discover random faults. As part of the design, a ***reliability analysis (unavailability analysis)*** is performed to evaluate if a proposed design configuration and testing frequency will achieve the required average PFD (i.e., SIL). This analysis is typically performed using simplified equations, Fault Tree Analysis or Markov models. In addition, ISA has compiled simplified reliability equations that can be used to perform the analysis. ISA technical reports TR84.00.02-2002, Parts 1 through 5, *Safety Instrumented Functions (SIF) – Safety Integrity Level (SIL) Evaluation Techniques*, provide detailed information on the use of these analysis techniques to calculate the average PFD for proposed SIF designs.

The unavailability analysis is used to determine the testing frequency for each SIF and the type of testing required (e.g., calibration, functional testing). Also, SIS designers are to ensure that the SIF designs include design features (e.g., bypass valves) that will allow the SIF to be tested as required. IEC 61511 also includes requirements for installation, commissioning, pre-startup acceptance testing, operations, MOC, and decommissioning (Reference 15-10). These requirements essentially outline a quality assurance system for SISs.

15.5   ASSET FAILURE AND ROOT CAUSE ANALYSES

Failure analysis is designed to help personnel understand how a failure occurred and to determine and implement needed corrections and improvements to assets. These corrections typically focus on redesigning the asset item, altering its ITPM plan and/or changing the process conditions. This analysis may result in (1) changes that affect only the asset item that failed or (2) changes that affect the failed asset item as well as similar asset items. Usually, this type of analysis is based on forensic engineering principles.

On the other hand, *root cause analysis* (RCA) is designed to discover why the failure occurred so that corrections and improvements can be made to the management systems that promote asset integrity (e.g., AIM program, operating procedures, engineering design practices, personnel training). Because RCA focuses on changes to management systems, the results tend to affect the asset item that failed, similar assets, and seemingly unrelated assets that are managed by the same management systems. For example, if a pressure vessel leaks because of corrosion and RCA finds that nondestructive testing (NDT) activities were not being performed as planned, the RCA might result in improvements in the ITPM program that not only affect pressure vessel inspections, but also instrument calibration, rotating equipment lubrication and so forth. The following subsections outline both failure analysis and RCA.

While the processes and results of failure analysis and RCA are distinctly different, they are quite often used together to analyze asset failures. For example, a failure analysis provides vital technical information for the RCA. Similarly, the RCA process and associated tools can be used to provide useful information (e.g., potential damage mechanism for an asset item, timeline of events) for the failure analysis.

15.5.1 Failure Analysis

Failure analysis processes focus on preserving, collecting, and analyzing evidence related to a failure. The analysis of the evidence begins with a macroscopic (i.e., broad) examination and progresses to a more microscopic (i.e., detailed) examination. Both nondestructive and destructive testing methods are sometimes used to determine the failure mechanisms that resulted in the failure. (See Appendix 6A for additional information on damage mechanisms.) Once the failure mechanisms are identified, the analysis team generates recommendations and corrective actions for reducing or eliminating the likelihood of recurrence of a specific failure.

Figure 15-4 displays an eight-step failure analysis process (Reference 15-12). Each step in the eight-step process is briefly described below.

Step 1 – Evaluate Conditions at the Site. Collect background information on the equipment and the process, as well as information related to the equipment conditions when the failure occurred (e.g., temperature, pressure, flow, operating mode).

Step 2 – Perform Preliminary Component Assessment. This typically involves performing a visual inspection. The visual inspection can uncover evidence of the type of failure that occurred.

Step 3 – Preserve "Fragile" Data Sources. The evidence and data collected needs to be preserved to ensure that they are not lost and do not degrade.

Step 4 – Perform Macroscopic Examination. Perform (1) a visual inspection of component surfaces, dimensions, etc., with appropriate visual inspection tools (e.g., low power magnification, low power microscopes), (2) NDT (e.g., radiography, eddy current testing), (3) nominal chemistry testing (e.g., moisture content of gearbox oil), and/or (4) basic asset tests (e.g., hardness testing).

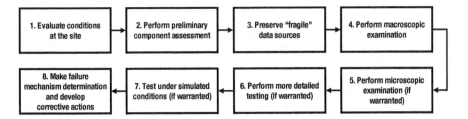

Figure 15-4. Sample failure analysis process

Step 5 – Perform Microscopic Examination (if warranted). For some failures, the preliminary visual inspection and macroscopic examination may not provide the information needed to determine the failure mechanism. In these cases, microscopic examination, using such tools as light microscopes, transmission electron microscopes, scanning electron microscopes, and energy-dispersive x-ray spectroscopy, may be needed to collect the needed information.

Step 6 – Perform More Detailed Testing (if warranted). Similar to Step 5, more detailed mechanical testing and/or chemical analysis may be needed to determine the failure mechanism. This testing typically focuses on determining the physical properties (e.g., metal hardness), compositions (e.g., chemical composition of lube oil), or other characteristics of material samples taken from the failed asset. This testing many times results in destroying or altering the sample; therefore, personnel need to verify that all visual examinations are completed before beginning such tests.

Step 7 – Test under Simulated Conditions (if warranted). Experiments to reproduce a specific failure in a controlled environment (based on reasoned hypotheses from the available data) may provide verification of the hypothesis and insights into ways to prevent subsequent failures.

Step 8 – Make the Failure Mechanism Determination and Develop Corrective Actions. Based on the facts generated by the data analysis, conclude which failure mechanisms were significant contributors. Then identify specific events/ characteristics that caused the failure mechanisms to occur, and develop corrective actions for eliminating the causes and/or reducing the likelihood of the causes resulting in the failure.

The supplemental materials accompanying this document contains additional resources for performing equipment failure analyses: (1) Additional detailed information on the analysis steps, (2) an equipment failure analysis checklist and (3) information on dominant failure mechanisms for typical plant equipment (e.g., tanks, vessels, pumps, piping). Chapter 8 of the CCPS book *Guidelines for Investigating Chemical Process Incidents, 2nd Edition* (Reference 15-12) also contains information on equipment failure analysis. In addition, the book *Understanding How Components Fail* (Reference 15-13) is a good source of information on equipment failure analysis.

API Recommended Practice 585, entitled *Pressure Equipment Integrity Incident Investigation* (Reference 15-14), describes how an effective investigation could be structured so an organization can learn from loss-of-containment incidents. It highlights the value in recognizing precursor events and promotes investigating them to determine the immediate, contributing and root causes. Although is focus is process pressure equipment in the refining and petrochemical industries, the principles in API RP 585 can also be applied to other types of equipment.

15.5.2 Root Cause Analysis

Root cause analysis (RCA) is a process designed to investigate and categorize the root causes (i.e., underlying causes leading to asset failures or human errors) of events with negative impacts such as adverse safety, health, environmental, quality, reliability and/or production effects. The term "event" is used to generically identify occurrences that produce, or have the potential to produce, these types of consequences. RCA is simply a tool designed to help identify not only WHAT and HOW an event occurred, but also WHY it happened. Only when investigators are able to determine why an event or failure occurred will they be able to specify workable corrective measures to prevent similar future events.

To effectively perform an RCA, a structured analysis process is needed. Many different RCA processes and tools are available. The following paragraphs discuss one approach, which includes the following four steps (Reference 15-15): (1) data collection, (2) data analysis, (3) root cause identification, and (4) recommendation generation and corrective action implementation. Each step is briefly described below:

Step 1 – Data Collection. The first analysis step is to gather four types of data:

1. *People* (e.g., witnesses, participants)

2. *Physical* (e.g., parts, chemical samples)

3. *Position* (e.g., location of people and physical evidence)

4. *Paper* (e.g., procedures, computer data).

Without complete information and an understanding of the event, the causal factors (CFs) and root causes associated with the event cannot be identified. The majority of time spent analyzing an event is spent gathering data. In addition, analysts may obtain information from other similar failures. This might include contacting personnel at other facilities that have experienced a similar failure and/or equipment vendor (or other outside) personnel who are knowledgeable about the failure.

Step 2 – Data Analysis. Organize the data and develop a model of how the event occurred. Two common methods for developing the model are causal factor (CF) charting and Fault Tree Analysis (FTA). CF charting/FTA provides a structure for investigators to organize and analyze the information gathered during the investigation and to identify gaps and deficiencies in knowledge as the investigation progresses. The CF chart is simply a sequence diagram that describes the events leading up to an occurrence, as well as the conditions surrounding these events. FTA is a Boolean logic tool to help model the combinations of human errors, asset failures, and external events that can produce the type of event being analyzed. Figures 15-4 and 15-5 (Reference 15-12) provide an example CF fault tree and CF chart, respectively.

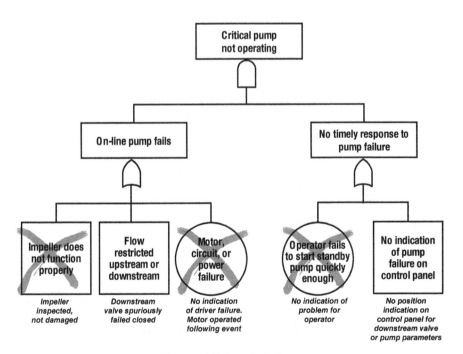

Figure 15-5. Sample fault tree

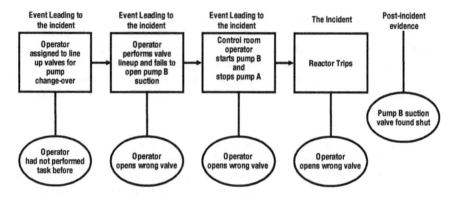

Figure 15-6. Sample causal factor chart

Step 3 – Root Cause Identification. After all of the CFs have been identified, the investigators begin root cause identification. Many approaches use root cause diagrams, charts or lists to aid the investigators in identifying the underlying reason(s) for each CF. Investigators use these tools to structure the reasoning process, which helps them answer questions about why particular CFs existed or occurred. The identification of root causes helps the investigator determine why the event occurred so that the problems surrounding the occurrence can be addressed. In addition, trending of the root causes of occurrences identified over a period of time can provide valuable insight concerning specific areas for improvement. As an added benefit of this process, RCA can be used to (1) help prevent the recurrence of specific events and (2) combine lessons learned from individual occurrences to identify major areas of weakness. This allows management to identify necessary actions to take before a seemingly unrelated incident or failure occurs.

Step 4 – Recommendation Generation and Implementation. The next step is the generation of recommendations. Following identification of the root cause(s) for a particular CF, the analyst can generate achievable recommendations for preventing its recurrence. The root cause analyst is often not responsible for developing appropriate corrective actions based on the RCA recommendations. Therefore, analysis reports need to provide the needed information for those who are implementing the recommendations (or developing reasonable alternatives). Once the RCA team has developed its recommendations, facility management can evaluate the recommendations to determine the corrective actions to be implemented. In developing corrective actions, management needs to ensure that the corrective actions are practical and that they will be effective. In addition, management needs to communicate rejected recommendations, and the basis for their rejection, to the RCA team. Facilities also need to ensure that a system is in place to track corrective actions to completion. When the entire occurrence has been charted, the investigators are better able to identify the major contributors to the incident. CFs are those contributors (human errors and/or component failures) that, if eliminated, would have either prevented the occurrence or reduced its severity. (Note: CFs are not the root causes of the event; rather, they are contributing causes to the event. In order to eliminate or reduce the likelihood of the event recurring, it is important that the root causes for CFs be identified and corrected.)

In many traditional analyses, the most visible CF is given all the attention. However, events are rarely caused by just one CF. They are often the result of a combination of contributors. Without identifying all of the CFs, the list of recommendations will likely not be complete. Consequently, the occurrence (and similar occurrences) may repeat because the organization did not learn all that it could from the original event.

The steps presented in this subsection outline one of many RCA processes that facilities can use to analyze asset failures. For more information on RCA, see Reference 15-12.

CHAPTER 15 REFERENCES

15-1   API 510, *Pressure Vessel Inspection Code: Maintenance Inspection, Rating, Repair and Alteration*, American Petroleum Institute, Washington, DC.

15-2   API 570, *Piping Inspection Code: Inspection, Repair, Alteration, and Re-rating of In-service Piping*, American Petroleum Institute, Washington, DC.

15-3   Montgomery, R. and W. Satterfield, "Applications of Risk-based Decision-making Tools for Process Equipment Maintenance," presented at ASME Pressure Vessel and Piping Conference, Cleveland, OH, 2002.

15-4   Center for Chemical Process Safety, *Guidelines for Developing Quantitative Safety Risk Criteria*, American Institute of Chemical Engineers, New York, NY, 2009.

15-5   Folk, T., "Risk Based Approach to Asset Integrity Success in Implementation," Process Plant Safety Symposium, American Institute of Chemical Engineers, New Orleans, LA, 2003.

15-6   SAE JA1011, *Evaluation Criteria for Reliability-Centered Maintenance (RCM) Processes*, SAE International, Warrendale, PA.

15-7   SAE JA1012, *A Guide to the Reliability-Centered Maintenance (RCM) Standard*, SAE International, Warrendale, PA.

15-8   Moubray, J., *Reliability-Centered Maintenance, Second Edition*, Industrial Press Inc., New York, NY, 1997.

15-9   Center for Chemical Process Safety, *Layer of Protection Analysis: Simplified Process Risk Assessment*, American Institute of Chemical Engineers, New York, NY, 2001.

15-10   AIChE Academy, "Advanced Concepts in Process Hazard Analysis," Course CH754, American Institute of Chemical Engineers, New York, NY, 2016.

15-11   IEC 61511, *Functional Safety: Safety Instrumented Systems for the Process Industry Sector - Part 1: Framework, Definitions, System, Hardware and Software Requirements*, International Electrotechnical Commission, Geneva, Switzerland.

15-12 Center for Chemical Process Safety, *Guidelines for Investigating Chemical Process Incidents, Second Edition*, American Institute of Chemical Engineers, New York, NY, 2003.

15-13 Wulpi, D., *Understanding How Components Fail, 3rd Edition*, ASM International, Materials Park, OH, 2013.

15-14 API Recommended Practice 585, *Pressure Equipment Integrity Incident Investigation*, American Petroleum Institute, Washington, DC.

15-15 ABSG Consulting Inc., *Root Cause Analysis Handbook, A Guide to Efficient and Effective Incident Investigation, Third Edition*, Rothstein Associates Inc., Brookfield, Connecticut, 2008.

Additional Chapter 15 Resources

API Recommended Practice 580, *Risk-Based Inspection*, American Petroleum Institute, Washington, DC.

API Recommended Practice 581, *Risk-Based Inspection Technology*, American Petroleum Institute, Washington, DC.

Smith, A., *Reliability Centered Maintenance*, McGraw-Hill Inc., New York, NY, 1993.

International Maritime Organization, *Guidelines for Formal Safety Assessment (FSA) for Use in the IMO Rule Making Process*, MSC/Circ. 1023–MEPC/Circ. 392, London, England, 2002.

International Society for Automation, *Safety Instrumented Functions (SIF) — Safety Integrity Level (SIL) Evaluation Techniques, Part 1: Introduction*, ISA-TR84.00.02, Research Triangle Park, NC.

International Society for Automation, *Safety Instrumented Functions (SIF) — Safety Integrity Level (SIL) Evaluation Techniques, Part 2: Determining the SIL of a SIF via Simplified Equations*, ISA-TR84.00.02, Research Triangle Park, NC.

International Society for Automation, *Safety Instrumented Functions (SIF) — Safety Integrity Level (SIL) Evaluation Techniques, Part 3: Determining the SIL of a SIF via Fault Tree Analysis*, ISA-TR84.00.02, Research Triangle Park, NC.

International Society for Automation, *Safety Instrumented Functions (SIF) — Safety Integrity Level (SIL) Evaluation Techniques, Part 4: Determining the SIL of a SIF via Markov Analysis*, ISA-TR84.00.02, Research Triangle Park, NC.

International Society for Automation, *Safety Instrumented Functions (SIF) — Safety Integrity Level (SIL) Evaluation Techniques, Part 5: Determining the PFD of Logic Solvers via Markov Analysis*, ISA-TR84.00.02, Research Triangle Park, NC.

ACRONYMS AND ABBREVIATIONS

ACCP	ASNT Central Certification Program
ACGIH	American Conference of Governmental Industrial Hygienists
AIChE	American Institute of Chemical Engineers
AI	Asset integrity
AIM	Asset integrity management
ALARP	As low as reasonably practicable
ANSI	American National Standards Institute
API	American Petroleum Institute
ASHRAE	American Society of Heating, Refrigerating, and Air-Conditioning Engineers
ASM	American Society for Metals (ASM International)
ASME	American Society of Mechanical Engineers
ASNT	American Society of Nondestructive Testing
ASTM	American Society of Testing and Materials (ASTM International)
AWS	American Welding Society
BPCS	Basic process control system
BPVC	Boiler and Pressure Vessel Code
CCD	Corrosion Control Document
CCPS	Center for Chemical Process Safety
CF	Causal factor
CFR	Code of Federal Regulations (U.S.)
CM	Condition monitoring
CML	Condition monitoring location
CMMS	Computerized maintenance management system
DCS	Distributed control system
DIERS	Design Institute for Emergency Relief Systems
DOT	U.S. Department of Transportation
E&I	Electrical and instrumentation
EHS	Environmental, health and safety

EPA	U.S. Environmental Protection Agency
ESD	Emergency shutdown
EPC	Engineering, procurement and construction
FAT	Factory acceptance test
FFS	Fitness for service
FM	Factory Mutual Insurance Company (FM Global)
FMEA	Failure Modes and Effects Analysis
FMECA	Failure Modes, Effects and Criticality Analysis
FRP	Fiber-reinforced plastic
FTA	Fault Tree Analysis
GHS	Globally Harmonized System of Classification and Labeling of Chemicals
HAZMAT	Hazardous material
HAZOP	Hazard and Operability [Study]
HIC	Hydrogen-induced cracking
HTHA	High-temperature hydrogen attack
HI	Hydraulic Institute
HIRA	Hazard identification and risk analysis
HVAC	Heating, ventilation and air conditioning
ICC	International Code Council
IDMS	Inspection Data Management System
IEC	International Electrotechnical Commission
IFC	International Fire Code
IIAR	International Institute of Ammonia Refrigeration
IMC	International Mechanical Code
IOW	Integrity operating window
IPL	Independent protection layer
ISA	International Society for Automation
ISO	International Organization for Standardization
ITP	Inspection and testing plan
ITPM	Inspection, testing and preventive maintenance
KPI	Key Performance Indicator
LOPA	Layer of Protection Analysis
LOPC	Loss of primary containment

MEL	Master equipment list
MI	Mechanical integrity
MOC	Management of change

NB	National Board (of Boiler and Pressure Vessel Inspectors)
NBBPVI	National Board of Boiler and Pressure Vessel Inspectors
NBIC	National Board Inspection Code
NDE	Nondestructive examination
NDT	Nondestructive testing
NEC	National Electric Code
NEP	National Emphasis Program
NFPA	National Fire Protection Association

OEE	Overall equipment effectiveness
OEM	Original equipment manufacturer
ORR	Operational readiness review
OSHA	Occupational Safety and Health Administration

P-F	Potential-to-functional failure
P&ID	Piping and instrumentation diagram
PFD	Probability of failure on demand *or* Process flow diagram
PHA	Process hazard analysis
PIMS	Pipeline integrity monitoring system
PM	Preventive maintenance
PMI	Positive material identification
PPE	Personal protective equipment
PRV	Pressure relief valve
PSM	Process safety management
PSSR	Pre-startup safety review
PSV	Pressure safety valve
PWHT	Post-weld heat treatment

QA	Quality assurance
QC	Quality control
QM	Quality management

RAGAGEP	Recognized and generally accepted good engineering practice
RBI	Risk-based inspection
RBPS	Risk-based process safety

RCA	Root Cause Analysis
RCM	Reliability-centered maintenance
RMP	Risk management program
ROI	Return on investment
RP	Recommended practice
RFID	Radio-frequency identification
SCBA	Self-contained breathing apparatus
SCE	Safety-critical equipment
SIF	Safety instrumented function
SIL	Safety integrity level
SIMOPS	Simultaneous operations
SIS	Safety instrumented system
SME	Subject matter expert
SOHIC	Stress oriented hydrogen induced cracking
TML	*See* CML
UL	Underwriters Laboratories Inc.
UMC	Uniform Mechanical Code
UPS	Uninterruptible power supply
U.S.	United States [of America]
USCG	United States Coast Guard
UT	Ultrasonic thickness

GLOSSARY

Acceptance criterion Technical basis used to determine whether equipment is deficient (e.g., when analyzing inspection, testing and preventive maintenance results).

Asset A process/facility involved in the use, storage, manufacturing, handling or transport of chemicals. It also refers to the equipment, such as vessels, piping systems, controls, safety systems, utilities, structures and other elements comprising such a process/facility.

Asset integrity The condition of an asset that is properly designed and installed in accordance with specifications and remains fit for purpose.

Asset integrity management (AIM) A process safety management system for ensuring the integrity of assets throughout their life cycle.

Causal factor (CF) A major unplanned, unintended contributor to an incident (a negative event or undesirable condition), that if eliminated would have either prevented the occurrence of the incident or reduced its severity or frequency. Also known as a *critical causal factor* or *contributing cause*.

Causal factor chart A sequence diagram that graphically depicts an incident from beginning to end; typically used to organize incident data and identify causal factors.

Certification Completion of the formal training and qualification requirements specified by applicable codes and standards.

Computerized maintenance management system (CMMS) Computer software for planning, scheduling and documenting maintenance activities. A typical CMMS includes work order generation, work instructions, parts and labor expenditure tracking, parts inventories and equipment histories.

Condition monitoring (CM) Observing, measuring and/or trending of indicators with respect to some independent parameter (usually time or cycles) to indicate the current and future ability of a structure, system or component to function within acceptance criteria.

Cost avoidance Return (often expressed in monetary terms) resulting from actions that prevent an incident from occurring.

Critical asset Asset, the malfunction or failure of which could (or is likely to) cause, contribute to, or fail to prevent or mitigate a major business impact or a major safety, environmental or security incident.

Critical operating parameter Process condition (e.g., flow rate, temperature) that can lead to an asset failure if limits are exceeded.

Criticality analysis Quantitative analysis of events and effects and their ranking in order of the seriousness of their consequences.

Damagelfailure mechanism A mechanical, chemical, physical or other process that results in equipment degradation. Identifying and inspecting for indications of damage mechanisms can be used to predict future failures.

Decision tree A logic tree used in reliability-centered maintenance (RCM) to help determine the correct type of maintenance (e.g., predictive, preventive) to perform to reduce the likelihood of equipment failures.

Deficiency A condition that does not meet acceptance criteria.

Equipment A piece of hardware that can be defined in terms of mechanical, electrical and/or instrumentation components contained within its boundaries.

Equipment class A grouping of individual equipment items with similar design and operation, such that facilities should perform similar ITPM activities on all of the items.

Facility The physical location where a management system activity is performed. In early life-cycle stages, a facility may be the company's central research laboratory or the engineering offices of a technology vendor. In later stages, the facility may be a typical chemical plant, storage terminal, distribution center, or corporate office. In the context of this document, a *facility* is a portion of or a complete plant, unit, site, complex or offshore platform or any combination thereof.

Failure Loss of ability to perform as required.

Failure analysis A systematic approach for analyzing asset failures to determine the failure mechanism(s) and the root cause(s) that resulted in the failure.

Failure mode A symptom or condition by which a failure is observed. A failure mode might be identified as loss of function, spurious operation (function without demand), an out-of-tolerance condition, or a simple physical characteristic such as a leak observed during inspection.

Failure Modes and Effects Analysis (FMEA) A hazard evaluation procedure in which all known failure modes of components or features of a system are considered in turn and undesired outcomes are noted.

Failure Modes, Effects and Criticality Analysis (FMECA) A variation of FMEA that includes an estimate of the significance of the consequence of a failure mode.

Fault tree A logic model that graphically portrays the combinations of failures that can lead to a specific main failure or accident of interest.

Fitness for service (FFS) A systematic approach for evaluating the current condition of an asset in order to determine if the equipment item is capable of operating at defined operating conditions (e.g., temperature, pressure).

Gantt chart A manner of depicting multiple, time-based project activities, usually on a bar chart with a horizontal time scale.

Hazard and Operability (HAZOP) Study A scenario-based hazard evaluation procedure in which a team uses a series of guide words to identify possible deviations from the intended design or operation of a process, then examines the potential consequences of the deviations and the adequacy of existing safeguards.

Hazard identification and risk analysis (HIRA) A systematic identification and evaluation of process hazards with the purpose of ensuring that sufficient safeguards are in place to manage the inherent risks. Also known as *process hazard analysis.*

Incident An event, or series of events, resulting in one or more undesirable consequences, such as harm to people, damage to the environment, or asset/business losses. Such events include fires, explosions, releases of toxic or otherwise harmful substances, and so forth.

Incipient failure An imperfection in the state or condition of hardware such that a degraded or catastrophic failure can be expected to result if corrective action is not taken.

Inspection A work activity designed to determine if ongoing work activities associated with operating and maintaining a facility comply with an established standard. Inspections normally provide immediate feedback to the persons in charge of the ongoing activities, but normally do not examine the management systems that help ensure that policies and procedures are followed.

Inspection, testing and preventive maintenance (ITPM) Scheduled proactive maintenance activities intended to (1) assess the current condition and/or rate of degradation of equipment, (2) test the operation/functionality of equipment, and/or (3) prevent equipment failure by restoring equipment condition.

Inspection, testing and preventive maintenance program (ITPM program) A management system that develops, maintains, monitors, and manages inspection, testing and preventive maintenance activities.

Inspection, testing and preventive maintenance plan (ITPM plan) List of recurring ITPM tasks and their associated schedule.

Inspection, testing and preventive maintenance task (ITPM task) An inspection, testing or preventive maintenance activity that is performed at some interval with the purpose of assessing the current condition of the asset in order to plan identified corrective activities to ensure ongoing asset integrity.

Integrity operating window (IOW) The parameters (i.e., safe upper and lower limits, run time) under which an asset can function without failure.

Layer of Protection Analysis (LOPA) An approach that analyzes one incident scenario (cause-consequence pair) at a time, using predefined values for the initiating event frequency, independent protection layer failure probabilities and consequence severity, in order to compare a scenario risk estimate to risk criteria for determining where additional risk reduction or more detailed analysis is needed.

Major incident See Table 5-1 and accompanying text.

Management of change (MOC) A process for evaluating and controlling modifications to facility design, operation, organization or activities - prior to implementation - to make certain that no new hazards are introduced and that the risk of existing hazards to employees, the public, or the environment is not unknowingly increased.

Metrics Leading and lagging measures of process safety management efficiency or performance. Metrics include predictive indicators, such as the number of improperly performed line-breaking activities during the reporting period, and outcome-oriented indicators, such as the number of incidents during the reporting period.

Nondestructive testing/examination (NDT/NDE) Evaluation of an equipment item with the intention of measuring an equipment parameter without damaging or destroying the equipment item.

Owner-user Person, plant, or corporation legally responsible for the safe operation of a pressure-retaining item (e.g., a pressure vessel).

Passive system A system in which failures are only revealed by testing or when a demand has occurred.

Performance measure A metric used to monitor or evaluate the operation of a program activity or management system.

Positive material identification (PMI) The determination of the materials of construction of an equipment item or component (e.g., piping).

Potential-to-functional failure interval (P-F interval) The average amount of time or number of cycles between a detectable potential failure condition and an actual functional failure.

Predictive maintenance An equipment maintenance strategy based on measuring the condition of equipment in order to assess whether it will fail during some future period, and then taking appropriate action to avoid the consequences of that failure.

Pre-startup safety review (PSSR) A management system for ensuring that new or modified processes are ready for startup by verifying that equipment is installed in a manner consistent with the design intent.

Preventive maintenance (PM) Maintenance that seeks to reduce the frequency and severity of unplanned shutdowns by establishing a fixed schedule of routine inspection and repairs.

Process hazard analysis (PHA) See *Hazard identification and risk analysis.*

Process safety information A compilation of chemical hazard, technology and equipment documentation needed to manage process safety.

Proof test The exercising of a passive system. The test is required to simulate the actual operation as much as is possible to be valid.

Quality assurance (QA) Activities performed to ensure that equipment is designed appropriately and to ensure that the design intent is not compromised, providing confidence throughout that a product or service will continually fulfill a defined need the equipment's entire life cycle.

Quality control (QC) Execution of a procedure or set of procedures intended to ensure that a design or manufactured product or performed service/activity adheres to a defined set of quality criteria or meets the requirements of the client or customer.

Quality management (QM) All the activities that an organization uses to direct, control and coordinate quality. These activities include formulating a quality policy, setting quality objectives and executing quality planning, quality assurance, quality control and quality improvement.

Recognized and generally accepted good engineering practice (RAGAGEP) This term, originally used by OSHA, stems from the selection and application of appropriate engineering, operating and maintenance knowledge when designing, operating and maintaining chemical facilities with the purpose of ensuring safety and preventing process safety incidents.

It involves the application of engineering, operating or maintenance activities derived from engineering knowledge and industry experience based upon the evaluation and analyses of appropriate internal and external standards, applicable codes, technical reports, guidance, or recommended practices or documents of a similar nature. RAGAGEP can be derived from singular or multiple sources and will vary based upon individual facility processes, materials, service, and other engineering considerations.

Reliability-centered maintenance (RCM) A systematic analysis approach for evaluating asset failure impacts on system performance and determining specific strategies for managing the identified asset failures. The failure management strategies may include preventive maintenance, predictive maintenance, inspections, testing and/or one-time changes such as design improvements or operational changes.

Remaining life An estimate, based on inspection results, of the time it will take for an equipment item to reach a defined retirement criterion such as minimum wall thickness.

Risk A measure of human injury, environmental damage, or economic loss in terms of both the incident likelihood and the magnitude of the loss or injury.

Risk analysis The estimation of scenario, process, facility and/or organizational risk by identifying potential incident scenarios, then evaluating and combining the expected frequency and impact of each scenario having a consequence of concern, then summing the scenario risks if necessary to obtain the total risk estimate for the level at which the risk analysis is being performed.

Risk-based inspection (RBI) A systematic approach for identifying credible failure mechanisms and using equipment failure consequences and likelihood to determine inspection strategies for equipment.

Root cause analysis (RCA) A formal investigation method that attempts to identify and address the management system failures that led to an incident. These root causes often are the causes, or potential causes, of other seemingly unrelated incidents. Identifies the underlying reasons the event was allowed to occur so that workable corrective actions can be implemented to help prevent recurrence of the event (or occurrence of similar events). In the context of asset integrity management, RCA is a formal investigation method used to identify and address the management system reasons that led to a failure.

Safeguards Any device, system, or action that either interrupts the chain of events following an initiating event or that mitigates the consequences. A safeguard can be an engineered system or an administrative control.

Safety-critical equipment or ***Safety-critical element (SCE)*** Equipment, the malfunction or failure of which could (or is likely to) cause or contribute to a major incident, or the purpose of which is to prevent a major incident or mitigate its effects.

Safety instrumented system (SIS) A separate and independent combination of sensors, logic solvers, final elements, and support systems that are designed and managed to achieve a specified safety integrity level. A SIS may implement one or more safety instrumented functions (SIFs).

Safety integrity level (SIL) Discrete level (one out of four) allocated to the SIF for specifying the safety integrity requirements to be achieved by the SIS.

Technical assurance The process for communicating that appropriate technology is being applied to process equipment.

Technical evaluation condition An equipment condition requiring further technical evaluation to determine suitability for continued service.

Testing Checking the operation/functionality of assets, including *proof testing*.

Verification activity A test, field observation or other activity used to ensure that personnel have acquired necessary skills and/or knowledge following training.

INDEX